Data Engineering with Databricks Cookbook

Build effective data and AI solutions using Apache Spark, Databricks, and Delta Lake

Pulkit Chadha

Data Engineering with Databricks Cookbook

Copyright © 2024 Packt Publishing

Group Product Manager: Apeksha Shetty

Publishing Product Manager: Deepesh Patel

Book Project Manager: Shambhavi Mishra

Senior Editor: Rohit Singh

Technical Editor: Kavyashree KS

Copy Editor: Safis Editing

Proofreaders: Safis Editing and Rohit Singh

Indexer: Manju Arasan

Production Designer: Alishon Mendonca

DevRel Marketing Executive: Nivedita Singh

First published: May 2024

Production reference: 1100524

Published by Packt Publishing Ltd.

Grosvenor House

11 St Paul's Square

Birmingham

B3 1RB, UK.

ISBN 978-1-83763-335-7

www.packtpub.com

To my wonderful wife, Latika Kapoor, thank you for always being there for me. Your support and belief in me have helped me achieve my dream of writing this book. You've motivated me every step of the way, and I couldn't have done it without you.

– Pulkit Chadha

Contributors

About the author

Pulkit Chadha is a seasoned technologist with over 15 years of experience in data engineering. His proficiency in crafting and refining data pipelines has been instrumental in driving success across diverse sectors such as healthcare, media and entertainment, hi-tech, and manufacturing. Pulkit's tailored data engineering solutions are designed to address the unique challenges and aspirations of each enterprise he collaborates with.

An alumnus of the University of Arizona, Pulkit holds a master's degree in management information systems along with multiple cloud certifications. His impactful career includes tenures at Dell Services, Adobe, and Databricks, shaping data-driven decision-making and business growth.

About the reviewers

Gaurav Chawla is a seasoned data scientist and machine learning engineer at JP Morgan Chase, and possesses more than a decade of expertise in machine learning and software engineering. His focus lies in specialized areas such as fraud detection and the comprehensive development of real-time machine learning models. Gaurav holds a master's degree in data science from Columbia University in the City of New York.

I express gratitude to my wife, Charika, for her constant support, and to our delightful son, Angad, who brings immense joy into our lives. I extend thanks to my parents for imparting valuable teachings and fostering a resilient foundation for my character.

Jaime Andres Salas is an exceptionally enthusiastic data professional. With a wealth of expertise spanning more than 6 years in data engineering and data management, he possesses extensive experience in designing and maintaining large-scale enterprise data platforms such as data warehouses, data lakehouses, and data pipeline solutions. Jaime Andres holds a bachelor's degree in electronic engineering from Espol and an MBA from UOC.

Throughout his professional journey, he has successfully undertaken significant big data and data engineering projects for a diverse range of industries, including retail, production, brewing, and insurance.

Mohit Raja Sudhera has over a decade of extensive experience in data and cloud engineering. He currently leads a team of talented engineers at a prominent and innovative healthcare provider globally.

His core competencies lie in designing and delivering scalable and robust data-intensive solutions utilizing highly performant systems such as Spark, Kafka, Snowflake, and Azure Databricks. Mohit spearheads the architecture and standardization of data-driven capabilities responsible for optimizing the performance and latency of reporting dashboards.

His dedication to self-education within the data engineering domain has afforded him invaluable exposure to crafting and executing data-intensive jobs/pipelines.

Table of Contents

2

Data Transformation and Data Manipulation with Apache Spark   35

3

Data Management with Delta Lake   65

4

Ingesting Streaming Data   99

5

Processing Streaming Data   139

6

Performance Tuning with Apache Spark   177

7

Performance Tuning in Delta Lake 219

Part 2 – Data Engineering Capabilities within Databricks

8

Orchestration and Scheduling Data Pipeline with Databricks Workflows 241

9

Building Data Pipelines with Delta Live Tables   281

10

Data Governance with Unity Catalog 321

11

Implementing DataOps and DevOps on Databricks 369

Preface

In the ever-evolving landscape of data, the need for robust, scalable, and efficient data engineering solutions has never been more pressing. *Data Engineering with Databricks Cookbook* is a testament to this necessity, offering a practical guide to mastering the intricacies of the **Databricks Lakehouse Platform**.

The evolving landscape of data engineering

In recent years, the field of data engineering has undergone a transformative shift, with the demand for efficient, scalable, and collaborative solutions reaching unprecedented heights. This book addresses this paradigm shift by delving into the intricacies of Apache Spark, a versatile engine for big data processing, Databricks, a collaborative and cloud-based platform, and Delta Lake, an open source storage layer that enhances the reliability and consistency of your data workflows.

A pragmatic approach to data engineering

This cookbook goes beyond being a compilation of recipes; it's a pragmatic guide aimed at empowering you to overcome real-world data engineering challenges. The recipes provided are designed to be practical, with step-by-step instructions, code snippets, and detailed explanations to facilitate a hands-on learning experience. Whether you are a seasoned data engineer or just embarking on your data journey, the book offers valuable insights and practical solutions to integrate these cutting-edge technologies seamlessly into your projects.

Key features

Some of the key objectives of this book are as follows:

- **In-depth recipes for the entire data engineering life cycle**: Navigate through a comprehensive set of recipes covering data extraction, transformation, loading, and effective management within a Lakehouse architecture

- **Practical learning**: Embrace a hands-on approach with detailed instructions, code examples, and explanations to ensure you gain practical expertise in applying these technologies to real-world scenarios

- **Best practices and optimization**: Benefit from industry best practices and expert tips to optimize your data engineering workflows, building scalable, efficient, and easy-to-maintain solutions

- **Real-world challenges and solutions**: Explore recipes addressing common challenges faced by data engineers in actual projects, providing practical insights for implementation

- **Collaboration and seamless integration**: Leverage the collaborative capabilities of Databricks and learn how to seamlessly integrate these technologies into your existing data infrastructure, fostering a more efficient and collaborative environment

Embark on a journey to master the art of data engineering with Apache Spark, Databricks, and Delta Lake. This cookbook is not just a guide; it's your companion in navigating the complexities of modern data engineering. Happy cooking!

Who this book is for

Data Engineering with Databricks Cookbook is crafted for a diverse audience, ranging from beginners to seasoned professionals in the field of data engineering. Here's who will benefit the most from this book:

- **Data engineers** who are looking to deepen their understanding of the Databricks Lakehouse Platform and want to efficiently process, manage, and analyze large datasets

- **Data scientists** who desire to leverage the power of Apache Spark for complex data analytics and machine learning tasks

- **Data analysts** who aim to transform data into actionable insights using Databricks and Delta Lake

- **Software developers** with an interest in data processing and analytics, who are looking to integrate Databricks into their software solutions

- **IT professionals** who manage big data infrastructure and are considering Databricks as a solution for their organization

Whether you're taking your first steps into big data or are an experienced practitioner, this book provides valuable knowledge and skills that can be applied to real-world data challenges. If you're passionate about data and eager to learn how to build scalable and effective data solutions, this book is for you.

What this book covers

Chapter 1, *Data Ingestion and Data Extraction with Apache Spark*, explores the fundamental processes of data ingestion and extraction using Apache Spark. From connecting to various data sources to efficiently extracting and loading data, you will gain hands-on experience in leveraging Apache Spark's capabilities for seamless data integration.

Chapter 2, *Data Transformation and Data Manipulation with Apache Spark*, delves into the transformative power of Apache Spark, focusing on data transformation and manipulation techniques. You will learn how to harness Spark's robust functionalities for reshaping and optimizing data, ensuring it aligns with specific business requirements and analytical needs.

Chapter 3, Data Management with Delta Lake, delves into Delta Lake, a critical component for effective data management. You will discover how to leverage Delta Lake's ACID transactions and versioning capabilities to ensure data reliability, consistency, and efficient management within the Lakehouse architecture.

Chapter 4, Ingesting Streaming Data, initiates the exploration of ingesting streaming data using Apache Spark. It covers the basics of streaming data ingestion, setting the stage for understanding real-time data processing and analysis.

Chapter 5, Processing Streaming Data, completes the exploration of streaming data by focusing on advanced techniques and best practices for processing real-time data with Apache Spark. You will gain insights into handling dynamic data streams and maintaining data integrity in dynamic, fast-paced environments.

Chapter 6, Performance Tuning with Apache Spark, delves into the intricacies of performance tuning in Apache Spark. From optimizing code to fine-tuning configurations, you will learn practical strategies to enhance the efficiency and speed of Spark applications, ensuring optimal performance for large-scale data processing.

Chapter 7, Performance Tuning in Delta Lake, builds upon performance tuning principles and focuses specifically on optimizing Delta Lake workflows. You will gain insights into techniques for improving the speed and efficiency of data transactions, making data management within the Lakehouse architecture more performant.

Chapter 8, Orchestration and Scheduling Data Pipeline with Databricks Workflows, guides you through the orchestration and scheduling of workflows in Databricks. From designing automated data pipelines to scheduling tasks efficiently, you will learn how to streamline your data engineering processes and ensure the timely execution of critical workflows.

Chapter 9, Building Data Pipelines with Delta Live Tables, helps you explore the innovative Delta Live Tables, showing how to build robust and dynamic data pipelines. The focus is on leveraging Delta Live Tables to simplify data pipeline development, enhance collaboration, and ensure data consistency in real time.

Chapter 10, Data Governance with Unity Catalog, introduces the concept of data governance using Unity Catalog in Databricks. You will discover how to implement effective data governance practices, including metadata management, data lineage tracking, and access control, to ensure data quality and compliance.

Chapter 11, Implementing DataOps and DevOps on Databricks, addresses the integration of DataOps and DevOps practices within the Databricks environment. You will learn how to implement collaborative and automated development and deployment processes, fostering a culture of continuous improvement and efficiency in data engineering workflows.

To get the most out of this book

To follow along with the examples in this chapter, you will need to have the following:

1. Install Docker Desktop as per the instructions at `https://www.docker.com/products/docker-desktop/`.

    If on Mac, make sure to choose the binary associated with your chip (Intel Chip or Apple Chip).

2. Install Git as per the instructions at `https://git-scm.com/book/en/v2/Getting-Started-Installing-Git`.

3. Clone the book's repository locally by running the following:

    ```
    git clone https://github.com/PacktPublishing/Data-Engineering-
    with-Databricks-Cookbook.git
    ```

4. Download and build the Docker images for the two nodes, one master spark cluster, and the JupyterLab notebook environment by running the following command in the cloned repository's root folder:

    ```
    $ sh build.sh
    ```

> **Note**
>
> This may take several minutes the first time since it has to download and install Spark and all other supporting libraries on the base images.

5. Start the local Apache Spark and JupyterLab notebook environment by running `docker-compose` from the root folder of the cloned repository:

    ```
    $ docker-compose up
    ```

This `docker-compose` file is creating a multi-container application that consists of the following services:

*   **ZooKeeper**: A service that provides coordination and configuration management for distributed systems.

*   **Kafka**: A service that provides a distributed streaming platform for publishing and subscribing to streams of data. It depends on ZooKeeper and uses port `9092`. It allows plaintext listeners and has some custom configuration options.

*   **JupyterLab**: A service that provides an interactive web-based environment for data science and machine learning. It uses ports `8888` and `4040` and shares a local volume with the other services. It has a custom image, which includes Spark 3.4.1.

- **spark-master**: A service that acts as the master node for a Spark cluster. It uses ports 8080 and 7077 and shares a local volume with the other services. It has a custom image, which includes Spark 3.4.1.

- **spark-worker-1 and spark-worker-2**: Two services that act as worker nodes for the Spark cluster. They depend on spark-master and use port 8081. They have custom images, which include Spark 3.4.1 and some environment variables to specify the worker cores and memory.

To run this docker-compose file, you need to have the following minimum system requirements:

- Docker Engine version 18.02.0+ and Docker Compose version 1.25.5+

- At least 6 GB of RAM and 10 GB of disk space

- A Linux, Mac, or Windows operating system that supports Docker

The system requirements in order to follow the recipes are as follows:

Software/Hardware covered in the book	OS requirements
Docker Engine version 18.02.0+	Windows, Mac OS X, and Linux (any)
Docker Compose version 1.25.5+	
Docker Desktop	
Git	

If you are using the digital version of this book, we advise you to type the code yourself or access the code via the GitHub repository (link available in the next section). Doing so will help you avoid any potential errors related to the copying and pasting of code.

Download the example code files

You can download the example code files for this book from GitHub at https://github.com/PacktPublishing/Data-Engineering-with-Databricks-Cookbook. In case there's an update to the code, it will be updated on the existing GitHub repository.

We also have other code bundles from our rich catalog of books and videos available at https://github.com/PacktPublishing/. Check them out!

Conventions used

There are a number of text conventions used throughout this book.

`Code in text`: Indicates code words in text, database table names, folder names, filenames, file extensions, pathnames, dummy URLs, user input, and Twitter handles. Here is an example: "Import the required libraries and create a `SparkSession` object."

A block of code is set as follows:

```
from pyspark.sql import SparkSession

spark = (SparkSession.builder
    .appName("read-csv-data")
    .master(«spark://spark-master:7077»)
    .config(«spark.executor.memory", "512m")
    .getOrCreate())

spark.sparkContext.setLogLevel("ERROR")
```

When we wish to draw your attention to a particular part of a code block, the relevant lines or items are set in bold:

```
builder = (SparkSession.builder
                .appName("optimize-delta-table")
                .master("spark://spark-master:7077")
                .config("spark.executor.memory", "512m")
```

Any command-line input or output is written as follows:

```
$ docker-compose stop
```

Bold: Indicates a new term, an important word, or words that you see onscreen. For example, words in menus or dialog boxes appear in the text like this. Here is an example: "The Spark UI also has a dedicated **Structured Streaming** tab, which shows detailed information about your streaming queries, such as input rate, processing rate, latency, state size, and event timeline."

> **Tips or important notes**
> Appear like this.

Sections

In this book, you will find several headings that appear frequently (*Getting ready, How to do it..., How it works..., There's more...,* and *See also*).

To give clear instructions on how to complete a recipe, use these sections as follows:

Getting ready

This section tells you what to expect in the recipe and describes how to set up any software or any preliminary settings required for the recipe.

How to do it...

This section contains the steps required to follow the recipe.

How it works...

This section usually consists of a detailed explanation of what happened in the previous section.

There's more...

This section consists of additional information about the recipe in order to make you more knowledgeable about the recipe.

See also

This section provides helpful links to other useful information for the recipe.

Get in touch

Feedback from our readers is always welcome.

General feedback: If you have questions about any aspect of this book, mention the book title in the subject of your message and email us at customercare@packtpub.com.

Errata: Although we have taken every care to ensure the accuracy of our content, mistakes do happen. If you have found a mistake in this book, we would be grateful if you would report this to us. Please visit www.packtpub.com/support/errata, select your book, click on the Errata Submission Form link, and enter the details.

Piracy: If you come across any illegal copies of our works in any form on the Internet, we would be grateful if you would provide us with the location address or website name. Please contact us at copyright@packt.com with a link to the material.

If you are interested in becoming an author: If there is a topic that you have expertise in and you are interested in either writing or contributing to a book, please visit authors.packtpub.com.

Share Your Thoughts

Once you've read *Data Engineering with Databricks Cookbook*, we'd love to hear your thoughts! Scan the QR code below to go straight to the Amazon review page for this book and share your feedback.

https://packt.link/r/1-837-63335-5

Your review is important to us and the tech community and will help us make sure we're delivering excellent quality content.

Download a free PDF copy of this book

Thanks for purchasing this book!

Do you like to read on the go but are unable to carry your print books everywhere?

Is your eBook purchase not compatible with the device of your choice?

Don't worry, now with every Packt book you get a DRM-free PDF version of that book at no cost.

Read anywhere, any place, on any device. Search, copy, and paste code from your favorite technical books directly into your application.

The perks don't stop there, you can get exclusive access to discounts, newsletters, and great free content in your inbox daily

Follow these simple steps to get the benefits:

1. Scan the QR code or visit the link below

https://packt.link/free-ebook/9781837633357

2. Submit your proof of purchase
3. That's it! We'll send your free PDF and other benefits to your email directly

Part 1 – Working with Apache Spark and Delta Lake

In this part, we will explore the essentials of data operations with Apache Spark and Delta Lake, covering data ingestion, extraction, transformation, and manipulation to align with business analytics. We will delve into Delta Lake for reliable data management with ACID transactions and versioning, and tackle streaming data ingestion and processing for real-time insights. This part concludes with performance tuning strategies for both Apache Spark and Delta Lake, ensuring efficient data processing within the Lakehouse architecture.

This part contains the following chapters:

- *Chapter 1, Data Ingestion and Data Extraction with Apache Spark*
- *Chapter 2, Data Transformation and Data Manipulation with Apache Spark*
- *Chapter 3, Data Management with Delta Lake*
- *Chapter 4, Ingesting Streaming Data*
- *Chapter 5, Processing Streaming Data*
- *Chapter 6, Performance Tuning with Apache Spark*
- *Chapter 7, Performance Tuning in Delta Lake*

1

Data Ingestion and Data Extraction with Apache Spark

Apache Spark is a powerful distributed computing framework that can handle large-scale data processing tasks. One of the most common tasks when working with data is loading it from various sources and writing it into various formats. In this hands-on chapter, you will learn how to load and write data files with Apache Spark using Python.

In this chapter, we're going to cover the following recipes:

- Reading CSV data with Apache Spark

- Reading JSON data with Apache Spark

- Reading Parquet data with Apache Spark

- Parsing XML data with Apache Spark

- Working with nested data structures in Apache Spark

- Processing text data in Apache Spark

- Writing data with Apache Spark

By the end of this chapter, you will have learned how to read, write, parse, and manipulate data in CSV, JSON, Parquet, and XML formats. You will have also learned how to analyze text data with **natural language processing** (NLP) and how to optimize data writing with buffering, compression, and partitioning.

Technical requirements

Before starting, make sure that your `docker-compose` images are up and running, and open the JupyterLab server running on the localhost (`http://127.0.0.1:8888/lab`). Also, ensure that you have cloned the Git repo for this book and have access to the notebook and data used in this chapter.

Remember to stop all services defined in the `docker-compose` file for this book when you are done running the code examples. You can do this by executing this command:

```
$ docker-compose stop
```

You can find the notebooks and data for this chapter at `https://github.com/PacktPublishing/Data-Engineering-with-Databricks-Cookbook/tree/main/Chapter01`.

Reading CSV data with Apache Spark

Reading CSV data is a common task in data engineering and analysis, and Apache Spark provides a powerful and efficient way to process such data. Apache Spark supports various file formats, including CSV, and it provides many options for reading and processing such data. In this recipe, we will learn how to read CSV data with Apache Spark using Python.

How to do it...

1. **Import libraries:** Import the required libraries and create a `SparkSession` object:

   ```
   from pyspark.sql import SparkSession

   spark = (SparkSession.builder
       .appName("read-csv-data")
       .master("spark://spark-master:7077")
       .config("spark.executor.memory", "512m")
       .getOrCreate())

   spark.sparkContext.setLogLevel("ERROR")
   ```

2. **Read the CSV data with an inferred schema:** Read the CSV file using the `read` method of `SparkSession`. In the following code, we specify the format of the file as `csv` using the `format` method of `SparkSession`. We then set the `header` option to `true` to indicate that the first row of the CSV file contains the column names. Finally, we specify the path to the CSV file to load it into a DataFrame using the `load` method:

   ```
   df = (spark.read.format("csv")
       .option("header", "true")
       .load("../data/netflix_titles.csv"))
   ```

> **Note**
>
> If your CSV file does not have a header row, you can set the header option to false, as shown:
>
> ```
> df = (spark.read.format("csv")
>     .option("header", "false")
>     .load("../data/netflix_titles.csv"))
> ```

3. **Display sample data in the DataFrame**: You can display the contents of the DataFrame using the show() method. This will display the first 20 rows of the DataFrame. If you want to display more or fewer rows, you can pass an integer argument to the show() method:

   ```
   # Display contents of DataFrame
   df.show()
   # Alternatively
   # df.show(10, truncate=False)
   ```

4. **Read the CSV data with an explicit schema**: First, you need to define the schema for your CSV file. You can do this using the StructType and StructField classes in Spark SQL. For example, if your CSV file has three columns, "name", "age", and "gender", and you want to specify that "name" is a string, "age" is an integer, and "gender" is a string, you could define the schema like this:

   ```
   from pyspark.sql.types import StructType, StructField,
   StringType, IntegerType, DateType

   schema = StructType([
       StructField("show_id", StringType(), True),
       StructField("type", StringType(), True),
       StructField("title", StringType(), True),
       StructField("director", StringType(), True),
       StructField("cast", StringType(), True),
       StructField("country", StringType(), True),
       StructField("date_added", DateType(), True),
       StructField("release_year", IntegerType(), True),
       StructField("rating", StringType(), True),
       StructField("duration", StringType(), True),
       StructField("listed_in", StringType(), True),
       StructField("description", StringType(), True)])
   ```

5. **Read the CSV file**: Next, you need to read the CSV file using the `spark.read.format("csv")` method and pass it in the schema as an argument. In this example, we're reading a CSV file located at `../data/netflix_titles.csv`, with the first row as the header, and using the previously defined schema:

```
df = (spark.read.format("csv")
    .option("header", "true")
    .schema(schema)
    .load("../data/netflix_titles.csv"))
```

6. You can display the contents of the DataFrame using the `show()` method. This will display the first five rows of the DataFrame. If you want to display more or fewer rows, you can change the n argument to the `show()` method:

```
df.show()
```

7. Finally, we need to stop the Spark session to release the resources used by Spark:

```
spark.stop()
```

Common issues faced while working with CSV data

Let's take a look at some issues frequently encountered when working with CSV data:

- **Issue**: Delimiter value is present within the data.

  Solution: You can use the `option("escapeQuotes", "true")` method while reading the CSV file to specify how to handle the delimiter value within the column values. In the following code, we specified that there are escape quotes for values that have the delimiter character:

  ```
  df = (spark.read.format("csv")
      .option("header", "true")
      .option("nullValue", "null")
      .option("escapeQuotes", "true")
      .schema(schema)
      .load("../data/netflix_titles.csv"))
  ```

- **Issue**: Null values and empty values are not handled correctly.

  Solution: You can use the `option("nullValue", "null")` method while reading the CSV file to specify how null values are represented in the CSV file. You can also use the `option("emptyValue", "")` method to specify how to handle empty values. In the following code, we specified the `"null"` value as the representation of null values and empty values to be `""` in the CSV file. You can specify any other string or character that represents null values in your CSV file:

  ```
  df = (spark.read.format("csv")
      .option("header", "true")
      .option("nullValue", "null")
  ```

```
        .option("emptyValues", "")
        .schema(schema)
        .load("../data/netflix_titles.csv"))
```

- **Issue**: Date formats are different and not handled correctly.

   Solution: You can use the `option("dateFormat", "LLLL d, y")` method while reading the CSV file to specify how `date` column values are represented in the CSV file. In the following code, we specified that `date` column values are formatted as `"LLLL d, y"`. You can specify any other datetime patterns for formatting and parsing dates in a CSV file:

```
df = (spark.read.format("csv")
        .option("header", "true")
        .option("nullValue", "null")
        .option("dateFormat", "LLLL d, y")
        .schema(schema)
        .load("../data/netflix_titles.csv"))
```

> **Note**
>
> When we used the `read` API in *step 2*, Apache Spark did not execute any jobs. This is because Apache Spark uses a lazy evaluation technique to delay the execution of transformations until an action is called. This allows Spark to optimize the execution plan and recover from failures. However, it can also lead to some challenges in debugging and troubleshooting. To work effectively with lazy evaluation, it's important to understand the distinction between transformations and actions and to consider the order and timing of transformations and actions in your code.

There's more...

Here are some additional details to help you become more knowledgeable about reading CSV data with Apache Spark:

- **Specifying options when reading CSV files**: In addition to specifying the `header` option, you can also specify other options when reading CSV files with Spark. For example, you can use the `delimiter` option to specify the delimiter used in the CSV file (for example, `option("delimiter", "|")` for a pipe-delimited file) or the `inferSchema` option to automatically infer the data types of the columns in the DataFrame (for example, `option("inferSchema", "true")`).

- **Handling missing or malformed data**: When reading CSV files with Spark, you may encounter missing or malformed data that can cause errors. To handle missing data, you can use the `nullValue` option to specify the value used to represent null values in the CSV file (for example, `option("nullValue", "NA")` for a file where `"NA"` represents null values). To handle malformed data, you can use the `mode` option to specify how to handle parsing errors (for example, `option("mode", "PERMISSIVE")` to ignore parsing errors and continue processing the file).

- **Working with large CSV files**: When working with large CSV files, you may run into memory and performance issues if you try to load the entire file into a DataFrame at once. To avoid this, you can use the `spark.read.csv()` method with the `maxColumns` and `maxCharsPerColumn` options to limit the number of columns and characters per column that Spark reads at a time. You can also use the `spark.readStream.csv()` method to read large CSV files as a stream, which allows you to process the data in real time as it is read from disk.

See also

- *CSV Files* – Spark 3.4.0 documentation: `https://spark.apache.org/docs/latest/sql-data-sources-csv.html`

- *Data Types* – Spark 3.4.0 documentation: `https://spark.apache.org/docs/latest/sql-ref-datatypes.html`

- *Datetime Patterns for Formatting and Parsing* – Spark 3.4.0 documentation: `https://spark.apache.org/docs/latest/sql-ref-datetime-pattern.html`

Reading JSON data with Apache Spark

In this recipe, we will learn how to ingest and load JSON data with Apache Spark. Finally, we will cover some common tasks in data engineering with JSON data.

How to do it...

1. **Import libraries**: Import the required libraries and create a `SparkSession` object:

```
from pyspark.sql import SparkSession
from pyspark.sql.functions import *

spark = (SparkSession.builder
    .appName("read-json-data")
    .master("spark://spark-master:7077")
    .config("spark.executor.memory", "512m")
    .getOrCreate())

spark.sparkContext.setLogLevel("ERROR")
```

2. **Load the JSON data into a Spark DataFrame**: The `read` method of the `SparkSession` object can be used to load JSON data from a file or a directory. The `multiLine` option is set to `true` to parse records that span multiple lines. We need to pass the path to the JSON file as a parameter:

```
df = (spark.read.format("json")
    .option("multiLine", "true")
    .load("../data/nobel_prizes.json"))
```

3. **View the schema of the DataFrame**: The `printSchema()` method of the DataFrame object can be used to display the schema of the JSON data. This will help us understand the structure of the data:

    ```
    df.printSchema()
    ```

4. **View the data in the DataFrame**: The `show()` method of the DataFrame object can be used to display the data in the JSON file. This will help us see what the data looks like and how it is structured:

    ```
    df.show()
    ```

5. **Flatten nested structures within the JSON**: If the JSON data has nested structures, we can use the `explode` function to simplify the data.

    In this example, we are flattening the `laureates` field, which is an array with `id`, `firstname`, `surname`/`lastname`, `share`, and `motivation` fields, into separate columns. The `explode` function will create a new row for each element in an array or map. We use the `withColumn` method to replace the existing `laureates` column. The method takes two arguments – the name of the new column and the expression that defines the values of the new column:

    ```
    df_flattened = (df
        .withColumn("laureates",explode(col("laureates")))
        .select(col("category"),
            col("year"),
            col("overallMotivation"),
            col("laureates.id"),
            col("laureates.firstname"),
            col("laureates.surname"),
            col("laureates.share"),
            col("laureates.motivation")
            )
        )


    df_flattened.show(truncate=False)
    ```

6. **Use a schema to enforce data types while reading data**: If we want to enforce data types on the JSON data, we can define a schema using the `StructType`, `StructField`, and data type classes from the `pyspark.sql.types` module. We can then pass the schema to the `read` method using the `schema` parameter:

    ```
    from pyspark.sql.types import StructType, StructField,
    StringType, IntegerType

    json_schema = StructType(
        [StructField('category', StringType(), True),
            StructField('laureates', ArrayType(StructType(
                [StructField('firstname', StringType(), True),
    ```

```
                    StructField('id', StringType(), True),
                    StructField('motivation', StringType(), True),
                    StructField('share', StringType(), True),
                    StructField('surname', StringType(), True)
                    ]), True), True),
            StructField('overallMotivation', StringType(), True),
            StructField('year', IntegerType(), True)])

    json_df_with_schema = (
        spark.read.format("json")
        .schema(json_schema)
        .option("multiLine", "true")
        .option("mode", "PERMISSIVE")
        .option("columnNameOfCorruptRecord", "corrupt_record")
        .load("../data/nobel_prizes.json"))
```

7. **Stop the SparkSession object**: We can use the `stop()` method of the `SparkSession` object to stop the Spark application and release the resources:

```
    spark.stop()
```

There's more...

You can also use options to specify how to read JSON data. For example, you can specify a schema to enforce the JSON data or options to handle corrupt data. Let's look at some in more detail.

The get_json_object() and json_tuple() functions

In Apache Spark, the `get_json_object()` and `json_tuple()` functions are used to extract values from JSON strings.

The `get_json_object()` function extracts a JSON object from a JSON string based on a specified JSON path expression and returns a string representation of the extracted object.

Here is an example code snippet that demonstrates how to use the `get_json_object()` function in Apache Spark. In this example, we create a DataFrame with a JSON string column. We use the `get_json_object()` function to extract the "name" field from the JSON string column based on the "`$.name`" JSON path expression. Finally, we cast the extracted value to a string using the `cast()` function:

```
from pyspark.sql.functions import get_json_object
from pyspark.sql.types import StringType

# create a DataFrame with a JSON string column
df = spark.createDataFrame([
```

```
    (1, '{"name": "Alice", "age": 25}'),
    (2, '{"name": "Bob", "age": 30}')
], ["id", "json_data"])

# extract the "name" field from the JSON string column
name_df = df.select(get_json_object("json_data", "$.name").
alias("name"))

# cast the extracted value to a string
name_str_df = name_df.withColumn("name_str", name_df["name"].
cast(StringType()))

name_str_df.show()
```

We get the following output:

```
+-----+--------+
| name|name_str|
+-----+--------+
|Alice|   Alice|
|  Bob|     Bob|
+-----+--------+
```

The json_tuple() function extracts multiple values from a JSON string based on a specified set of JSON path expressions and returns a tuple of the extracted values.

Here is an example code snippet that demonstrates how to use the json_tuple() function in Apache Spark. In this example, we create a DataFrame with a JSON string column. We use the json_tuple() function to extract the "name" and "age" fields from the JSON string column based on the specified "name" and "age" JSON path expressions. Finally, we use the alias() function to rename the "name" and "age" extracted columns:

```
from pyspark.sql.functions import json_tuple

# create a DataFrame with a JSON string column
df = spark.createDataFrame([
    (1, '{"name": "Alice", "age": 25}'),
    (2, '{"name": "Bob", "age": 30}')
], ["id", "json_data"])

# extract the "name" and "age" fields from the JSON string column
name_age_df = df.select(json_tuple("json_data", "name", "age").
alias("name", "age"))

name_age_df.show()
```

We get the following output:

```
+-----+---+
| name|age|
+-----+---+
|Alice| 25|
|  Bob| 30|
+-----+---+
```

Handling corrupt data

If the JSON data has corrupt records, we can use the `option` method to set the `mode` parameter to `"PERMISSIVE"`. This will cause Spark to set fields with corrupt data to null and store the corrupt record in a new column called `corrupt_record`, which can be used to investigate and handle the errors:

```
json_df_with_schema = (spark.read.format("json")
    .schema(json_schema)
    .option("multiLine", "true")
    .option("mode", "PERMISSIVE")
    .option("columnNameOfCorruptRecord", "corrupt_record")
    .load("../data/nobel_prizes.json"))
```

The flatten() and collect_list() functions

While working with JSON data, you may need to transform an array of arrays into a single array by merging all the elements of the inner arrays. It returns a new DataFrame with a single column that contains the merged array.

Here is an example code snippet that demonstrates how to use the `flatten()` function in Apache Spark. In this example, we create a DataFrame with an array of arrays column. Then, we use the `collect_list()` function to group all the arrays together into a single array column. Finally, we use the `flatten()` function to merge all the elements of the inner arrays into a single array column:

```
from pyspark.sql.functions import flatten, collect_list

# create a DataFrame with an array of arrays column
df = spark.createDataFrame([
    (1, [[1, 2], [3, 4], [5, 6]]),
    (2, [[7, 8], [9, 10], [11, 12]])
], ["id", "data"])
# use collect_list() function to group by specified columns
collect_df = df.select(collect_list("data").alias("data"))
collect_df.show(truncate=False)
```

We get the following output:

```
+----------------------------------------------------+
|data                                                |
+----------------------------------------------------+
|[[[7, 8], [9, 10], [11, 12]], [[1, 2], [3, 4], [5, 6]]]|
+----------------------------------------------------+
```

Now, we will merge the inner array elements:

```
# use flatten() function to merge all the elements of the inner
arrays
flattened_df = collect_df.select(flatten("data").alias("merged_data"))
flattened_df.show(truncate=False)
```

We get the following output:

```
+---------------------------------------------------+
|merged_data                                        |
+---------------------------------------------------+
|[[1, 2], [3, 4], [5, 6], [7, 8], [9, 10], [11, 12]]|
+---------------------------------------------------+
```

> **Note**
>
> The flatten() function only works with array columns. If you have a nested structure with multiple levels of arrays, you can use the explode() function to flatten the structure before using the flatten() function.

Here is an example that demonstrates how to use the flatten() function with a nested array structure. In this example, we create a DataFrame with a nested array column. We first use the explode() function to flatten the outermost array, and then use collect_list() to group all the inner arrays together. Finally, we use flatten() to merge all the elements of the inner arrays into a single array column:

```
from pyspark.sql.functions import explode, flatten, collect_list

# create a DataFrame with nested array column
df = spark.createDataFrame([
    (1, [[[1, 2], [3, 4]], [[5, 6], [7, 8]]]),
    (2, [[[9, 10], [11, 12]], [[13, 14], [15, 16]]])
], ["id", "data"])

# explode the outermost array to flatten the structure
exploded_df = df.select(col("id"),explode("data").alias("inner_data"))
```

```
# use collect_list() to group all the inner arrays together
grouped_df = exploded_df.groupBy("id").agg(collect_list("inner_data").
alias("merged_data"))

# use flatten() to merge all the elements of the inner arrays
flattened_df = grouped_df.select(flatten("merged_data").alias("final_
data"))

flattened_df.show(truncate=False)
```

We get the following output:

```
+-----------------------------------------------+
| final_data                                    |
+-----------------------------------------------+
| [[1, 2], [3, 4], [5, 6], [7, 8]]              |
| [[9, 10], [11, 12], [13, 14], [15, 16]]       |
+-----------------------------------------------+
```

See also

- *JSON Files* – Spark 3.4.0 documentation: https://spark.apache.org/docs/latest/sql-data-sources-json.html

- *Data Types* – Spark 3.4.0 documentation: https://spark.apache.org/docs/latest/sql-ref-datatypes.html

- *Datetime Patterns for Formatting and Parsing* – Spark 3.4.0 documentation: https://spark.apache.org/docs/latest/sql-ref-datetime-pattern.html

Reading Parquet data with Apache Spark

Apache Parquet is a columnar storage format designed to handle large datasets. It is optimized for the efficient compression and encoding of complex data types. Apache Spark, on the other hand, is a fast and general-purpose cluster computing system that is designed for large-scale data processing.

In this recipe, we will explore how to read Parquet data with Apache Spark using Python.

How to do it...

1. **Import libraries**: Import the required libraries and create a SparkSession object:

```
from pyspark.sql import SparkSession

spark = (SparkSession.builder
```

```
    .appName("read-parquet-data")
    .master("spark://spark-master:7077")
    .config("spark.executor.memory", "512m")
    .getOrCreate())

spark.sparkContext.setLogLevel("ERROR")
```

2. **Load the Parquet data**: We use the `spark.read.format("parquet")` method to load the Parquet data into a Spark DataFrame, as in this example:

```
df = (spark.read.format("parquet")
    .load("../data/recipes.parquet")
```

3. **View the schema of the DataFrame**: The `printSchema()` method of the DataFrame object can be used to display the schema of the JSON data. This will help us understand the structure of the data:

```
df.printSchema(
```

4. **View the data in the DataFrame**: The `show()` method of the DataFrame object can be used to display the data. This will help us see what the data looks like and how it is structured:

```
df.show()
```

5. **Stop the SparkSession object**: We can use the `stop()` method of the `SparkSession` object to stop the Spark application and release the resources:

```
spark.stop()
```

Common scenarios encountered while working with Parquet data

Let's quickly review a couple of challenges when you are handling Parquet data.

Reading partitioned data

To load partitioned Parquet files with Apache Spark, you can use the `spark.read` API and specify the path to the directory containing the Parquet files. Spark will automatically recognize the directory as a partitioned dataset and load the data accordingly.

For example, the `data/partitioned_recipes` path contains recipes partitioned by recipe category, and we can load the data as follows:

```
df_partitioned = (spark.read.format("parquet")
    .load("../data/partioned_recipes"))
df_partitioned.printSchema()
```

You can also read only a few partitions by loading a wildcard file path as shown:

```
df_partitioned = (spark.read.format("parquet")
    .load("../data/partitioned_recipes/DatePublished=2020-01*"))
df_partitioned.printSchema()
```

Schema merging

By default, if you try to read Parquet files that have different schemas, Spark will not merge the schemas as `spark.sql.parquet.mergeSchema` is set to `false`. However, if you set the `mergeSchema` option to `true`, Spark will merge the schemas of the files and try to create a unified schema that can be used to read all the files.

In this example, `mergeSchema` is set to `true` to enable schema merging. Spark will try to merge the schemas of all files in the directory specified by path. If the schemas are compatible, Spark will successfully create a unified schema for all files.

When we read all partitions, Spark inferred the `ReviewCount` column but did not infer the `Images` column in the inferred schema. When we read just a few partitions (that is, `2020-01*`), Spark inferred the schema to have the `Images` column but did not infer the `ReviewCount` column. This is because the `mergeSchema` option was set to `false` by default. After setting `mergeSchema` to `true`, both `ReviewCount` and `Images` are in the inferred schema.

Here's an example code snippet:

```
df_merged_schema = (spark.read.format("parquet")
    .option("mergeSchema", "true")
    .load("../data/partitioned_recipes"))
```

> **Note**
>
> Schema merging can be an expensive operation, especially if the files have different and complex schemas. In some cases, schema merging may not be possible, in which case Spark will still throw an error. Additionally, schema merging may not always result in a schema that is optimal for your use case. In such cases, you may want to manually handle schema evolution to ensure that the schema is suitable for your needs.

See also

- *Parquet Files* – Spark 3.4.0 documentation: https://spark.apache.org/docs/latest/sql-data-sources-parquet.html

- *Data Types* – Spark 3.4.0 documentation: https://spark.apache.org/docs/latest/sql-ref-datatypes.html

- *Datetime Patterns for Formatting and Parsing* – Spark 3.4.0 documentation: https://spark.apache.org/docs/latest/sql-ref-datetime-pattern.html

Parsing XML data with Apache Spark

Reading XML data is a common task in big data processing, and Apache Spark provides several options for reading and processing XML data. In this recipe, we will explore how to read XML data with Apache Spark using the built-in XML data source. We will also cover some common issues faced while working with JSON data and how to solve them. Finally, we will cover some common tasks in data engineering with JSON data.

Note

We also need to install the `spark-xml` package on our cluster. The `spark-xml` package is a third-party library for Apache Spark released by Databricks. The package enables the processing of XML data in Spark applications and provides the ability to read and write XML files using the Spark DataFrame API, which makes it easy to integrate with other Spark components and perform complex data analysis tasks. We can install the package by running the following command:

`$SPARK_HOME/bin/spark-shell –packages com.databricks:spark-xml_2.12:0.16.0****`

We have already installed this package to the Docker images we are running locally.

How to do it...

1. **Import libraries**: Import the required libraries and create a `SparkSession` object. By adding `.config('spark.jars.packages', 'com.databricks:spark-xml_2.12:0.16.0')`, we are adding the JAR file needed to work with XML data:

    ```
    from pyspark.sql import SparkSession
    from pyspark.sql.functions import *

    spark = (SparkSession.builder
        .appName("read-xml-data")
        .config('spark.jars.packages', 'com.databricks:spark-xml_2.12:0.16.0')
        .master("spark://spark-master:7077")
        .config("spark.executor.memory", "512m")
        .getOrCreate())

    spark.sparkContext.setLogLevel("ERROR")
    ```

2. **Define the XML file path and create a DataFrame**: We define the path to the XML file that we want to read and create a DataFrame using the `spark.read.format()` method with the `"xml"` argument. We also set the `rowTag` option to the name of the XML element we want to use as the row tag. To work with more complex XML files that contain multiple nested

elements, we can use the `option("rootTag", "tagname")` function to specify the XML element that should be treated as the root of the DataFrame:

```
# Read XML file into a DataFrame
df = (spark.read.format("com.databricks.spark.xml")
    .option("rowTag", "row")
    .load("..//ta/nobel_prizes.xml"))
```

3. **Display the DataFrame**: We can use the `show()` method to display the contents of the DataFrame in a tabular format:

```
df.show()
```

4. **Access data from the DataFrame**: We can use the `select()` method to choose specific columns from the DataFrame. We pass the names of the columns as arguments to the method, separated by commas:

```
df.select("category", "year").show()
```

If you want to access an element in complex types such as arrays, maps, and structs, we can use the `getItem` method, which extracts a single element or value. It takes a single argument that specifies the index of the element to extract from the complex type.

For example, in our DataFrame, the `laureates` column is an array of objects, and we can use `getItem` to extract the first element of the array with `col("laureates").getItem(0)`.

Furthermore, if we want to access data from nested elements in the XML file, we can use the `.` (dot) operator to access child elements, as in this example:

```
(df.select("category", "year"
    , col("laureates").getItem(0).id).show())
```

This will select the `category` and `year` fields from the `laureates` array element and get the first item's `id` value.

5. **Flatten nested structures**: If the JSON data has nested structures, we can use the `explode` function to simplify the data.

In this example, we are flattening the `laureates` field, which is an array with `id`, `firstname`, `surname/lastname`, `share`, and `motivation` fields, into separate columns. The `explode` function will create a new row for each element in an array or a map. We use the `withColumn` method to replace the existing `laureates` column. The method takes two arguments – the name of the new column and an expression that defines the values of the new column:

```
df_flattened = (
    df
    .withColumn("laureates",explode(col("laureates")))
    .select(col("category"),
    col("year"),
```

```
        col("overallMotivation"),
        col("laureates.id"),
        col("laureates.firstname"),
        col("laureates.surname"),
        col("laureates.share"),
        col("laureates.motivation")))

    df_flattened.show(truncate=False)
```

6. **Use a schema to enforce data types**: If we want to enforce data types on the JSON data, we can define a schema using the StructType, StructField, and data type classes from the pyspark.sql.types module. We can then pass the schema to the read method using the schema parameter:

```
    from pyspark.sql.types import StructType, StructField,
    StringType, IntegerType

    schema = StructType(
        [StructField('category', StringType(), True),
            StructField('laureates', ArrayType(StructType(
                [StructField('firstname', StringType(), True),
                StructField('id', StringType(), True),
                StructField('motivation', StringType(), True),
                StructField('share', StringType(), True),
                StructField('surname', StringType(), True)]),
            True), True),
            StructField('overallMotivation', StringType(), True),
            StructField('year', IntegerType(), True)])

    # Read XML file into a DataFrame
    df_with_schema = (spark.read.format("com.databricks.spark.xml")
        .schema(schema)
        .option("rowTag", "row")
        .load("../data/nobel_prizes.xml"))

    df_with_schema.show()
```

7. **Stop the SparkSession object**: We can use the stop() method of the SparkSession object to stop the Spark application and release the resources:

```
    spark.stop()
```

There's more...

Apache Spark provides various other options that can be used to customize the behavior of the XML reader. These include the following:

- `excludeAttribute`: A comma-separated list of attribute names that should be excluded from the DataFrame

- `inferSchema`: Specifies whether the schema should be inferred from the data or not

- `ignoreSurroundingSpaces`: Specifies whether surrounding spaces should be ignored or not

- `mode`: Specifies the behavior of the reader when encountering corrupt records or parsing errors

See also

- *XML Data Source for Apache Spark 3.x* documentation at `https://github.com/databricks/spark-xml`

- *pyspark.sql.Column.getItem* documentation at `https://spark.apache.org/docs/latest/api/python/reference/pyspark.sql/api/pyspark.sql.Column.getItem.html`

Working with nested data structures in Apache Spark

In this recipe, we will walk you through the step-by-step process of handling nested data structures such as arrays, maps, and so on with Apache Spark. This recipe will equip you with the essential knowledge and practical skills needed to work with complex data types using Apache Spark's distributed computing capabilities.

How to do it...

1. **Import libraries**: Import the required libraries and create a `SparkSession` object: `SparkSession` is a unified entry point for Spark applications. It provides a simplified way to interact with various Spark functionalities, such as **resilient distributed datasets (RDDs)**, DataFrames, datasets, SQL queries, streaming, and more. You can create a `SparkSession` object using the `builder` method, which allows you to configure the application name, master URL, and other options. We will also define `SparkContext`, which is the entry point to any Spark functionality. It represents the connection to a Spark cluster and is responsible for coordinating and distributing operations on that cluster:

   ```
   from pyspark.sql import SparkSession
   from pyspark.sql.functions import *
   ```

```
spark = (SparkSession.builder
    .appName("nested-dataframe")
    .master("spark://spark-master:7077")
    .config("spark.executor.memory", "512m")
    .getOrCreate())

spark.sparkContext.setLogLevel("ERROR")
```

2. **Load the data**: We use the `spark.read.format("json")` method to load the JSON data into a Spark DataFrame, as in this example:

```
df = (spark.read.format("json")
    .option("multiLine", "true")
    .load("../data/Stanford Question Answering Dataset.json"))
```

3. **Explode nested data using the explode function**: If we want to explode the `paragraphs` column and create a new row for each paragraph, we can use the `explode` function. Furthermore, in order to extract data from the nested data structure, we can use dot notation. For instance, to extract `context` and `qas` of each title, we can use the following code:

```
df_exploded = (
    df.select("title",
        explode("paragraphs").alias("paragraphs"))
    .select("title",
        col("paragraphs.context").alias ("context"),
        explode(col("paragraphs.qas")).alias("questions")))

df_exploded.show()
```

4. **Get unique values with an array**: We can also use **higher-order functions** (**HOFs**) to manipulate nested columns in place. For example, we can use the `array_distinct` function to de-duplicate the `answers` array in the `qas` column to only have distinct answers in the array, as shown in the following code:

```
df_array_distinct = (
df_exploded.select("title","context"
                            ,col("questions.id").
alias("question_id")
                            ,col("questions.question").
alias("question_text")
                            ,array_distinct("questions.
answers").alias("answers")))

df_array_distinct.show()
```

5. **Stop the SparkSession object**: We can use the `stop()` method of the `SparkSession` object to stop the Spark application and release the resources:

```
spark.stop()
```

Common issues and considerations

While working with nested data structures, you might run into the following common issues.

A large number of rows with explode

The `explode()` function may result in a large number of rows, which can be inefficient to process.

If possible, try to avoid using the `explode` function if you don't need to "flatten" the data. If you do need to use `explode`, consider using it on a subset of the data or aggregating the data before exploding it to reduce the number of resulting rows.

Dot notation can be tricky to use with deeply nested data structures

Use the `getItem` function instead of dot notation to extract nested fields. The `getItem` function takes an index or a key as an argument and returns the corresponding element from an array or a map. For example, to extract the value of the first answer for each question using `getItem`, we can use the following code. In this example, we use the `getItem` function to extract the first element from the `answers` array, and then use the `getField` function to extract the `text` field from the resulting struct:

```
(df_array_distinct
    .select("title","context","question_text",
        col("answers").getItem(0).getField("text"))
    .show())
```

Nested data with null values

Nested data can contain missing or null values, which can cause errors while trying to extract data.

Use the `isNull` and `isNotNull` functions to filter out rows with missing or null values. For example, to filter out all rows where the value field of the first contact is null, we can use the following code. In this example, we use the `getItem` function to extract the first element from the `answers` array, and then use the `getField` function to extract `textfield` from the resulting struct. We then use the `isNotNull` function to filter out rows where `textfield` is null:

```
(df_array_distinct
.filter(col("answers").getItem(0).getField("text").isNotNull()).show())
```

There's more...

In addition to the explode and filter functions, PySpark provides a wide range of functions that can be used to work with nested data structures, including array_contains, map_keys, map_values, explode_outer, and many others. Let's review each.

The array_contains function

The array_contains function is a built-in function in Spark that allows you to check whether an array contains a specified element. The function takes two arguments: the array to check and the element to search for. It returns a Boolean value indicating whether the array contains the specified element.

Here's an example to help illustrate how the array_contains function works. Let's say we have a DataFrame containing an array of strings for each row, and we want to check whether the "apple" string appears in the array for each row. In this code, we're using the select method of the DataFrame to select the fruits column, as well as a new column called contains_apple that uses the array_contains function to check whether each row's fruits array contains the "apple" string. The alias method is used to give the new column a more descriptive name.

As we can see, the array_contains function correctly identifies that the first and third rows contain the "apple" string in their fruits arrays, while the second row does not:

```
from pyspark.sql.functions import array_contains

df = spark.createDataFrame(
    [(["apple", "orange", "banana"],),
        (["grape", "kiwi", "melon"],),
        (["pear", "apple", "pineapple"],)],
    ["fruits"])

(df.select("fruits"
                , array_contains("fruits", "apple")
                .alias("contains_apple"))
  .show(truncate=False))
```

When we run this code, we get the following output:

```
+-----------------------------+----------------+
|fruits                       | contains_apple |
+-----------------------------+----------------+
|[apple, orange, banana]      | true           |
|[grape, kiwi, melon]         | false          |
|[pear, apple, pineapple]     | true           |
+-----------------------------+----------------+
```

The map_keys and map_values functions

The `map_keys` function in Apache Spark is used to extract the keys from a map column in a DataFrame. It takes a single argument, which is the name of the map column from which to extract the keys. The function returns an array of all the keys in the map column.

The `map_values` function in Apache Spark is used to extract the values from a map column in a DataFrame. It takes a single argument, which is the name of the map column from which to extract the values. The function returns an array of all the values in the map column.

Here's an example to demonstrate how the `map_keys` and `map_values` functions work in Spark. Let's say we have a DataFrame with a single column called `user_info`, which contains a map column with various user information, such as name, age, and email address:

```
from pyspark.sql.functions import map_keys
from pyspark.sql import SparkSession

spark = SparkSession.builder.appName("map_keys_example").getOrCreate()

data = [
    {"user_info": {"name": "Alice", "age": 28, "email": "alice@example.com"}},
    {"user_info": {"name": "Bob", "age": 35, "email": "bob@example.com"}},
    {"user_info": {"name": "Charlie", "age": 42, "email": "charlie@example.com"}}
]

df = spark.createDataFrame(data)
df.show(truncate=False)
```

Now, let's use the `map_keys` function to extract the keys and the `map_values` function to extract the values from the `user_info` map column:

```
(df.select("user_info",
    map_keys("user_info").alias("user_info_keys"),
    map_values("user_info").alias("user_info_values"))
    .show(truncate=False))
```

This will produce the following output:

```
+-----------------------------------------------------------+----------
--------+--------------------------------------+
|user_info                                                  |user_info_keys      |user_info_val-
ues                                         |
+-----------------------------------------------------------+----------
--------+--------------------------------------+
```

```
|{name -> Alice, email -> alice@example.com, age -> 28}     |[name,
email, age]|[Alice, alice@example.com, 28]     |
|{name -> Bob, email -> bob@example.com, age -> 35}         |[name,
email, age]|[Bob, bob@example.com, 35]         |
|{name -> Charlie, email -> charlie@example.com, age -> 42}|[name,
email, age]|[Charlie, charlie@example.com, 42]|
+------------------------------------------------------------+----------
--------+--------------------------------+
```

The explode_outer function

The `explode_outer` function works similarly to the `explode` function but with an important difference: if the array or map column is null, the `explode_outer` function will still return a row for that null column, whereas the `explode` function will not.

Here's an example to demonstrate how the `explode_outer` function works in Spark. As we can see, the `explode_outer` function has expanded each element of the `words` array into its own row in the output DataFrame. In addition, the function has also created a row for the null `words` column, which the `explode` function would have skipped over:

```
data = [
    {"words": ["hello", "world"]},
    {"words": ["foo", "bar", "baz"]},
    {"words": None}
]

df = spark.createDataFrame(data)
(df.select(explode_outer("words").alias("word")).show(truncate=False))
```

We get the following output:

```
+-----+
|word |
+-----+
|hello|
|world|
|foo  |
|bar  |
|baz  |
|null |
+-----+
```

See also

- *JSON Files* – Spark 3.4.0 documentation: `https://spark.apache.org/docs/latest/sql-data-sources-json.html#working-with-nested-data-using-higher-order-functions`

- *Functions* – PySpark 3.4.0 documentation: `https://spark.apache.org/docs/latest/api/python/reference/pyspark.sql/functions.html`

Processing text data in Apache Spark

In this recipe, we will walk you through the step-by-step process of leveraging the power of Spark to handle and manipulate textual information efficiently. This recipe will equip you with the essential knowledge and practical skills needed to tackle text-based challenges using Apache Spark's distributed computing capabilities.

How to do it...

1. **Import libraries**: Import the required libraries and create a `SparkSession` object:

    ```
    from pyspark.sql import SparkSession
    from pyspark.sql.functions import *

    spark = (SparkSession.builder
        .appName("text-processing")
        .master("spark://spark-master:7077")
        .config("spark.executor.memory", "512m")
        .getOrCreate())

    spark.sparkContext.setLogLevel("ERROR")
    ```

2. **Load the data**: We use the `spark.read.format("csv")` method to load the CSV data into a Spark DataFrame, as in this example:

    ```
    df = (spark.read.format("csv")
        .option("header",True)
        .option("multiLine", "true")
        .load("../data/Reviews.csv"))
    ```

3. **Explore the DataFrame**: Once we have loaded the text data into a DataFrame, we can explore the data using the `show()` method. The following code displays the first 10 rows of the DataFrame:

    ```
    df.show(10, truncate=False)
    ```

4. **Clean text data with regular expressions**: You can use regular expressions to match and extract patterns in text data using the `regexp_replace()` function in Apache Spark. In

this example, the `[^a-zA-Z]` regular expression matches all non-alphabetical characters, and the `regexp_replace()` function replaces them with an empty string. We are also chaining the `. +` regular expression that matches multiple spaces and replaces it with a single space with the `regexp_replace()` function. The resulting DataFrame is overwritten to the `Text` column with text data that has been cleaned of non-alphabetical characters:

```
# Apply regular expression to remove all non-alphabetic
characters
df_clean = (df
    .withColumn("Text", regexp_replace("Text", "[^a-zA-Z ]",
"")) 
    .withColumn("Text", regexp_replace("Text", " +", " ")))

df_clean.show()
```

5. **Tokenize the text data**: Tokenization is the process of breaking down text data into smaller units, such as words or phrases. To tokenize the text data, we can split the text into words using the `split()` function. We can apply this function to the `Text` column of the DataFrame and store the result in a new column called `"words"`. Alternatively, we can use the built-in tokenizer function in the MLlib library of Apache Spark.

The following code tokenizes the text data:

```
df_with_words = (df_clean.withColumn("words", split(df_clean.
Text, "\\\s+")))
df_with_words.show()

#Alternatively
from pyspark.ml.feature import Tokenizer

# Tokenize the text data
tokenizer = Tokenizer(inputCol='Text', outputCol='words')
df_with_words = tokenizer.transform(df_clean)
df_with_words.show()
```

6. **Remove stop words**: Stop words are common words that do not carry much meaning, such as "the," "and," and "in." We can remove these stop words using the `StopWordsRemover` transformer in Apache Spark. The following code removes stop words from the `"words"` column:

```
from pyspark.ml.feature import StopWordsRemover

remover = StopWordsRemover(inputCol="words",
outputCol="filtered_words")
df_stop_words_removed = remover.transform(df_with_words)

df_stop_words_removed.show()
```

7. **Compute the word frequency**: To compute the word frequency, we can use the `explode()` and `groupBy()` functions in Apache Spark. This code first uses the `explode()` function to transform the `filtered_words` column into a new column called `word`, which contains one word per row. It then uses the `groupBy()` function to group the rows based on the `word` column and computes the count of each group, which is stored in a new column called `count`. Finally, it sorts the results in descending order of the count and displays the first 10 rows:

```
df_exploded = (df_stop_words_removed
    .select(explode(df_stop_words_removed.filtered_words).
alias("word")))
word_count = (df_exploded
                        .groupBy("word")
                        .count()
                        .orderBy("count", ascending=False))
word_count.show(n=100)
```

8. **Convert text data into numerical features**: **Machine learning** (ML) algorithms work with numerical data, so we need to convert the text data into numerical features. MLlib is Apache Spark's scalable ML library, with APIs in Java, Scala, Python, and R. It provides common learning algorithms and utilities, such as classification, regression, clustering, collaborative filtering, feature extraction, and pipelines. It also leverages Spark's distributed computing framework to handle large-scale data processing and analysis. One way to do this is by using the `CountVectorizer` function in the MLlib library, which creates a **bag-of-words** (**BoW**) representation of the text data:

```
from pyspark.ml.feature import CountVectorizer

# Convert the text data into numerical features
vectorizer = CountVectorizer(inputCol='filtered_words',
outputCol='features')
vectorized_data = vectorizer.fit(df_stop_words_removed).
transform(df_stop_words_removed)
vectorized_data.show(10, truncate=False)
```

9. **Save the processed data**: Finally, the processed data can be saved in any desired format, such as JSON or Parquet:

```
(vectorized_data.repartition(1)
    .write.mode("overwrite")
    .json("../data/data_lake/reviews_vectorized.json"))
```

10. **Stop the SparkSession object**: We can use the `stop()` method of the `SparkSession` object to stop the Spark application and release the resources:

```
spark.stop()
```

There's more...

Apache Spark provides many other features for processing text data, which we'll discuss next.

Using the regexp_extract() function

This function is used to extract substrings from a string that match a specified regular expression pattern. In this example, the `\\bq\\w*` regular expression matches all words that start with the letter q, and the `regexp_extract` function extracts them from the text data. The resulting DataFrame will contain a new `"q_words"` column with the extracted words:

```
from pyspark.sql.functions import regexp_extract

df_q_words = (vectorized_data
    .withColumn("q_words", regexp_extract("text", "\\\\\\\\
bq\\\\\\\\\w*", 0)))
df_q_words.show()
```

Using the rlike() function

This function is used to test whether a string matches a specified regular expression pattern. In this example, the `rlike()` function is used to test whether the text data contains the word `quick`. The resulting DataFrame will contain a new `"contains_qood"` column with a Boolean value indicating whether the text data contains the word `good`:

```
df_good_word = (vectorized_data
    .withColumn("contains_qood", expr("text rlike 'quick'")))
df_good_word.show()
```

Customizing stop words

The built-in list of stop words in the `StopWordsRemover()` function may not be suitable for all types of text data. We may need to customize the list of stop words based on our specific needs.

To customize the `StopWordsRemover()` function in Apache Spark, you can follow these steps:

1. Create an instance of the `StopWordsRemover` class and specify the stop words to be removed as a list. In the example, `input_col` and `output_col` refer to the input and output columns of the `StopWordsRemover()` function, respectively:

    ```
    custom_stopwords =  ["/><br", "-", "/>I","/>The"]
    stopwords_remover = StopWordsRemover(inputCol="words",
        outputCol="filtered_words",
        stopWords=custom_stopwords)
    ```

```
df_stop_words_removed = stopwords_remover.transform(df_with_
words)
```

```
df_stop_words_removed.show()
```

2. Use the `transform()` method of the `StopWordsRemover` instance to apply the stop word removal to the text data. By specifying a custom list of stop words, you can tailor the `StopWordsRemover()` function to your specific use case and improve the accuracy of downstream analysis:

```
data_with_stopwords_removed = stopwords_remover.transform(data)
```

> **Note**
>
> You can also use the `setStopWords()` method of the `StopWordsRemover` instance to modify the list of stop words after instantiation:
>
> ```
> custom_stopwords = ["br", "get", "im","ive"]
> ```
>
> ```
> stopwords_remover = StopWordsRemover(inputCol="words",
> outputCol="filtered_words")
> stopwords_remover.setStopWords(custom_stopwords)
> ```
>
> ```
> df_stop_words_removed = stopwords_remover.transform(df_with_words)
> ```
>
> ```
> df_stop_words_removed.show()
> ```

See also

- *MLlib: Main Guide* – Spark 3.4.0 documentation: `https://spark.apache.org/docs/latest/ml-guide.html`
- *Tokenizer* – PySpark 3.4.0 documentation: `https://spark.apache.org/docs/latest/api/python/reference/api/pyspark.ml.feature.Tokenizer.html`
- *StopWordsRemover* – PySpark 3.4.0 documentation: `https://spark.apache.org/docs/latest/api/python/reference/api/pyspark.ml.feature.StopWordsRemover.html`

Writing data with Apache Spark

In this recipe, we will walk you through the step-by-step process of leveraging the power of Spark to write data in various formats. This recipe will equip you with the essential knowledge and practical skills needed to write data using Apache Spark's distributed computing capabilities.

How to do it...

1. **Import libraries**: Import the required libraries and create a `SparkSession` object:

```
from pyspark.sql import SparkSession

spark = (SparkSession.builder
    .appName("write-data")
    .master("spark://spark-master:7077")
    .config("spark.executor.memory", "512m")
    .getOrCreate())

spark.sparkContext.setLogLevel("ERROR")
```

2. Read a CSV file using the `read` method of `SparkSession`:

```
from pyspark.sql.types import StructType, StructField,
StringType, IntegerType, DateType

df = (spark.read.format("csv")
    .option("header", "true")
    .option("nullValue", "null")
    .option("dateFormat", "LLLL d, y")
    .load("../data/netflix_titles.csv"))
```

3. **Write the CSV data**: Once we have the data in a Spark DataFrame, we can write it to a CSV file using the `write` method of the DataFrame object. We need to specify the path of the CSV file and the delimiter to be used in the CSV file:

```
(df.write.format("csv")
    .option("header", "true")
    .mode("overwrite")
    .option("delimiter", ",")
    .save("../data/data_lake/netflix_csv_data"))
```

In the previous code, we have specified the `header` and `delimiter` options as we want to write the header and a comma delimiter to the CSV file.

> **Note**
>
> The mode parameter controls what happens if the data or table exists. The four modes are the following:
>
> - overwrite: Replaces the old data with the new one but drops indexes and constraints
> - append: Adds new rows to the old data without changing or deleting it
> - ignore: Skips the write if the data or table exists, avoiding duplicates
> - error or errorifexists: Fails the write if the data or table exists, preventing overwriting or appending

4. **Write the DataFrame to JSON format**: Now that we have created the DataFrame, we can write in a specific format. The write method on the DataFrame returns a DataFrameWriter object that provides methods to write data in various formats. We set the write mode to "overwrite" to overwrite any existing file at the given path. The json method is used to write the DataFrame in JSON format. Here is the sample code:

```
(df.write.format("json")
    .mode("overwrite")
    .save("../data/data_lake/netflix_json_data"))
```

5. **Write the DataFrame to Parquet format**: Now that we have created the DataFrame, we can write it to Parquet format. The write method on the DataFrame returns a DataFrameWriter object that provides methods to write data in various formats. We set the write mode to "overwrite" to overwrite any existing file at the given path. The parquet method is used to write the DataFrame in Parquet format. Here is the sample code:

```
(df.write.format("parquet")
    .mode("overwrite")
    .save("../data/data_lake/netflix_parquet_data"))
```

6. **Stop the Spark session**: Finally, we need to stop the Spark session to release the resources used by Spark:

```
spark.stop()
```

There's more...

Apache Spark provides many other features for writing data, which we'll discuss next.

Writing compressed data

We can also write the data in a compressed format such as GZIP or BZIP2. To write the data in a compressed format, we need to specify the compression codec in the `save` method of the DataFrame object:

```
(df.write.format("csv")
    .mode("overwrite")
    .option("header", "true")
    .option("delimiter", ",")
    .option("codec", "org.apache.hadoop.io.compress.GzipCodec")
    .save("../data/data_lake/netflix_csv_data.gz"))
```

> **Note**
>
> In the previous code, we specified the compression codec as `org.apache.hadoop.io.compress.GzipCodec` to write the CSV data in a GZIP compressed format. Similarly, we can use `org.apache.hadoop.io.compress.BZip2Codec` to write the CSV data in a BZIP2 compressed format.

Specifying the number of partitions

Partitioning in Apache Spark is a way to split data into multiple parts across a cluster so that each part can be processed in parallel by different nodes. Partitioning is important for optimizing the performance of Spark applications as it affects the amount of data shuffling, the load balancing across nodes, and the level of fault tolerance. We can also specify the number of partitions to be used while writing the data. The number of partitions determines the number of files created while writing the data. We can specify the number of partitions in the `repartition` method of the DataFrame object:

```
(df.repartition(4)
    .write.format("csv")
    .mode("overwrite")
    .option("header", "true")
    .option("delimiter", ",")
    .save("../data/data_lake/netflix_csv_data_4_part"))
```

> **Note**
>
> In the previous code, we specified the number of partitions as 4 to create four files while writing the data.

Using coalesce() to reduce the number of partitions

If you have a large DataFrame object with a high number of partitions, it can slow down the writing process. You can use the `coalesce()` method to reduce the number of partitions before writing the data to a file. Here's an example of how to use the `coalesce` method:

```
(df.coalesce(1)
    .write.format("csv")
    .mode("overwrite")
    .option("header", "true")
    .option("delimiter", ",")
    .save("../data/data_lake/netflix_csv_data_whole"))
```

Using partitionBy() to write partitions based on a column

Use the `partitionBy()` property of the `DataFrameWriter` class to partition data based on a column while writing the CSV data. If you want to partition the output CSV data based on a specific column, you can use the `partitionBy()` property of the `DataFrameWriter` class. Here's an example of how to use the `partitionBy()` property:

```
# partition the CSV data by the 'release_year' column
(df.write.format('csv')
    .option('header', 'true')
    .option('delimiter', ',')
    .mode('overwrite')
    .partitionBy('release_year')
    .save("../data/data_lake/netflix_csv_data_partitioned"))
```

See also

- *JSON Files –*Spark 3.4.0 documentation on reading and writing data: `https://spark.apache.org/docs/latest/sql-data-sources-json.html`

- *PySpark Overview* – PySpark documentation: `https://spark.apache.org/docs/latest/api/python/index.html`

- *Apache Parquet documentation*: `https://parquet.apache.org/docs/`

2

Data Transformation and Data Manipulation with Apache Spark

Apache Spark is a powerful distributed computing framework that can handle large-scale data processing tasks. One of the most common tasks when working with data is loading it from various sources and writing it into various formats. In this hands-on chapter, you will gain a comprehensive understanding of how to transform and manipulate data using Apache Spark.

In this chapter, we're going to cover the following main recipes:

- Applying basic transformations to data with Apache Spark
- Filtering data with Apache Spark
- Performing joins with Apache Spark
- Performing aggregations with Apache Spark
- Using window functions with Apache Spark
- Writing custom UDFs in Apache Spark
- Handling null values with Apache Spark

By the end of this chapter, you will have learned how to use Apache Spark to perform various data manipulation tasks such as applying basic transformations, filtering your data, and using window functions to perform time-based or rank-based calculations on your data. Additionally, you will learn how to write custom **user-defined functions** (**UDFs**) to apply your own logic to your data.

Technical requirements

Before starting, make sure that your docker-compose images are up and running and open JupyterLab's server running on localhost (http://127.0.0.1:8888/lab). Also, ensure that you have cloned the Git repo for this book and have access to the notebook and data used in this chapter.

Remember to stop all services defined in the docker-compose file for this book when you are done running the code examples. You can do this by executing the following command:

```
$ docker-compose stop
```

You can find the notebooks for this chapter at https://github.com/PacktPublishing/Data-Engineering-with-Databricks-Cookbook/tree/main/Chapter02.

Applying basic transformations to data with Apache Spark

In this recipe, we will discuss the basics of Apache Spark. We will use Python as our primary programming language and the PySpark API to perform basic transformations on a dataset of Nobel Prize winners.

How to do it...

1. **Import the libraries**: Import the required libraries and create a SparkSession object:

    ```
    from pyspark.sql import SparkSession
    from pyspark.sql.functions import transform, col, concat, lit

    spark = (SparkSession.builder
        .appName("basic-transformations")
        .master("spark://spark-master:7077")
        .config("spark.executor.memory", "512m")
        .getOrCreate())

    spark.sparkContext.setLogLevel("ERROR")
    ```

2. **Read file**: Read the nobel_prizes.json file using the read method of SparkSession:

    ```
    df = (spark.read.format("json")
        .option("multiLine", "true")
        .load("../data/nobel_prizes.json"))
    ```

3. **Apply transformations to array columns**: In Apache Spark, the transform function can be used on array columns to apply a user-defined function to each element of the array and return a new array column. The transform function takes two arguments: the array column to transform and the user-defined function to apply to each element of the array.

In the previous example, we have a DataFrame where the `laureates` column is an array column. We use the `transform` function to apply a transformation function to each element of the `laureates` column and create a new array column, `laureates_full_name`, which contains the result of applying the concatenated full name function to each element of the original `laureates` column.

Here's an example of how to use the `transform` function to apply a user-defined function to each element of an array column in Apache Spark:

```
df_transformed = (
    df.select("category",
        "overallMotivation",
        "year",
        "laureates",
        transform(col("laureates"),
            lambda x: concat(x.firstname,lit(" "), x.surname))
        .alias("laureates_full_name")))

df_transformed.show()
```

> **Note**
>
> The user-defined function passed to the `transform` function must take a single argument of the same type as the elements in the array column. In our example, the function takes an object argument because the `laureates` column contains object values. Also, if you have multiple array columns in your DataFrame, you can use the `arrays_zip` function to combine multiple array columns into a single array column before applying the `transform` function.

4. **Drop duplicates from the DataFrame**: In Apache Spark, you can use the `dropDuplicates` function to remove duplicates from a DataFrame based on one or more columns.

    In this example, we have a DataFrame with three columns: `category`, `overallMotivation`, and `year`. The resulting DataFrame, `df_deduped`, has one row for each unique combination of values in the Name and Age columns:

    ```
    df_deduped = df.dropDuplicates(["category","overallMotivation",
        "year"])
    df_deduped.show()
    ```

5. **Sort DataFrame by applying the orderBy function**: In Apache Spark, you can sort a DataFrame using the `orderBy` or `sort` functions. These functions allow you to specify one or more columns to sort by and the sorting order (ascending or descending).

In our example, we use the `orderBy` function to sort the DataFrame by the `year` column in ascending order. The resulting DataFrame, `df_sorted`, has rows sorted by age in ascending order:

```
df_sorted = df.orderBy("year")
df_sorted.show()
```

You can also sort multiple columns by passing a list of column names to the `orderBy` function. In the example, we use the `orderBy` function to sort the DataFrame by the `year` column in descending order, then by the `category` column in ascending order. The resulting DataFrame, `df_sorted`, has rows sorted by age in ascending order, and for rows with the same year, the rows are sorted by name in ascending order:

```
# Sort by year in descending order, then by category in
ascending order
df_sorted = df.orderBy(["year", "category"], ascending=[False,
True])
df_sorted.show()
```

Alternatively, you can use the `sort` function to achieve the same result. In this example, we use the `sort` function to sort the DataFrame by the Age column in ascending order, then by the Name column in descending order. The resulting DataFrame, `df_sorted`, has rows sorted by age in ascending order, and for rows with the same age, the rows are sorted by name in descending order:

```
# Sort by Age in ascending order, then by Name in descending
order
df_sorted = df.sort(["year", "category"], ascending=[False,
True])
df_sorted.show()
```

> **Note**
>
> Both the `orderBy` and `sort` functions return a new sorted DataFrame, and the original DataFrame remains unchanged.

6. **Rename columns**: In Apache Spark, you can rename columns of a DataFrame using the `withColumnRenamed` function. This function allows you to specify the current name of a column and the new name that you want to give it.

   In the example, we use the `withColumnRenamed` function to rename the Age column to Years. The resulting DataFrame, `df_renamed`, has the same data as the original DataFrame, but with the Age column renamed to Years.

Here's an example of how to use the `withColumnRenamed` function to rename columns of a DataFrame:

```
df_renamed = df.withColumnRenamed("category", "Topic")
df_renamed.show()
```

7. **You can also use the select function to rename multiple columns at once**: In this example, we use the `selectExpr` function to rename both columns. The resulting DataFrame, `df_renamed`, has the same data as the original DataFrame, but with the `Name` column renamed to `FirstName` and the `Age` column renamed to `Years`:

```
df_renamed = (
    df.selectExpr("category as Topic",
        "year as Year",
        "overallMotivation as Motivation"))
df_renamed.show(5)
```

The output will be as follows:

```
+----------+----+--------------------+
|     Topic|Year|          Motivation|
+----------+----+--------------------+
| chemistry|2022|                null|
| economics|2022|                null|
|literature|2022|                null|
|      peace|2022|                null|
|   physics|2022|                null|
+----------+----+--------------------+
```

8. **Stop the Spark session**: Finally, we need to stop the Spark session to release the resources used by Spark:

```
Spark.stop()
```

There's more…

Apache Spark provides many other transformation functions that can be used to process data. These include `filter`, `groupBy`, `reduceByKey`, `flatMap`, and many more. These functions can be used to perform complex data processing operations on large datasets and will be covered in the following recipes.

See also

- Apache Spark documentation: `https://spark.apache.org/docs/latest/`
- Apache Spark programming guide: `https://spark.apache.org/docs/latest/programming-guide.html`

Filtering data with Apache Spark

In this recipe, we will discuss how to filter data in Apache Spark. We will use Python as our primary programming language and the PySpark API to perform various filtering options with Apache Spark on datasets of Netflix titles and Nobel Prize winners.

How to do it...

1. **Import the libraries**: Import the required libraries and create a `SparkSession` object:

    ```
    from pyspark.sql import SparkSession
    from pyspark.sql.functions import array_contains, col, explode

    spark = (SparkSession
        .builder
        .appName("filter-data")
        .master("spark://spark-master:7077")
        .config("spark.executor.memory", "512m")
        .getOrCreate())

    spark.sparkContext.setLogLevel("ERROR")
    ```

2. **Read file**: Read the `netflix_titles.csv` file using the `read` method of `SparkSession`:

    ```
    df = (spark.read.format("csv")
        .option("header", "true")
        .option("nullValue", "null")
        .option("dateFormat", "LLLL d, y")
        .load("../data/netflix_titles.csv"))
    ```

3. **Filter the DataFrame**: Now that we have created a DataFrame, we can filter the data based on certain conditions. In this example, we will filter the DataFrame to only include titles released after 2020. As you can see, the resulting DataFrame only contains the rows where `release_year` is greater than 2020:

    ```
    filtered_df = df.filter(col("release_year") > 2020)
    filtered_df.show()
    ```

4. **Combine multiple conditions**: We can also combine multiple conditions using the & (and) and | (or) operators. In this example, we will filter the DataFrame to only include titles from the United States and whose `release_year` is greater than 2020:

    ```
    filtered_df = (
        df.filter(
            (col("country") == "United States")
    ```

```
            & (col("release_year") > 2020)))
filtered_df.show()
```

5. **Filter based on a list of values**: We can also filter the DataFrame based on a list of values using the `isin` method. In this example, we will filter the DataFrame to only include titles from the US, the UK, and India:

```
filtered_df = (
    df.filter(
        col("country")
        .isin(["United States", "United Kingdom",   "India"])))
filtered_df.show()
```

The output is as follows:

show_id	type	title	director	cast	country	date_added	release_year	rating	duration	listed_in	description
s10	Movie	The Starling	Theodore Melfi	Melissa McCarthy,...	United States	September 24, 2021	2021	PG-13	104 min	Comedies, Dramas	A woman adjusting...
s16	TV Show	Dear White People	null	Logan Browning, B...	United States	September 22, 2021	2021	TV-MA	4 Seasons	TV Comedies, TV D...	"Students of colo...
s41	TV Show	He-Man and the Ma...	null	Yuri Lowenthal, K...	United States	September 16, 2021	2021	TV-Y7	1 Season	Kids' TV, TV Sci-...	Mighty teen Adam ...
s56	TV Show	Nailed It	null	Nicole Byer, Jacq...	United States	September 15, 2021	2021	TV-PG	6 Seasons	Reality TV	Home bakers with ...
s82	Movie	Kate	Cedric Nicolas-Tr...	Mary Elizabeth Wi...	United States	September 10, 2021	2021	R	106 min	Action & Adventure	Slipped a fatal p...
s83	TV Show	Lucifer	null	Tom Ellis, Lauren...	United States	September 10, 2021	2021	TV-14	6 Seasons	Crime TV Shows, T...	Bored with being ...
s98	TV Show	Kid Cosmic	null	Jack Fisher, Tom ...	United States	September 7, 2021	2021	TV-Y7	2 Seasons	Kids' TV, TV Come...	A boy's superhero...
s102	Movie	Untold: Breaking ...	Chapman Way, Macl...	null	United States	September 7, 2021	2021	TV-MA	80 min	Documentaries, Sp...	Under pressure to...
s107	TV Show	Bunk'd	null	Peyton List, Kara...	United States	September 5, 2021	2021	TV-G	5 Seasons	Kids' TV, TV Come...	"The Ross sibling...
s126	TV Show	Q-Force	null	Sean Hayes, Wanda...	United States	September 2, 2021	2021	TV-MA	1 Season	Crime TV Shows, T...	A gay superspy an...
s185	TV Show	Sparking Joy	null	Marie Kondo	United States	August 31, 2021	2021	TV-PG	1 Season	Reality TV	In this reality s...
s221	TV Show	Family Reunion	null	Loretta Devine, T...	United States	August 26, 2021	2021	TV-PG	4 Seasons	Kids' TV, TV Come...	When the McKellan...
s232	Movie	The Water Man	David Oyelowo	David Oyelowo, Ro...	United States	August 25, 2021	2021	PG	92 min	Children & Family...	Desperate to save...
s242	TV Show	Manifest	null	Melissa Roxburgh,...	United States	August 21, 2021	2021	TV-14	3 Seasons	TV Dramas, TV Mys...	When a plane myst...
s248	Movie	Sweet Girl	Brian Andrew Mendoza	Jason Momoa, Isab...	United States	August 20, 2021	2021	R	110 min	Action & Adventur...	He lost the love ...
s255	TV Show	Go! Go! Cory Carson	null	Alan C. Lim, Paul...	United States	August 18, 2021	2021	TV-Y	5 Seasons	Kids' TV	Beep, beep – go, ...
s272	TV Show	Brand New Cherry ...	null	Rosa Salazar, Eri...	United States	August 13, 2021	2021	TV-MA	1 Season	TV Dramas, TV Hor...	A filmmaker heads...
s273	TV Show	Fast & Furious Sp...	null	Tyler Posey, Char...	United States	August 13, 2021	2021	TV-Y7	5 Seasons	Kids' TV	A government agen...
s275	TV Show	Grace and Frankie	null	Jane Fonda, Lily ...	United States	August 13, 2021	2021	TV-MA	7 Seasons	TV Comedies, TV D...	They're not frien...
s281	TV Show	Bake Squad	null	null	United States	August 11, 2021	2021	TV-PG	1 Season	Reality TV	Expert bakers ele...

only showing top 20 rows

Figure 2.1 – Output of records filtered by country

6. **Stop the Spark session**: Finally, we need to stop the Spark session to release the resources used by Spark:

```
spark.stop()
```

There's more...

In this section, we will be looking at how the filter method works with string columns, data range conditions, arrays, and maps.

Filtering on string

To filter on string substrings in Apache Spark, you can use the `like` or `rlike` functions. The `like` function allows you to filter based on a pattern match using wildcard characters, while the `rlike` function allows you to filter based on a regular expression match.

Here's an example of how to use the `like` function to filter a DataFrame based on a substring match. In this example, we use the `like` function with the `col` function to filter the DataFrame based on a substring match for `Crime` in the `listed_in` column. The `%` character is used as a wildcard to match any number of characters before or after the `app` substring:

```
# filter the DataFrame based on a substring match
filtered_df = df.filter(col("listed_in").like("%Crime%"))

# display the filtered DataFrame
filtered_df.show()
```

Next, let's see an example of how to use the `rlike` function to filter a DataFrame based on a regular expression match. In this example, we use the `rlike` function with the `col` function to filter the DataFrame based on a regular expression match for either `Crime` or `Thrillers` in the `listed_in` column using the `(Crime|Thrillers)` regular expression pattern:

```
# filter the DataFrame based on a regular expression match
filtered_df = df.filter(col("listed_in").rlike("(Crime|Thrillers)"))

# display the filtered DataFrame
filtered_df.show()
```

Filtering on data ranges

To filter on data ranges with Apache Spark, you can use comparison operators such as >, >=, <, <=, or `between` to compare column values against specific ranges.

Here's an example of how to filter a DataFrame based on a range of dates. In this example, we will use the >= and <= comparison operators with the `col` function to filter the DataFrame based on whether the `date_added` column falls within the range from `2021-02-01` to `2021-03-01`:

```
# filter the DataFrame based on a date range
filtered_df = df.filter((col("date_added") >= "2021-02-01") &
(col("date_added") <= "2021-03-01"))

# display the filtered DataFrame
filtered_df.show()
```

Alternatively, we can use the `between` function with the `col` function to filter the DataFrame based on whether the `date_added` column falls within the range from `2021-02-01` to `2021-03-01`:

```
# filter the DataFrame based on a date range
filtered_df = df.filter((col("date_added").
between("2021-02-01","2021-03-01")))
```

```
# display the filtered DataFrame
filtered_df.show()
```

Filtering on arrays

To filter on array columns in Apache Spark, you can use the `array_contains` function or use the `explode` function to convert the array column into multiple rows and then apply filter operations.

Here's an example of how to use the `array_contains` function to filter a DataFrame based on a value in an array column. In this example, we use the `array_contains` function with the `col` function to filter the DataFrame based on whether the `RecipeIngredientParts` array contains the value `apple`:

```
from pyspark.sql.functions import array_contains, col

# Read parquet file into a DataFrame
df_recipes = (spark.read
    .format("parquet")
    .load("../data/recipes.parquet"))

# filter the DataFrame based on a value in the array column
filtered_df = df_recipes.filter(array_
contains(col("RecipeIngredientParts"),
    "apple"))

# display the filtered DataFrame
filtered_df.show()
```

Filtering on map columns

To filter on map columns in Apache Spark, you can use the `getItem` method to access specific keys and values in the map. Here's an example.

Suppose you have a DataFrame with a column called `laureates` that contains a map. To filter the DataFrame based on the values of the map, you can use the `getItem` method. In this example, we use it to access the value associated with the `firstname` and `surname` keys in the `laureates` column. We then filter the DataFrame based on whether the value is equal to `Albert` for first name and `Einstein` for surname. The resulting filtered DataFrame contains only one row from the original DataFrame:

```
# Read JSON file into a DataFrame
df_nobel_prizes = (spark.read
    .format("json")
    .option("multiLine", "true")
```

```
    .load("../data/nobel_prizes.json"))

df_nobel_prizes_exploded = (
    df_nobel_prizes
    .withColumn("laureates",explode(col("laureates")))
    .select(col("category"),
        col("year"),
        col("overallMotivation"),
        col("laureates")))

filtered_df = (
    df_nobel_prizes_exploded
    .filter(
        (col("laureates").getItem("firstname") == "Albert")
        & (col("laureates").getItem("surname") == "Einstein")))

filtered_df.show(truncate=False)
```

See also

- Apache Spark documentation: https://spark.apache.org/docs/latest/programming-guide.html

- PySpark documentation: https://spark.apache.org/docs/latest/api/python/index.html

Performing joins with Apache Spark

In this recipe, we will discuss how to perform joins between DataFrames in Apache Spark. We will use Python as our primary programming language and the PySpark API and go over the various types of joins we can perform in Apache Spark.

How to do it...

1. **Import the libraries**: Import the required libraries and create a SparkSession object:

```
from pyspark.sql import SparkSession
from pyspark.sql.functions import broadcast

spark = (SparkSession.builder
    .appName("perform-joins")
    .master("spark://spark-master:7077")
```

```
        .config("spark.executor.memory", "512m")
        .getOrCreate())

    spark.sparkContext.setLogLevel("ERROR")
```

2. **Create the DataFrame**: We will start by creating the datasets for cars, customers, transactions, and fraud:

```
cards_df = (spark.read.format("csv")
    .option("header", "true")
    .option("nullValue", "null")
    .load("../data/Credit Card/CardBase.csv"))

customers_df = (spark.read.format("csv")
    .option("header", "true")
    .option("nullValue", "null")
    .load("../data/Credit Card/CustomerBase.csv"))

transactions_df = (spark.read.format("csv")
    .option("header", "true")
    .option("nullValue", "null")
    .load("../data/Credit Card/TransactionBase.csv"))

fraud_df = (spark.read.format("csv")
    .option("header", "true")
    .option("nullValue", "null")
    .load("../data/Credit Card/FraudBase.csv"))
```

3. **Perform an inner join**: In the inner join example, only the rows that have matching values in both datasets are returned. We will be joining the cards dataset to the customer data since all cards being associated with a customer performing an inner join is the right approach. To perform an inner join, we will call the `join()` function with the two datasets and specify the join type as `inner`:

```
customer_cards_df = (
    cards_df.join(customers_df.
    on='Cust_ID',
    how='inner'))
customer_cards_df.show()
```

4. **Perform a left outer join**: In the left outer join example, rows in the left datasets are kept in the datasets regardless of their matching with the right dataset. We will be joining the transactions dataset to the frauds dataset. Since not all transactions are frauds, performing a left outer join is the right approach. To perform a left outer join, we will call the `join()` function with the two datasets and specify the join type as `left_outer`:

```
joined_transactions_df = (
    transactions_df.join(fraud_df,
        on='Transaction_ID',
        how='left_outer'))
joined_transactions_df.show()
```

5. **Perform joins with complex conditions**: Sometimes, we may need to perform joins on complex conditions that involve multiple columns and operators. In such cases, it can be useful to break down the condition into simpler parts and use the & and | operators to combine them.

   In this example, we want to identify which customers have had a fraudulent transaction. To perform this, we will apply an inner join between the `Customer_cards_df` and `joined_transactions_df` DataFrames. The joins condition needs to compare the `Card_Number` column in `Customer_cards_df` to the `Credit_Card_ID` column in `joined_transactions_df` (note the column name mismatch). We also need to apply an `isNotNull` condition on the `Fraud_Flag` column in `joined_transactions_df`. We do this by passing a join expression to the `join()` method as shown:

```
joinExpr = (
    (customer_cards_df["Card_Number"] == joined_transactions_
df["Credit_Card_ID"])
    & (joined_transactions_df["Fraud_Flag"].isNotNull()))

customer_with_fraud_df = (
    Customer_cards_df.join(joined_transactions_df,
        on=joinExpr,
        how='inner'))
customer_with_fraud_df.show()
```

6. **Stop the Spark session**: Finally, we need to stop the Spark session to release the resources used by Spark:

```
spark.stop()
```

There's more...

In addition to the inner join and left outer join, Apache Spark also supports other types of joins. We will explore some of these in this section. We will also look at broadcast joins. Broadcast joins are useful in Apache Spark for improving join performance.

Right outer join

To perform a right outer join, we will call the `join()` function with the two datasets and specify the join type as `right_outer`. In the right outer join example, all the rows from the right dataset and matching rows from the left dataset are returned. If there are no matching rows in the left dataset, then the values in the left columns will be null:

```
right_join = df1.join(df2, on='Name', how='right_outer')
right_join.show()
```

Full outer join

To perform a full outer join, we will call the `join()` function with the two datasets and specify the join type as `outer`. In the full outer join example, all the rows from both datasets are returned. If there are no matching rows in either dataset, then the values in the corresponding columns will be null:

```
full_join = df1.join(df2, on='Name', how='outer')
full_join.show()
```

Cross join

To perform a cross join, we will call the `crossJoin()` function with the two datasets. In the cross join example, all the rows from both datasets are combined in all possible ways. This is also known as a Cartesian product:

```
cross_join = df1.crossJoin(df2)
cross_join.show()
```

Broadcast join

When joining large datasets, Spark may need to shuffle data across the network, which can be slow and resource-intensive. However, if one of the datasets is small enough to fit in memory, we can use the broadcast join to speed up the join operation and avoid shuffling. In a broadcast join, Spark broadcasts the small dataset to all worker nodes, which can then be used to perform the join locally. To use the broadcast join, we can call the `broadcast()` function on the smaller dataset.

In this example, we perform a broadcast join on the Name and Gender columns. Since the df2 dataset is small enough to fit in memory, we broadcast it to all the worker nodes using the `broadcast()` function. The resulting DataFrame has only the rows that have matching values in both datasets:

```
broadcast_join = df1.join(broadcast(df2), ["Name", "Gender"], "inner")
broadcast_join.show()
```

Multiple join conditions

If we want to join datasets on multiple conditions, we can specify them as a list of column names using the `on()` method.

In this example, we perform an inner join on the Name, Gender, and Age columns. The resulting DataFrame has only the rows that have matching values in both datasets for all three columns:

```
multi_join = df1.join(df2, on=['Name', 'Gender', 'Age'], how='inner')
multi_join.show()
```

See also

- *pyspark.sql.DataFrame.join* – The Spark documentation on joins: https://spark.apache.
  org/docs/3.1.2/api/python/reference/api/pyspark.sql.DataFrame.
  join.html

Performing aggregations with Apache Spark

In this recipe, we will discuss how to perform aggregations on DataFrames in Apache Spark. We will use Python as our primary programming language and the PySpark API and go over the various techniques for aggregating your data in Apache Spark.

How to do it...

1. **Import the libraries**: Import the required libraries and create a SparkSession object:

    ```
    from pyspark.sql import SparkSession
    from pyspark.sql.functions import col, max, count, min, approx_
    count_distinct
    from pyspark.sql.types import StructType, StructField,
    StringType, IntegerType, DateType

    spark = (SparkSession.builder
        .appName("perform-aagregations")
        .master("spark://spark-master:7077")
        .config("spark.executor.memory", "512m")
        .getOrCreate())

    spark.sparkContext.setLogLevel("ERROR")
    ```

2. **Read file**: Read the netfix_titles.csv file using the read method of SparkSession:

    ```
    df = (spark.read.format("csv")
        .option("header", "true")
        .option("nullValue", "null")
        .option("dateFormat", "LLLL d, y")
        .load("../data/netflix_titles.csv"))
    ```

3. **Perform aggregations**: In Apache Spark, the `groupBy` operation is used to group data in a distributed collection (DataFrame) based on one or more keys. It allows you to perform aggregations or transformations on the grouped data. The `groupBy` operation requires specifying one or more keys based on which the grouping will occur. These keys can be column names in a DataFrame. Here's an example using a DataFrame:

    ```
    # Group the data by a column
    grouped_df = df.groupBy("country")
    ```

> **Note**
>
> The `groupBy` operation in Spark is a transformation, which means it is lazily evaluated. It builds a logical plan but doesn't execute until an action operation (such as collect, show, or write) is triggered. Also, the `groupBy` function will return a `GroupedData` object and not a DataFrame.

4. **Perform aggregations or transformations**: After the `groupBy` operation, you can apply various aggregation or transformation functions to the grouped data. These functions include `count`, `sum`, `avg`, `min`, `max`, and so on. You can chain multiple operations together.

    Here's an example that calculates the count of rows for each group:

    ```
    # Count the number of rows in each group
    count_df = grouped_df.count()
    count_df.show()
    ```

    You can also apply custom aggregation functions using the `agg` function:

    ```
    # Apply custom aggregation using max
    max_release_df = grouped_df.agg(max(col("date_added")))
    max_release_df.show()
    ```

5. **Apply multiple aggregations**: In this example, we will calculate the total sales amount for each product. To do this, we will use the `groupBy` and `agg` methods of the DataFrame API. Here, we use the `groupBy` method to group the data by country. Then, we use the `agg` method to perform an aggregation to get the count of the `date_added` column using the `max` and `min` functions:

    ```
    release_date_gouped_df = (
        df.groupBy("country")
        .agg(
            count("show_id").alias("NumberOfReleases"),
            max("date_added").alias("LastReleaseDate"),
            min("date_added").alias("FirstReleaseDate")))
    release_date_gouped_df.show(3)
    ```

The output is as follows:

```
+----------------+----------------+-------------------+-------------------+
|         country|NumberOfReleases|     LastReleaseDate|    FirstReleaseDate|
+----------------+----------------+-------------------+-------------------+
|            null|             831|  September 9, 2021|  December 14, 2018|
|  Ama K. Abebrese|               1| Kobina Amissah Sam| Kobina Amissah Sam|
|      Aziz Ansari|               1|        Carla Gallo|        Carla Gallo|
+----------------+----------------+-------------------+-------------------+
only showing top 3 rows
```

Figure 2.2 – DataFrame output after multiple aggregates

6. **Stop the Spark session**: Finally, we need to stop the Spark session to release the resources used by Spark:

```
spark.stop()
```

There's more...

Aggregations are a powerful tool for summarizing data in Spark. In addition to the sum function used in this example, Spark provides many other aggregation functions, such as avg, min, max, count, and more. You can use these functions in combination with the groupBy method to perform complex aggregations on your data.

In addition to the aggregation functions provided by Spark, there are several other techniques you can use to summarize your data. Here are a few examples.

Pivot tables

Pivot tables allow you to summarize your data by grouping it along multiple dimensions.

In Apache Spark, you can use the pivot function along with groupBy to transform a DataFrame by converting the values in one column into multiple columns. This operation is particularly useful when you want to perform a cross-tabulation or when you want to reshape your data.

The pivot function takes three parameters: the pivot column, the values column, and the optional list of distinct values that will become the new columns.

In this example, we are grouping the DataFrame by the country column. Then, we apply the pivot function to the type column. The distinct values in the type column will become the new columns in the resulting DataFrame. Finally, we count the number of records using the count function.

The resulting DataFrame will have one row per distinct value in the `country` column, and each type will have its own column with the count as values. Any missing values will be represented as null in the resulting DataFrame:

```
pivot_table = df.groupBy("country").pivot("type").agg(count("show_
id"))
pivot_table.show()
```

> **Note**
>
> The pivot operation requires the shuffling of data, so it can be an expensive operation, especially on large datasets. Additionally, the number of distinct values in the pivot column should be reasonably small, as the resulting DataFrame will have one column for each distinct value.

Approximate aggregations

Approximate functions in Apache Spark provide a way to efficiently calculate aggregations on large datasets with reduced computational cost. They are designed to trade off accuracy for performance when the exact result is not strictly required. Here are a few reasons why you might choose to use approximate functions in Apache Spark:

- **Performance optimization**: Approximate aggregations can significantly improve query performance and reduce computational resources compared to exact aggregations, especially when dealing with large datasets. By sacrificing a small amount of accuracy, you can achieve substantial gains in query execution time.

- **Scalability**: Approximate functions are particularly useful when working with massive datasets that cannot fit into memory. Instead of processing the entire dataset, approximate aggregations can estimate results based on statistical techniques and a smaller subset of the data. This enables you to scale your computations to larger datasets without overwhelming your system's resources.

- **Trade-off between accuracy and cost**: Approximate functions allow you to control the trade-off between accuracy and computational cost. You can adjust the parameters such as **relative standard deviation** (**RSD**) to determine the level of approximation error you are willing to accept. This flexibility enables you to find the right balance for your specific use case.

It's important to note that the choice between approximate and exact aggregations depends on your specific requirements. If you need precise results, exact aggregations should be used. However, when approximate results are acceptable and performance is a priority, approximate functions can be a valuable tool in your Spark workflows.

Apache Spark provides several approximate functions that you can use for different types of aggregations. Here is a list of some commonly used approximate functions in Apache Spark:

- `approxQuantile()`: Computes the approximate quantiles of a numeric column in a DataFrame. It returns an array of approximate quantile values at the specified probabilities.

  To calculate approximate quantiles, we use the `approxQuantile()` function and pass the column name, the list of desired quantiles (0.25, 0.5, and 0.75 in this case), and the RSD value. The RSD determines the maximum approximation error allowed. In this example, we set it to `0.1`.

- `approxCountDistinct()`: Computes the approximate count of distinct values in a column. It takes a column as input and returns an approximate count of distinct values based on the provided RSD value.

  To calculate the approximate distinct count, we use the `approxCountDistinct()` function and pass the column name along with the desired RSD value.

Here's an example code snippet demonstrating the usage of these functions:

```
# Define a Schema
schema = StructType([
    StructField("Id", IntegerType(), True),
    StructField("ProductId", StringType(), True),
    StructField("UserId", StringType(), True),
    StructField("ProfileName", StringType(), True),
    StructField("HelpfulnessNumerator", StringType(), True),
    StructField("HelpfulnessDenominator", StringType(), True),
    StructField("Score", IntegerType(), True),
    StructField("Time", StringType(), True),
    StructField("Summary", StringType(), True),
    StructField("Text", StringType(), True)])

review_df = (spark.read.format("csv")
    .option("header",True)
    .schema(schema)
    .load("../data/Reviews.csv"))

# Approximate quantile calculation
quantiles = review_df.approxQuantile("Score", [0.25, 0.5, 0.75], 0.1)
print("Approximate Quantiles:", quantiles)

# Approximate distinct count calculation
approx_distinct_count = review_df.select(
    approx_count_distinct("ProductId",
        rsd=0.1).alias("approx_distinct_count"))
approx_distinct_count.show()
```

> **Note**
>
> Approximate aggregations provide faster results compared to exact aggregations but with a controlled amount of approximation error. The RSD value determines the trade-off between accuracy and performance. Adjust the RSD value based on your specific requirements.

See also

- Apache Spark documentation on aggregation: `https://spark.apache.org/docs/latest/sql-getting-started.html#aggregations`
- PySpark API documentation on DataFrames: `https://spark.apache.org/docs/latest/api/python/reference/pyspark.sql/api/pyspark.sql.DataFrame.html`

Using window functions with Apache Spark

In this recipe, we will discuss how to apply window functions to DataFrames in Apache Spark. We will use Python as our primary programming language and the PySpark API.

How to do it...

1. **Import the libraries**: Import the required libraries and create a `SparkSession` object:

    ```
    from pyspark.sql import SparkSession
    from pyspark.sql.functions import col, row_number, lead, lag,
    count, avg

    spark = (SparkSession.builder
        .appName("apply-window-functions")
        .master("spark://spark-master:7077")
        .config("spark.executor.memory", "512m")
        .getOrCreate())

    spark.sparkContext.setLogLevel("ERROR")
    ```

2. **Read file**: Read the `netflix_titles.csv` file using the `read` method of `SparkSession`:

    ```
    df = (spark.read
        .format("csv")
        .option("header", "true")
        .option("nullValue", "null")
        .option("dateFormat", "LLLL d, y")
        .load("../data/netflix_titles.csv"))
    ```

3.   **Define a window specification**: Before using any window function, we need to define a window specification that specifies how to partition and order the data for computation.

This window specification partitions the data by the `country` column and orders each partition by the `date_added` column in ascending order:

```
from pyspark.sql.window import Window
window_spec = Window.partitionBy("country").orderBy("date_
added")
```

4.   **Use the row_number window function**: The `row_number` window function in Apache Spark assigns a unique sequential integer to each row within a partition of a DataFrame or dataset. It is often used for tasks such as generating row identifiers or creating rankings based on certain criteria.

Here's an overview of how `row_number` works:

*   **Partitioning**: The data is divided into partitions based on a specified partitioning scheme. Each partition is processed independently, and row numbers are assigned within each partition.

*   **Ordering**: Within each partition, the rows are ordered based on one or more columns specified in the `ORDER BY` clause. The ordering determines the sequence in which the row numbers are assigned.

*   **Row numbering**: The `row_number` function is applied to each row in the partition, assigning a unique integer value. The first row in each partition receives the number 1, the second row receives 2, and so on.

In the example, `df` is the DataFrame on which we want to apply the `row_number` function with the window specification defined in the previous step. The resulting DataFrame will contain an additional column called `row_number` that holds the assigned row numbers. Each row number represents the position of the row within its partition:

```
# Assign row numbers within each partition
result = df.withColumn("row_number", row_number().over(window_
spec))
result.select("title","country","date_added","row_number").
show()
```

> **Note**
>
> `row_number` is a non-deterministic function, meaning that the order of rows within a partition may vary across different executions of the same query. If you require a stable order, you can use other functions such as `rank` or `dense_rank` instead.

5.   **Use the lead and lag window functions**: The `lead` and `lag` window functions in Apache Spark allow you to access the values of subsequent or previous rows within a partition. They are useful for tasks such as computing the difference between consecutive values or comparing values across adjacent rows.

The `lead` function retrieves the value of a column from the next row within the partition. It takes two arguments: the column to retrieve and the number of rows to look ahead. If the specified number of rows exceeds the partition boundary, it returns `null`.

The `lag` function retrieves the value of a column from the previous row within the partition. It also takes two arguments: the column to retrieve and the number of rows to look back. If the specified number of rows exceeds the partition boundary, it returns `null`.

In this example, we use `lead("date_added").over(windowSpec)` to add a new `lead_date_added` column that retrieves the value of the `date_added` column from the next row within each partition. Similarly, we use `lag("date_added").over(windowSpec)` to add a new `lag_date_added` column that retrieves the value of the `value` column from the previous row within each partition.

The output will show the original `category` and `value` columns along with the `lead_value` column containing the value from the next row and the `lag_value` column containing the value from the previous row within each partition:

```
# Add lead column
df = df.withColumn("lead_date_added", lead("date_added").
over(window_spec))
# Add lag column
df = df.withColumn("lag_date_added", lag("date_added").
over(window_spec))

df.select("title","country","date_added","lead_date_added","lag_
date_added").show(3)
```

The output is as follows:

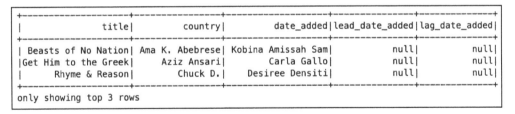

```
+------------------+-----------------+------------------+---------------+--------------+
|             title|          country|        date_added|lead_date_added|lag_date_added|
+------------------+-----------------+------------------+---------------+--------------+
| Beasts of No Nation| Ama K. Abebrese| Kobina Amissah Sam|           null|          null|
|Get Him to the Greek|     Aziz Ansari|        Carla Gallo|           null|          null|
|       Rhyme & Reason|        Chuck D.|    Desiree Densiti|           null|          null|
+------------------+-----------------+------------------+---------------+--------------+
only showing top 3 rows
```

Figure 2.3 – Output from lead and lag window functions

6. **Close the SparkSession**: After we have finished working with Spark, we need to close `SparkSession` to release the resources:

```
spark.stop()
```

There's more...

Spark supports many more window functions, such as `rank`, `dense_rank`, `percent_rank`, `cume_dist`, `first_value`, `last_value`, and `nth_value`. In this section, we will look at how we can nest window functions and window frames.

Nested window functions

PySpark allows us to nest window functions; that is, we can use the output of one window function as the input for another. For example, we can calculate the running total of a column using the `sum` window function and then use the lead window function to calculate the difference between the current row and the next row. Here's an example:

```
from pyspark.sql.functions import sum, lead
from pyspark.sql.window import Window

window_spec = Window.partitionBy("country").orderBy("release_year")
df = df.withColumn("running_total", count("show_id").over(window_
spec))
df = df.withColumn("next_running_total", lead("running_total").
over(window_spec))
df = df.withColumn("diff", df["next_running_total"] - df["running_
total"])
(df.filter(df.next_running_total.isNotNull())
.select("show_id","country","release_year","running_total","next_
running_total","diff")
.show())
```

Window frames

Window functions allow you to perform calculations over a sliding window of data, rather than the entire dataset. This can be useful for computing running totals, moving averages, and more.

We can use window frames to specify a more specific range of rows within a window. The window frame specifies a start and end point within the window, and the aggregation function is only computed for the rows within this frame.

In this example, we have a DataFrame with two columns: `id` and `value`. We create a window specification using the `orderBy` function to order the rows by `id`. The `rowsBetween(-2, 0)` parameter specifies the range of rows to consider for the rolling average, which includes the current row and two preceding rows. We then calculate the rolling average using the `avg` function and the `over` method with the window specification. Finally, we add the rolling average as a new column to the DataFrame using the `withColumn` method:

```
data = [(1, 10), (2, 15), (3, 20), (4, 25), (5, 30)]
df = spark.createDataFrame(data, ["id", "value"])
```

```
windowSpec = Window.orderBy("id").rowsBetween(-2, 0)
df = df.withColumn("rolling_avg", avg(df["value"]).over(windowSpec))

df.show()
```

The output will be as follows:

```
+---+-----+-----------+
| id|value|rolling_avg|
+---+-----+-----------+
|  1|   10|       10.0|
|  2|   15|       12.5|
|  3|   20|       15.0|
|  4|   25|       20.0|
|  5|   30|       25.0|
+---+-----+-----------+
```

Writing custom UDFs in Apache Spark

In this recipe, we will discuss how to write custom **UDFs** in Apache Spark. Writing UDFs in Apache Spark provides the flexibility and expressiveness necessary to perform custom data transformations and computations. UDFs enable you to extend the capabilities of Spark's built-in functions, integrate with external libraries, and achieve specific data processing requirements in a scalable and distributed manner. We will use Python as our primary programming language and the PySpark API.

How to do it...

1. **Import the libraries**: Import the required libraries and create a `SparkSession` object:

```
from pyspark.sql import SparkSession
from pyspark.sql.functions import explode, col

spark = (SparkSession.builder
            .appName("write-udfs")
            .master("spark://spark-master:7077")
            .config("spark.executor.memory", "512m")
            .getOrCreate())

spark.sparkContext.setLogLevel("ERROR")
```

2. **Read file**: Read the `nobel_prizes.json` file using the `read` method of `SparkSession`:

```
df = (spark.read.format("json")
    .option("multiLine", "true")
    .load("../data/nobel_prizes.json"))

df_flattened = (
    df
    .withColumn("laureates",explode(col("laureates")))
    .select(col("category"),
        col("year"),
        col("overallMotivation"),
        col("laureates.id"),
        col("laureates.firstname"),
        col("laureates.surname"),
        col("laureates.share"),
        col("laureates.motivation"))
    .filter(col("laureates.firstname").isNotNull() &
col("laureates.surname").isNotNull()))
```

3. **Define a Python function that will be used as a UDF, which extends the functionality of Spark's built-in functions**: You can define a UDF in Python as a regular Python function. In this example, we define a function named `concat` that takes two parameters, `first_name` and `last_name`, and returns their concatenation:

```
def concat(first_name, last_name):
    return first_name + " " + last_name
```

4. **Register the function as a UDF using the udf function from pyspark.sql.functions**: Once you have defined the UDF, you need to register it with Spark using the `udf` function from `pyspark.sql.functions`. The `udf` function takes the UDF as an argument and returns a UDF object that can be used to apply the UDF on a DataFrame:

```
from pyspark.sql.functions import udf
concat_udf = udf(concat)
```

5. **Set return UDF type:** You can also specify the return type of the UDF explicitly by passing the return type as the second argument to the `udf` function:

```
from pyspark.sql.types import StringType
concat_udf = udf(concat, StringType())
```

6. **Apply UDF**: To apply the UDF on a DataFrame, you can use the `withColumn` method, which adds a new column to the DataFrame with the UDF's output. The `withColumn` method takes two arguments – the name of the new column and the UDF object:

```
df_flattened = df_flattened.withColumn("full_name",
    concat_udf(df_flattened["firstname"],
    df_flattened["surname"]))
```

7. **Finally, show the DataFrame to see the result**: You can use the `show` method to display the DataFrame. The `show` method takes an optional integer argument that specifies the number of rows to display:

```
df_flattened.show()
```

8. **Stop the Spark session**: Finally, we need to stop the Spark session to release the resources used by Spark:

```
spark.stop()
```

There's more...

In Apache Spark, you can register the UDF using the `spark.udf.register` function. This allows you to make the UDF accessible by name in Spark SQL queries. Here's an example of how to use the `spark.udf.register` function in Python.

In this example, we start by creating a `SparkSession` object using `SparkSession.builder.getOrCreate()`. Then, we define a UDF called `square_udf`, which squares the input. Next, we register the UDF using `spark.udf.register`. The first argument is the name we want to assign to the UDF (`square` in this case), the second argument is the function itself (`square_udf`), and the third argument is the return type of the UDF (`IntegerType()` in this case).

After registering the UDF, we create a DataFrame called `df` with a single column named `num`. We then create a temporary view called `numbers` using `createOrReplaceTempView`, which allows us to reference the DataFrame in SQL queries.

Finally, we use the registered UDF square in a SQL query to select the num column and apply the UDF to it. The result is stored in the resulting DataFrame and displayed using `show()`:

```
from pyspark.sql.functions import udf
from pyspark.sql.types import IntegerType

# Define a UDF
def square_udf(x):
    return x ** 2

# Register the UDF
```

```
spark.udf.register("square", square_udf, IntegerType())

# Create a DataFrame
df = spark.createDataFrame([(1,), (2,), (3,), (4,), (5,)], ["num"])

# Use the registered UDF in a SQL query
df.createOrReplaceTempView("numbers")
result = spark.sql("SELECT num, square(num) AS square_num FROM
numbers")

# Show the result
result.show()
```

See also

- *Functions* – Spark 3.4.0 documentation: `https://spark.apache.org/docs/3.4.0/sql-ref-functions.html#udfs-user-defined-functions`

- *Scalar UDFs* – Spark 3.4.0 documentation: `https://spark.apache.org/docs/3.4.0/sql-ref-functions-udf-scalar.html`

- *pyspark.sql.functions.pandas_udf* – PySpark 3.1.2 documentation: `https://spark.apache.org/docs/3.1.2/api/python/reference/api/pyspark.sql.functions.pandas_udf.html`

Handling null values with Apache Spark

Handling null values is an essential part of data processing in Apache Spark. Null values are missing or unknown values in a dataset that can affect the analysis and modeling process. Apache Spark provides multiple ways to handle null values to ensure data quality and data integrity. In this recipe, we will discuss how to handle null values in Apache Spark using Python.

How to do it...

1. **Import the libraries**: Import the required libraries and create a SparkSession object:

    ```
    from pyspark.sql import SparkSession
    from pyspark.sql.functions import explode, col, when

    spark = (SparkSession.builder
        .appName("handle-nulls")
        .master("spark://spark-master:7077")
        .config("spark.executor.memory", "512m")
    ```

```
        .getOrCreate())

    spark.sparkContext.setLogLevel("ERROR")
```

2. **Read file**: Read the nobel_prizes.json file using the read method of SparkSession:

```
df = (spark.read.format("json")
    .option("multiLine", "true")
    .load("../data/nobel_prizes.json"))

df_flattened = (
    df
    .withColumn("laureates",explode(col("laureates")))
    .select(col("category"),
        col("year"),
        col("overallMotivation"),
        col("laureates.id"),
        col("laureates.firstname"),
        col("laureates.surname"),
        col("laureates.share"),
        col("laureates.motivation")))
```

3. **Drop null values**: Dropping null values means removing rows that contain null values from the DataFrame. We can use the dropna() function to drop null values. The code will drop the rows that contain null values and display the resulting DataFrame on the console:

```
# Dropping rows with null values
df_dropna = df_flattened.dropna()

# Displaying the DataFrame after dropping null values
df_dropna.show()
```

4. **Fill null values**: Filling null values means replacing null values with a specific value or with the mean or median of the column. We can use the fillna() function to fill null values. The code will replace all null values with N/A and display the resulting DataFrame on the console:

```
# Filling null values with a specific value
df_fillna = df_flattened.fillna("N/A")

# Displaying the DataFrame after filling null values
df_fillna.show()
```

5. **Replace null values**: Replacing null values means replacing null values with a specific value based on some conditions. We can use the when() and otherwise() functions to replace null values. The code will replace null values in the age column with 0 and null values in the gender column with Unknown based on conditions specified and display the resulting DataFrame on the console:

```
# Replacing null values based on conditions
df_replace = (
    df_flattened.withColumn("category",
        when(col("category").isNull(),
        "").otherwise(col("category")))
    .withColumn("overallMotivation",
        when(col("overallMotivation").isNull(),
        "").otherwise(col("overallMotivation")))
    .withColumn("firstname", when(col("firstname").isNull(),
        "").otherwise(col("firstname")))
    .withColumn("surname", when(col("surname").isNull(),
        "").otherwise(col("surname")))
    .withColumn("year", when(col("year").isNull(), 9999).
otherwise(col("year"))))

# Displaying the DataFrame after replacing null values
df_replace.show()
```

6. **Stop the Spark session**: Finally, we need to stop the Spark session to release the resources used by Spark:

```
spark.stop()
```

There's more...

Apart from the methods mentioned in this guide, Apache Spark provides many more ways to handle null values, such as imputing null values using machine learning models. We can also use third-party libraries such as pandas and numpy to handle null values in Apache Spark.

Handling null values in UDFs

When working with UDFs, we may encounter null values in the input data. To handle these null values, we can use the isNull method to check for null values and na.fill to replace null values with default values. For example, suppose we have a UDF that takes a column as input and returns a new column with some transformation applied:

```
from pyspark.sql import SparkSession
from pyspark.sql.functions import udf
from pyspark.sql.types import StringType
```

```
# Sample DataFrame with null values
data = [("John", 25), ("Alice", None), ("Bob", 30)]
df = spark.createDataFrame(data, ["name", "age"])

# Define a UDF to handle null values
def process_name(name):
    if name is None:
        return "Unknown"
    else:
        return name.upper()

# Register the UDF
process_name_udf = udf(process_name, StringType())

# Apply the UDF to the DataFrame
df_with_processed_names = df.withColumn("processed_name", process_
name_udf(df["name"]))

# Show the resulting DataFrame
df_with_processed_names.show()
```

Handling null values in machine learning pipelines

When building machine learning models with Spark, we may encounter null values in the input data. To handle these null values, we can use the Imputer transformer in Spark's `ml.feature` module. The Imputer transformer replaces null values with the mean (default), median, or mode of the column. Here is an example:

```
from pyspark.sql import SparkSession
from pyspark.ml.feature import Imputer

# Create a sample DataFrame with missing values
data = [
    (1, 2.0),
    (2, None),
    (3, 5.0),
    (4, None),
    (5, 7.0)
]
df = spark.createDataFrame(data, ["id", "value"])

# Create an instance of Imputer and specify the input/output columns
imputer = Imputer(inputCols=["value"], outputCols=["imputed_value"])
```

```
# Fit the imputer to the data and transform the DataFrame
imputer_model = imputer.fit(df)
imputed_df = imputer_model.transform(df)

# Show the resulting DataFrame
imputed_df.show()
```

See also

- Apache Spark documentation on handling missing data: `https://spark.apache.org/docs/latest/sql-data-sources.html`

- pandas documentation on handling missing data: `https://pandas.pydata.org/docs/user_guide/missing_data.html`

- NumPy documentation on handling missing data: `https://numpy.org/doc/1.14/neps/missing-data.html`

3

Data Management with Delta Lake

Delta Lake is an open source storage layer that enables building a lakehouse architecture with various compute engines and APIs. It provides features such as **atomicity, consistency, isolation, and durability (ACID)** transactions, scalable metadata, time travel, schema evolution, and **data manipulation language (DML)** operations. It is compatible with Apache Spark and other query engines.

This chapter provides a comprehensive overview of how to manage and optimize Delta tables using Apache Spark. It covers topics such as creating Delta tables, querying and analyzing them, optimizing them for better performance and cost-effectiveness, managing table metadata, migrating data to Delta Lake, and versioning Delta tables using time travel and table versioning. Additionally, this chapter explains how to perform incremental loads of data into Delta tables, including deduplication, writing change data, and reading change data feeds.

In this chapter, we'll cover the following recipes:

- Creating a Delta Lake table
- Reading a Delta Lake table
- Updating data in a Delta Lake table
- Merging data into Delta tables
- Change data capture in Delta Lake
- Optimizing Delta Lake tables
- Versioning and time travel for Delta Lake tables
- Managing Delta Lake tables

By the end of this chapter, you will have a solid understanding of how to manage and optimize Delta Lake tables using Spark and Delta Lake APIs.

Technical requirements

Before starting, make sure that your `docker-compose` images are up and running, and open the JupyterLab server running on the localhost (`127.0.0.1:8888/lab`). Also, ensure that you have cloned the Git repo for this book and have access to the notebook and data used in this chapter.

Remember to stop all services defined in the `docker-compose` file for this book when you are done running the code examples. You can do this by executing this command:

```
$ docker-compose stop
```

You can find the notebooks and data for this chapter at `https://github.com/PacktPublishing/Data-Engineering-with-Databricks-Cookbook/tree/main/Chapter03`.

Creating a Delta Lake table

In this recipe, we will explore the process of creating a table in Delta Lake, a powerful open source data lake storage format. Delta Lake provides features such as transactional capabilities, schema evolution, time travel, and concurrency control, making it an ideal choice for scalable and reliable data management. We will guide you through the hands-on steps, explain the underlying concepts, and address common issues you may encounter along the way. This recipe will give you a clear understanding of how to create a Delta Lake table and leverage its advanced capabilities in your data workflows. Let's get started!

How to do it...

1. **Import the required libraries**: Start by importing the necessary libraries for working with Delta Lake. In this case, we need the `delta` module and the `SparkSession` class from the `pyspark.sql` module:

    ```
    from delta import configure_spark_with_delta_pip, DeltaTable
    from pyspark.sql import SparkSession
    ```

2. **Create a SparkSession object**: To interact with Spark and Delta Lake, you need to create a `SparkSession` object:

    ```
    builder = (SparkSession.builder
        .appName("create-delta-table")
        .master("spark://spark-master:7077")
        .config("spark.executor.memory", "512m")
        .config("spark.sql.extensions",
            "io.delta.sql.DeltaSparkSessionExtension")
    ```

```
    .config("spark.sql.catalog.spark_catalog", "org.apache.
spark.sql.delta.catalog.DeltaCatalog"))

spark = configure_spark_with_delta_pip(builder).getOrCreate()
spark.sparkContext.setLogLevel("ERROR")
```

3. **Create a Delta table**: Before we load data, we will need to create a table in the Delta catalog. This can be done with the CREATE OR REPLACE TABLE command. You can specify the table name, which is default.netflix_titles, and also specify the data format as DELTA and the location of the table data, which in our case is /opt/workspace/data/delta_lake/netflix_titles:

```
%%sparksql
CREATE OR REPLACE TABLE default.netflix_titles (
    show_id STRING,
    type STRING,
    title STRING,
    director STRING,
    cast STRING,
    country STRING,
    date_added STRING,
    release_year STRING,
    rating STRING,
    duration STRING,
    listed_in STRING,
    description STRING
) USING DELTA LOCATION '/opt/workspace/data/delta_lake/netflix_
titles';
```

4. **Read your data**: Once the table is created, you need data to populate it. We will create a DataFrame with CSV data using the read method of the spark object:

```
# Read CSV file into a DataFrame
df = (spark.read
    .format("csv")
    .option("header", "true")
    .load("../data/netflix_titles.csv"))
```

5. **Write the data to Delta Lake**: Once you have the data, you can write it to Delta Lake using the write method on the DataFrame. Specify the format as delta and choose a mode for writing the data. In this case, we use the overwrite mode to overwrite any existing table at the specified location in your storage system where you want to create the table. Make sure you have the necessary permissions to write data to that location.

The `format("delta")` method sets the format to Delta Lake. The `mode("overwrite")` method determines the behavior when writing the data. The `"overwrite"` mode replaces any existing table with the same name at the specified location. If you want to append the data or perform other actions, you can choose a different mode such as `append` or `ignore`. Finally, the `saveAsTable` method is used to write the data to the specified table:

```
df.write.format("delta").mode("overwrite").saveAsTable("default.
netflix_titles")
```

6. **Query the Delta table**: Now that we have loaded our table with data, we can simply start querying it with `sparksql`:

```
%%sparksql
SELECT * FROM default.netflix_titles LIMIT 3;
```

You will see the following output:

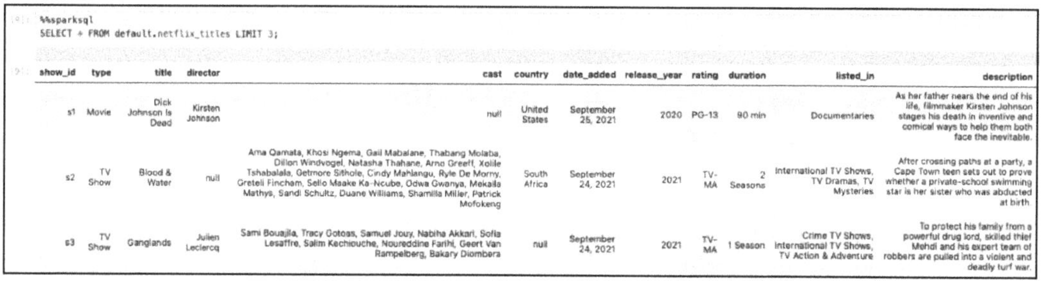

```
%%sparksql
SELECT * FROM default.netflix_titles LIMIT 3;
```

show_id	type	title	director	cast	country	date_added	release_year	rating	duration	listed_in	description
s1	Movie	Dick Johnson Is Dead	Kirsten Johnson	null	United States	September 25, 2021	2020	PG-13	90 min	Documentaries	As her father nears the end of his life, filmmaker Kirsten Johnson stages his death in inventive and comical ways to help them both face the inevitable.
s2	TV Show	Blood & Water	null	Ama Qamata, Khosi Ngema, Gail Mabalane, Thabang Molaba, Dillon Windvogel, Natasha Thahane, Arno Greeff, Xolile Tshabalala, Getmore Sithole, Cindy Mahlangu, Ryle De Morny, Greteli Fincham, Sello Maake Ka-Ncube, Odwa Gwanya, Mekaila Mathys, Sandi Schultz, Duane Williams, Shamilla Miller, Patrick Mofokeng	South Africa	September 24, 2021	2021	TV-MA	2 Seasons	International TV Shows, TV Dramas, TV Mysteries	After crossing paths at a party, a Cape Town teen sets out to prove whether a private-school swimming star is her sister who was abducted at birth.
s3	TV Show	Ganglands	Julien Leclercq	Sami Bouajila, Tracy Gotoas, Samuel Jouy, Nabiha Akkari, Sofia Lesaffre, Salim Kechiouche, Noureddine Farihi, Geert Van Rampelberg, Bakary Diombera	null	September 24, 2021	2021	TV-MA	1 Season	Crime TV Shows, International TV Shows, TV Action & Adventure	To protect his family from a powerful drug lord, skilled thief Mehdi and his expert team of robbers are pulled into a violent and deadly turf war.

Figure 3.1 – An overview of query result of the Delta table just created

7. **Stop the Spark session**: Finally, we need to stop the Spark session to release the resources used by Spark:

```
spark.stop()
```

There's more...

When creating Delta tables, there are additional topics you may encounter:

- **Partitioning**: Delta Lake supports partitioning, which improves query performance by organizing data based on certain columns. You can partition Delta Lake tables based on one or more columns, such as date, region, or any other relevant attribute. Partitioning can significantly enhance query performance, especially when dealing with large datasets. You can specify the partition columns while creating the table or use the `partitionBy` method when writing the data.

- **Schema evolution**: Delta Lake allows schema evolution, which means you can evolve the structure of your Delta tables over time. You can add, drop, or modify columns without affecting existing data. Delta Lake handles schema changes seamlessly, ensuring backward compatibility and data integrity. You can leverage schema evolution features such as the `mergeSchema` option and the `alter` command to modify the schema of your Delta Lake tables.

- **Time travel**: Delta Lake provides time travel capabilities, allowing you to query previous versions of your Delta tables based on timestamps or version numbers. This feature is useful for auditing, debugging, or reverting to specific points in time. You can use the `versionAsOf` or `timestampAsOf` options when querying Delta Lake tables to access their historical states. We will cover this topic later in the *Versioning and time travel for Delta Lake tables* recipe.

- **Concurrency control**: Delta Lake provides built-in concurrency control mechanisms to handle concurrent read and write operations. This ensures that data consistency is maintained even when multiple processes or users are accessing the same Delta Lake table simultaneously. Delta Lake uses optimistic concurrency control and transactional isolation to provide ACID guarantees.

- **Optimizations**: Delta Lake incorporates several performance optimizations to improve query performance and data processing. It leverages advanced indexing techniques, predicate pushdown, data skipping, and caching mechanisms to accelerate query execution. Delta Lake also supports Z-ordering, which improves data locality and reduces I/O during query execution.

See also

- **Delta Lake documentation**: The official Delta Lake documentation provides in-depth information on various topics, including table creation, reading and writing data, schema evolution, and more. You can refer to it for comprehensive details and advanced usage: `https://docs.delta.io/latest/index.html`.

- **Apache Spark documentation**: If you want to explore more about Spark and its capabilities, the official Apache Spark documentation is a valuable resource: `https://spark.apache.org/docs/latest/`.

- **Delta Lake GitHub repository**: The Delta Lake GitHub repository contains additional examples, tutorials, and community resources that can further enhance your understanding: `https://github.com/delta-io/delta`.

Reading a Delta Lake table

Reading tables in Delta Lake is a common task in data processing and analysis. Delta Lake provides ACID transactional capabilities and schema enforcement, making it a powerful data storage solution for big data workloads. In this hands-on recipe, we will explore how to read a table in Delta Lake using Python.

How to do it...

1. **Import the required libraries**: We'll start by importing the necessary libraries for working with Delta Lake. In this case, we need the `delta` module and the `SparkSession` class from the `pyspark.sql` module:

```
from delta import configure_spark_with_delta_pip, DeltaTable
from pyspark.sql import SparkSession
```

2. **Create a SparkSession object**: To interact with Spark and Delta Lake, you need to create a `SparkSession` object:

```
builder = (SparkSession.builder
    .appName("read-delta-table")
    .master("spark://spark-master:7077")
    .config("spark.executor.memory", "512m")
    .config("spark.sql.extensions",
        "io.delta.sql.DeltaSparkSessionExtension")
    .config("spark.sql.catalog.spark_catalog",
        "org.apache.spark.sql.delta.catalog.DeltaCatalog"))

spark = configure_spark_with_delta_pip(builder).getOrCreate()
spark.sparkContext.setLogLevel("ERROR")
```

3. **Read the Delta Lake table**: To read a Delta Lake table, we need to specify the format as `delta` and provide a path to the Delta Lake table:

```
# For PySpark:
df = spark.read.format("delta").load("/opt/workspace/data/delta_
lake/netflix_titles")
```

4. **Explore the data**: Once we have the DataFrame, we can explore the data. The `printSchema()` method displays the schema of the DataFrame, including column names and data types. The `show()` method displays a sample of the data in a tabular format. By default, it shows the first 20 rows. You can customize the number of rows to display by specifying an argument, such as `df.show(5)`, to show only 10 rows:

```
# Display the DataFrame schema
df.printSchema()
# Show a sample of the data
df.show(5)
```

Alternatively, we can also use SQL to explore Delta tables:

```
%%sparksql
SELECT * FROM delta.'/opt/workspace/data/delta_lake/netflix_
titles' LIMIT 3;
```

This will result in the following output:

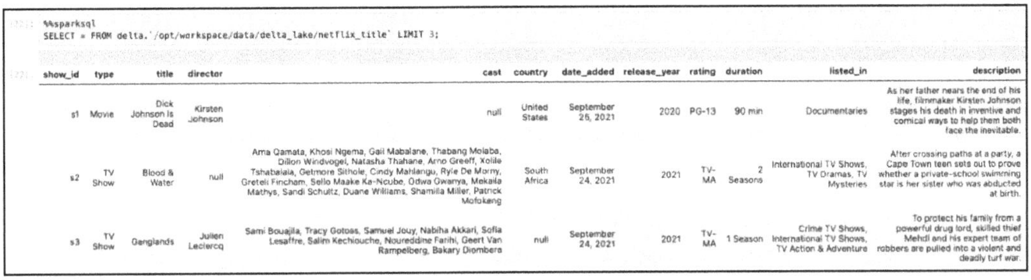

```
%%sparksql
SELECT * FROM delta.`/opt/workspace/data/delta_lake/netflix_title` LIMIT 3;
```

show_id	type	title	director	cast	country	date_added	release_year	rating	duration	listed_in	description
s1	Movie	Dick Johnson Is Dead	Kirsten Johnson	null	United States	September 25, 2021	2020	PG-13	90 min	Documentaries	As her father nears the end of his life, filmmaker Kirsten Johnson stages his death in inventive and comical ways to help them both face the inevitable.
s2	TV Show	Blood & Water	null	Ama Qamata, Khosi Ngema, Gail Mabalane, Thabang Molaba, Dillon Windvogel, Natasha Thahane, Arno Greeff, Xolile Tshabalala, Getmore Sithole, Cindy Mahlangu, Ryle De Morny, Greteli Fincham, Sello Maake Ka-Ncube, Odwa Gwanya, Mekaila Mathys, Sandi Schultz, Duane Williams, Shamilla Miller, Patrick Mofokeng	South Africa	September 24, 2021	2021	TV-MA	2 Seasons	International TV Shows, TV Dramas, TV Mysteries	After crossing paths at a party, a Cape Town teen sets out to prove whether a private-school swimming star is her sister who was abducted at birth.
s3	TV Show	Ganglands	Julien Leclercq	Sami Bouajila, Tracy Gotoas, Samuel Jouy, Nabiha Akkari, Sofia Lesaffre, Salim Kechiouche, Noureddine Farihi, Geert Van Rampelberg, Bakary Diombera	null	September 24, 2021	2021	TV-MA	1 Season	Crime TV Shows, International TV Shows, TV Action & Adventure	To protect his family from a powerful drug lord, skilled thief Mehdi and his expert team of robbers are pulled into a violent and deadly turf war.

Figure 3.2 – Query result of the Delta table

5. **Retrieve the Delta table history**: You can use the `DESCRIBE HISTORY` command to retrieve information about operations performed on a Delta table, including details about the user, timestamp, and other relevant metadata. This command returns operations in reverse chronological order, providing a historical view of the table's modifications. By default, Delta Lake retains the table history for a period of 30 days:

```
%%sparksql
DESCRIBE HISTORY '/opt/workspace/data/delta_lake/netflix_titles'
```

This will result in the following output:

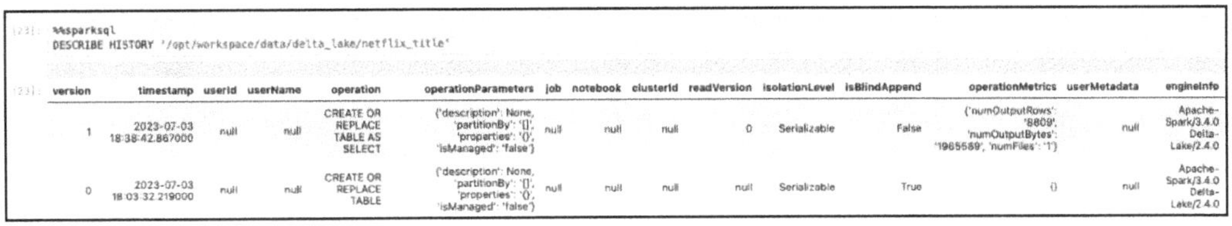

```
%%sparksql
DESCRIBE HISTORY '/opt/workspace/data/delta_lake/netflix_title'
```

version	timestamp	userId	userName	operation	operationParameters	job	notebook	clusterId	readVersion	isolationLevel	isBlindAppend	operationMetrics	userMetadata	engineInfo
1	2023-07-03 18:38:42.867000	null	null	CREATE OR REPLACE TABLE AS SELECT	{'description': None, 'partitionBy': '[]', 'properties': '{}', 'isManaged': 'false'}	null	null	null	0	Serializable	False	{'numOutputRows': '8809', 'numOutputBytes': '1965589', 'numFiles': '1'}	null	Apache-Spark/3.4.0 Delta-Lake/2.4.0
0	2023-07-03 18:03:32.219000	null	null	CREATE OR REPLACE TABLE	{'description': None, 'partitionBy': '[]', 'properties': '{}', 'isManaged': 'false'}	null	null	null	null	Serializable	True	{}	null	Apache-Spark/3.4.0 Delta-Lake/2.4.0

Figure 3.3 – Describing the history of the Delta table

> **Note**
>
> The retention period for the table history can be configured. If you need to retain the history for a longer period, you can modify the Delta Lake configuration to increase the retention duration. This can be done by setting the `delta.logRetentionDuration` and `delta.deletedFileRetentionDuration` table properties on the Delta table.
>
> For example, the following code sets the properties so that you can query any version of the table that was created in the past 60 days and can restore any files that were deleted in the past 14 days by rolling back the table version:
>
> ```
> %%sparksql
> ALTER TABLE delta.`/opt/workspace/data/delta_lake/netflix_titles`
> SET TBLPROPERTIES (
>   delta.logRetentionDuration = 'interval 60 days',
>   delta.deletedFileRetentionDuration = 'interval 14 days'
> );
> ```

6. **Stop the Spark session**: Finally, we need to stop the Spark session to release the resources used by Spark:

    ```
    spark.stop()
    ```

There's more...

Delta Lake provides various techniques to optimize query performance. Some of the optimization features include ZORDER, indexing, and caching. These features can significantly improve query execution time. For example, you can use ZORDER to organize data based on specific columns to accelerate data skipping during filtering. Indexing can be applied to columns that are frequently used for filtering or joining, further improving query performance. Caching data in memory can also speed up subsequent queries. Refer to the Delta Lake documentation for more details on performance optimization techniques.

See also

- Delta Lake documentation: `https://docs.delta.io/latest/index.html`
- PySpark documentation: `https://spark.apache.org/docs/latest/api/python/index.html`
- Delta table Python library documentation: `https://docs.delta.io/0.4.0/api/python/index.html`
- Data retention on a Delta table: `https://docs.delta.io/latest/delta-batch.html#-data-retention`

Updating data in a Delta Lake table

In this recipe, we will learn how to update data in a table in Delta Lake using Python and SQL. Delta Lake is an open source storage layer providing ACID transactions, schema enforcement, and data versioning for big data workloads. Updating data allows us to modify existing records or insert new records into a Delta Lake table efficiently.

How to do it...

1. **Import the required libraries**: Start by importing the necessary libraries for working with Delta Lake. In this case, we need the `delta` module and the `SparkSession` class from the `pyspark.sql` module:

   ```
   from delta import configure_spark_with_delta_pip, DeltaTable
   from pyspark.sql import SparkSession
   from pyspark.sql.functions import expr, lit
   ```

2. **Create a SparkSession object**: To interact with Spark and Delta Lake, you need to create a `SparkSession` object:

   ```
   builder = (SparkSession.builder
       .appName("upsert-delta-table")
       .master("spark://spark-master:7077")
       .config("spark.executor.memory", "512m")
       .config("spark.sql.extensions",
           "io.delta.sql.DeltaSparkSessionExtension")
       .config("spark.sql.catalog.spark_catalog",
           "org.apache.spark.sql.delta.catalog.DeltaCatalog"))

   spark = configure_spark_with_delta_pip(builder).getOrCreate()
   spark.sparkContext.setLogLevel("ERROR")
   ```

3. **Read the Delta Lake table**: We will need to create a `DeltaTable` object that represents a Delta Lake table stored in the specified path:

   ```
   # For PySpark:
   deltaTable = DeltaTable.forPath(spark, "/opt/workspace/data/
   delta_lake/netflix_titles")
   ```

4. **Explore the data**: We use the `toDF()` method to return a DataFrame representation of the Delta table and use the `show(5)` method to print the first five rows of the DataFrame to the console:

   ```
   deltaTable.toDF().show(5)
   ```

5. **Update records where director is null**: We will use the `update()` method, which takes two arguments: `condition` and `set`. The `condition` argument specifies a predicate that filters the rows to be updated. The `set` argument specifies a map of column names and values to be assigned to the updated rows:

```
# Update director to not have nulls
deltaTable.update(
  condition = expr("director IS NULL"),
  set = { "director": lit("") })
```

Alternatively, we can also write SQL commands to update Delta tables:

```
%%sparksql
UPDATE delta.'/opt/workspace/data/delta_lake/netflix_titles' SET
director = "" WHERE director IS NULL;
```

This will result in the following output:

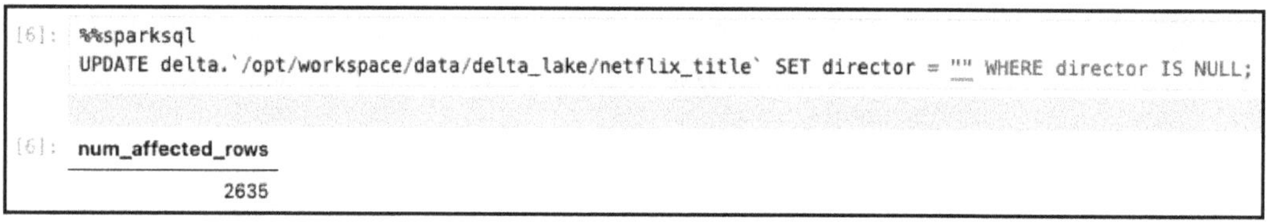

```
[6]: %%sparksql
     UPDATE delta.`/opt/workspace/data/delta_lake/netflix_title` SET director = "" WHERE director IS NULL;

[6]: num_affected_rows
               2635
```

Figure 3.4 – Getting the number of rows affected after an update is performed on a Delta table

6. **Stop the Spark session**: Finally, we need to stop the Spark session to release the resources used by Spark:

```
spark.stop()
```

See also

- Delta Lake documentation: https://docs.delta.io/latest/index.html
- Apache Spark documentation: https://spark.apache.org/docs/latest/index.html

Merging data into Delta tables

In this hands-on recipe, we will learn how to merge data into a table using Delta Lake in Python. It provides powerful capabilities to perform updates, deletes, and merges on data stored in a Delta table, making it an excellent choice for managing big data workloads with data integrity and reliability.

How to do it...

1. **Import the required libraries**: Start by importing the necessary libraries for working with Delta Lake. In this case, we need the `delta` module and the `SparkSession` class from the `pyspark.sql` module:

```
from delta import configure_spark_with_delta_pip, DeltaTable
from pyspark.sql import SparkSession
```

2. **Create a SparkSession object**: To interact with Spark and Delta Lake, you need to create a `SparkSession` object:

```
builder = (SparkSession.builder
    .appName("read-delta-table")
    .master("spark://spark-master:7077")
    .config("spark.executor.memory", "512m")
    .config("spark.sql.extensions",
        "io.delta.sql.DeltaSparkSessionExtension")
    .config("spark.sql.catalog.spark_catalog",
        "org.apache.spark.sql.delta.catalog.DeltaCatalog"))

spark = configure_spark_with_delta_pip(builder).getOrCreate()
spark.sparkContext.setLogLevel("ERROR")
```

3. **Create a Delta table**: Before we load data, we will need to create a table in the Delta catalog. This can be done with the `CREATE OR REPLACE TABLE` command. You can specify the table name, which is `default.movie_and_show_titles`, and also specify the data format as `DELTA` and the location of the table data, which in our case is `/opt/workspace/data/delta_lake/movie_and_show_titles`:

```
%%sparksql
CREATE OR REPLACE TABLE default.movie_and_show_titles (
    show_id STRING,
    type STRING,
    title STRING,
    director STRING,
    cast STRING,
    country STRING,
    date_added STRING,
    release_year STRING,
    rating STRING,
    duration STRING,
    listed_in STRING,
    description STRING
) USING DELTA LOCATION '/opt/workspace/data/delta_lake/movie_
and_show_titles';
```

4. **Create a DeltaTable object for our target table**: This code creates a `DeltaTable` object that represents a Delta Lake table stored in the specified path, and then converts it into a `DataFrame` object and displays the first five rows of the DataFrame:

```
# For PySpark:
deltaTable_titles = DeltaTable.forPath(spark,
    "/opt/workspace/data/delta_lake/movie_and_show_titles")

deltaTable_titles.toDF().show(5) # Delta table is empty at first
```

5. **Read the Delta Lake table**: To read a Delta Lake table, we need to specify the format as `delta` and provide a path to the Delta Lake table. For PySpark, we use the `spark.read.format("delta")` syntax and load the table using `.load("/opt/workspace/data/delta_lake/netflix_titles")`. Merge operations only allow a single update to each row. The reason for this is that Delta Lake employs a technique called **optimistic concurrency control (OCC)** to protect data quality and avoid clashes between simultaneous transactions. When more than one row is updated with the same criterion, there is a chance that another transaction might change one of those rows before the `MERGE` action is finalized, causing a write conflict and a transaction failure. This means that you cannot have duplicate records in the source; that is, you cannot have more than one row in the source updating a single row in the target. To avoid this, we will drop duplicates from our source. Since there is no numeric identifier, we will consider a row duplicate if it has the same type, title, director, and `date_added` value:

```
# For PySpark:
df_netflix = spark.read.format("delta").load("/opt/workspace/
data/delta_lake/netflix_titles")
df_netflix_deduped = df_netflix.dropDuplicates(["type", "title",
    "director", "date_added"])
```

6. **Merge the source DataFrame with the target DeltaTable**: This code performs a merge operation between `deltaTable_titles` and `df_netflix_deduped`, in order to update and insert records based on certain matching conditions. The `whenMatchedUpdate()` function specifies the updates to be performed on matched records. In this case, the following columns will be updated with values from the `updates` DataFrame. The `whenNotMatchedInsert()` function specifies the values to be inserted for unmatched records:

```
(deltaTable_titles.alias('movie_and_show_titles')
 .merge(df_netflix_deduped.alias('updates')
          ,"""lower(movie_and_show_titles.type) =
lower(updates.type)
              AND lower(movie_and_show_titles.title) =
lower(updates.title)
              AND lower(movie_and_show_titles.director) =
lower(updates.director)
              AND movie_and_show_titles.date_added = updates.
date_added""")
```

```
.whenMatchedUpdate(set ={
    "show_id": "updates.show_id",
    "type": "updates.type",
    "title" : "updates.title",
    "director" : "updates.director",
    "cast" : "updates.cast",
    "country" : "updates.country",
    "date_added" : "updates.date_added",
    "release_year" : "updates.release_year",
    "rating" : "updates.rating",
    "duration" : "updates.duration",
    "listed_in" : "updates.listed_in",
    "description" : "updates.description"})
.whenNotMatchedInsert(values = {
    "show_id": "updates.show_id",
    "type": "updates.type",
    "title" : "updates.title",
    "director" : "updates.director",
    "cast" : "updates.cast",
    "country" : "updates.country",
    "date_added" : "updates.date_added",
    "release_year" : "updates.release_year",
    "rating" : "updates.rating",
    "duration" : "updates.duration",
    "listed_in" : "updates.listed_in",
    "description" : "updates.description"})  .execute())
```

7. **Retrieve the Delta table history**: Describe the history of the Delta table to view statistics from the merge operation:

```
%%sparksql
DESCRIBE HISTORY "/opt/workspace/data/delta_lake/movie_and_show_
titles"
```

This will result in the following output:

Figure 3.5 – Describing the history of the Delta table to see the merge operation listed

8. **Read an additional CSV dataset:** In the code, we specify the format of the file as `"csv"` using the `format` method of SparkSession. We then set the `"header"` option to `"true"` to indicate that the first row of the CSV file contains the column names. Finally, we specify the path to the CSV file to load it into a DataFrame using the `load` method. As stated earlier, merge operations only allow a single update to each row. To avoid errors, we will drop duplicates of our source based on the same type and title:

```
# Read CSV file into a DataFrame
df_titles = (spark.read
        .format("csv")
        .option("header", "true")
        .load("../data/titles.csv"))
df_titles_deduped = df_titles.dropDuplicates(["type", "title"])
```

9. **Merge the source DataFrame with the target Delta table:** This code performs a merge operation between `deltaTable_titles` and `df_titles_deduped` in order to update and insert records based on certain matching conditions. The `whenMatchedUpdate()` function specifies the updates to be performed on matched records. In this case, the following columns will be updated with values from the `updates` DataFrame. The `whenNotMatchedInsert()` function specifies the values to be inserted for unmatched records. We also modify the following statement to map the appropriate source columns to the target columns:

```
(deltaTable_titles.alias('movie_and_show_titles')
 .merge(df_titles_deduped.alias('updates')
        ,"""lower(movie_and_show_titles.type) =
lower(updates.type)
            AND lower(movie_and_show_titles.title) =
lower(updates.title)
            AND movie_and_show_titles.release_year = updates.
release_year""")
 .whenMatchedUpdate(set ={
     "show_id" : "updates.id",
     "type" : "updates.type",
     "title" : "updates.title",
     "country" : "updates.production_countries",
     "release_year" : "updates.release_year",
     "rating" : "updates.age_certification",
     "duration" : "updates.runtime",
     "listed_in" : "updates.genres",
     "description" : "updates.description"})
 .whenNotMatchedInsert(values = {
     "show_id" : "updates.id",
     "type" : "updates.type",
     "title" : "updates.title",
     "country" : "updates.production_countries",
```

```
                "release_year" : "updates.release_year",
                "rating" : "updates.age_certification",
                "duration" : "updates.runtime",
                "listed_in" : "updates.genres",
                "description" : "updates.description"})
        .execute())
```

Alternatively, we can also write the MERGE INTO command in SQL to perform a merge operation:

```
%%sparksql
MERGE INTO default.movie_and_show_titles
USING titles_deduped
ON lower(default.movie_and_show_titles.type) = lower(titles_
deduped.type)
      AND lower(default.movie_and_show_titles.title) =
lower(titles_deduped.title)
      AND default.movie_and_show_titles.release_year = titles_
deduped.release_year
WHEN MATCHED THEN
  UPDATE SET
      show_id = titles_deduped.id,
      type = titles_deduped.type,
      title = titles_deduped.title,
      country = titles_deduped.production_countries,
      release_year = titles_deduped.release_year,
      rating = titles_deduped.age_certification,
      duration = titles_deduped.runtime,
      listed_in = titles_deduped.genres,
      description = titles_deduped.description
WHEN NOT MATCHED
  THEN INSERT (
      show_id,
      type,
      title,
      country,
      release_year,
      rating,
      duration,
      listed_in,
      description
  )
  VALUES (
      titles_deduped.id,
      titles_deduped.type,
      titles_deduped.title,
```

```
            titles_deduped.production_countries,
            titles_deduped.release_year,
            titles_deduped.age_certification,
            titles_deduped.runtime,
            titles_deduped.genres,
            titles_deduped.description
        )
```

This will result in the following output:

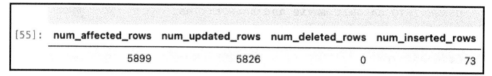

Figure 3.6 – Getting the number of rows impacted by the merge operation

10. **Retrieve the Delta table history**: Describe the history of the Delta table to view statistics from the merge operation:

```
%%sparksql
DESCRIBE HISTORY "/opt/workspace/data/delta_lake/movie_and_show_
titles"
```

This will result in the following output:

Figure 3.7 – Getting the history of the Delta table and noting the merge operation logged

11. **Stop the Spark session**: Finally, we need to stop the Spark session to release the resources used by Spark:

```
spark.stop()
```

There's more...

When merging data into Delta tables, there are additional topics you may have to consider. Here are some of them, along with code examples in Python to illustrate the scenarios:

- **Schema mismatch**: One common issue is when the schema of the new data does not match the schema of the Delta table. This can happen if the column names or types are different. Let's consider an example where the column names in the new data DataFrame do not match the Delta table's schema.

  In the code, we compare the schemas of the new data DataFrame (`new_data_df`) and the Delta table (`deltaTable`) using the `schema` attribute. If the schemas don't match, we can align the column names by selecting only the columns present in the Delta table's schema. This ensures that the column names in the new data match the Delta table's schema:

  ```
  # New data with different column names
  new_data_df = spark.read.format("csv").option("header", "true").
  load("path/to/new_data.csv")

  # Delta table schema
  delta_table_schema = deltaTable.toDF().schema

  # Check for schema mismatch
  if not new_data_df.schema == delta_table_schema:
      # Handle schema mismatch: align column names
      new_data_df = new_data_df.select(*delta_table_schema.
  fieldNames())
  ```

- **Handling deletes**: During the merge process, you may encounter scenarios where you want to delete specific rows from the Delta table based on certain conditions. Delta Lake provides the `whenMatchedDelete()` action to handle such cases. Here's an example that demonstrates deleting rows using this action:

  ```
  deltaTable.alias("target").merge(
      new_data_df.alias("source"),
      merge_condition
  ).whenMatchedDelete().whenMatchedUpdateAll().
  whenNotMatchedInsertAll().execute()
  ```

See also

- **Spark documentation**: Refer to the Apache Spark documentation for more information on Spark APIs and concepts: https://spark.apache.org/docs/latest/

- **Delta Lake GitHub repository**: Visit the Delta Lake GitHub repository for additional resources and examples: https://github.com/delta-io/delta

- **Partitioning and bucketing**: Consider partitioning and bucketing your Delta table to optimize performance and enable efficient data pruning. You can find more details in the Delta Lake documentation: `https://docs.delta.io/latest/best-practices.html#partitioning`

- **Delta Lake documentation**: Explore the Delta Lake documentation to learn more about advanced features and best practices for incremental loading: `https://docs.delta.io/latest/index.html`

Change data capture in Delta Lake

Change data capture (**CDC**) is a technique to capture and process the changes made to a data source, such as a database or a file system. CDC can be useful for various scenarios, such as data synchronization, replication, auditing, and analytics.

Delta Lake supports CDC through a feature called **change data feed** (**CDF**), which allows Delta tables to track row-level changes between versions of a Delta table. When enabled on a Delta table, the runtime records "change events" for all data written into the table.

In this recipe, we will learn how to apply CDC to a table using Delta Lake in Python.

How to do it...

1. **Import the required libraries**: Start by importing the necessary libraries for working with Delta Lake. In this case, we need the `delta` module and the `SparkSession` class from the `pyspark.sql` module:

```
from delta import configure_spark_with_delta_pip, DeltaTable
from pyspark.sql import SparkSession
```

2. **Create a SparkSession object**: To interact with Spark and Delta Lake, you need to create a `SparkSession` object:

```
builder = (SparkSession.builder
                .appName("read-delta-table")
                .master("spark://spark-master:7077")
                .config("spark.executor.memory", "512m")
                .config("spark.sql.extensions", "io.delta.sql.
DeltaSparkSessionExtension")
                .config("spark.sql.catalog.spark_catalog", "org.
apache.spark.sql.delta.catalog.DeltaCatalog"))

spark = configure_spark_with_delta_pip(builder).getOrCreate()
spark.sparkContext.setLogLevel("ERROR")
```

3. **Create a Delta table with CDF enabled**: We can enable CDF on a table while creating it by applying the `delta.enableChangeDataFeed` table property. In the following code, we create a table with the name `default.movie_and_show_titles_cdf`:

```
%%sparksql
CREATE OR REPLACE TABLE default.movie_and_show_titles_cdf (
      show_id STRING,
      type STRING,
      title STRING,
      director STRING,
      cast STRING,
      country STRING,
      date_added STRING,
      release_year STRING,
      rating STRING,
      duration STRING,
      listed_in STRING,
      description STRING
) USING DELTA LOCATION '/opt/workspace/data/delta_lake/movie_
and_show_titles_cdf'
TBLPROPERTIES (delta.enableChangeDataFeed = true, medallionLevel
= 'bronze');
```

Alternatively, you can alter an existing table to enable capturing CDF from that point onward:

```
%%sparksql
ALTER TABLE delta.`/opt/workspace/data/delta_lake/movie_and_
show_titles` SET TBLPROPERTIES (delta.enableChangeDataFeed =
true)
```

4. Write data into the `bronze` table:

```
# Read CSV file into a DataFrame
df = (spark.read
        .format("csv")
        .option("header", "true")
        .load("../data/netflix_titles.csv"));

df.write.format("delta").mode("append").saveAsTable("default.
movie_and_show_titles_cdf")
```

5. **Create a silver Delta table**: We can enable CDF on a table while creating it by applying the `delta.enableChangeDataFeed` table property. We will also add some custom user-defined metadata to this table. This will be `updatedFromTable`, which contains the upstream table name, and `updatedFromTableVersion`, which contains the last version of the source table propagated to this table successfully processed (`-1` means the table has never been propagated):

```
%%sparksql
CREATE OR REPLACE TABLE default.movie_and_show_titles_cleansed (
        show_id STRING,
        type STRING,
        title STRING,
        director STRING,
        cast STRING,
        country STRING,
        date_added STRING,
        release_year STRING,
        rating STRING,
        duration STRING,
        listed_in STRING,
        description STRING
) USING DELTA LOCATION '/opt/workspace/data/delta_lake/movie_
and_show_titles_cleansed'
TBLPROPERTIES (delta.enableChangeDataFeed = true, medallionLevel
= 'silver'
, updatedFromTable= 'default.movie_and_show_titles_cdf',
updatedFromTableVersion=-1);
```

6. **Get the last updated version**: Get the value of the last updated version from the `silver` table:

```
#get the value of the last updated version from the silver table
lastUpdateVersion = int(spark.sql("SHOW TBLPROPERTIES default.
movie_and_show_titles_cleansed ('updatedFromTableVersion')").
first()["value"])+1
lastUpdateVersion
```

7. **Get the latest version**: Get the value of the latest version from the `silver` table:

```
#get the value of the latest version from the silver table
latestVersion = spark.sql("DESCRIBE HISTORY default.movie_and_
show_titles_cdf").first()["version"]
latestVersion
```

8. **Create a temporary view**: Create a temporary view of changes to the `bronze` table since the last load of the `silver` table:

```
%%sparksql
CREATE OR REPLACE TEMPORARY VIEW bronzeTable_latest_version as
SELECT * FROM (
    SELECT *,
            RANK() OVER (
            PARTITION BY (lower(type), lower(title),
lower(director), date_added)
            ORDER BY _commit_version DESC) as rank
    FROM table_changes('default.movie_and_show_titles_
cdf',{lastUpdateVersion},{latestVersion})
        WHERE type IS NOT NULL AND title IS NOT NULL AND director
IS NOT NULL AND  _change_type !='update_preimage'
)
WHERE rank = 1;
```

9. **Merge change data into the silver table**: This code executes a `MERGE  INTO` statement that performs a CDC operation on a Delta Lake table. We specify three possible actions based on the match between the source and target rows:

   - If the source row matches the target row and `_change_type` is either `update_postimage` or `update_preimage`, then update all columns of the target row with the corresponding columns of the source row. This action handles updates in the source table.

   - If the source row matches the target row and `_change_type` is `delete`, then delete the matching target row. This action handles deletions in the source table.

   - If the source row does not match any target row and `_change_type` is `insert`, then insert all columns of the source row into the target table. This action handles insertions in the source table:

```
%%sparksql
MERGE INTO default.movie_and_show_titles_cleansed t
USING bronzeTable_latest_version s
ON lower(t.type) = lower(s.type)
AND lower(t.title) = lower(s.title)
AND lower(t.director) = lower(s.director)
AND t.date_added = s.date_added
WHEN MATCHED AND s._change_type='update_postimage' OR s._change_
type='update_postimage' THEN UPDATE SET *
WHEN MATCHED AND s._change_type='delete' THEN DELETE
WHEN NOT MATCHED AND s._change_type='insert' THEN INSERT *
```

This will result in the following output:

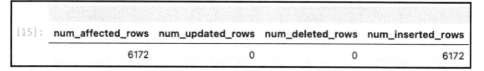

	num_affected_rows	num_updated_rows	num_deleted_rows	num_inserted_rows
[15]:	6172	0	0	6172

Figure 3.8 – Getting the number of rows impacted by MERGE INTO

10. **Alter the Delta table**: Update the user-defined metadata value for updatedFromTableVersion:

```
%%sparksql
ALTER TABLE default.movie_and_show_titles_cleansed SET
TBLPROPERTIES(updatedFromTableVersion = {latestVersion});
```

11. **Drop the temporary view**: We will use the DROP VIEW statement, which takes the view name or a path as an argument and deletes the view definition from the catalog:

```
%%sparksql
DROP VIEW bronzeTable_latest_version
```

12. **Simulate data changes**: Update the bronze table by deleting a few roles and updating a few rows:

```
%%sparksql
DELETE FROM default.movie_and_show_titles_cdf WHERE country is
NULL
 %%sparksql
UPDATE default.movie_and_show_titles_cdf SET director = '' WHERE
director is NULL
```

13. **Update the downstream table incrementally**: Execute *steps 6, 7, 8, 9, 10*, and *11* again to propagate only the changes to the silver table. This will result in the following output:

	num_affected_rows	num_updated_rows	num_deleted_rows	num_inserted_rows
[26]:	422	0	422	0

Figure 3.9 – Getting the number of rows impacted by the update command

14. **Stop the Spark session**: Finally, we need to stop the Spark session to release the resources used by Spark:

```
spark.stop()
```

See also

- *Table deletes, updates, and merges* – `https://docs.delta.io/latest/delta-update.html`

- *Table properties and table options* – `https://docs.databricks.com/sql/language-manual/sql-ref-syntax-ddl-tblproperties.html`

- *Upsert into a Delta Lake table using merge* – `https://docs.databricks.com/delta/merge.html`

- *How to Simplify CDC With Delta Lake's Change Data Feed* – `https://www.databricks.com/blog/2021/06/09/how-to-simplify-cdc-with-delta-lakes-change-data-feed.html`

- *Change data feed* – `https://docs.delta.io/2.0.0/delta-change-data-feed.html`

Optimizing Delta Lake tables

To optimize Delta Lake tables, we will focus on improving performance and reducing storage space. We'll cover various techniques and strategies that can be applied to enhance the efficiency of Delta Lake tables.

How to do it...

1. **Import the required libraries**: Start by importing the necessary libraries for working with Delta Lake. In this case, we need the `delta` module and the `SparkSession` class from the `pyspark.sql` module:

   ```
   from delta import configure_spark_with_delta_pip, DeltaTable
   from pyspark.sql import SparkSession
   ```

2. **Create a SparkSession object**: To interact with Spark and Delta Lake, you need to create a `SparkSession` object:

   ```
   builder = (SparkSession.builder
                   .appName("optimize-delta-table")
                   .master("spark://spark-master:7077")
                   .config("spark.executor.memory", "512m")
                   .config("spark.sql.extensions", "io.delta.sql.
   DeltaSparkSessionExtension")
                   .config("spark.sql.catalog.spark_catalog", "org.
   apache.spark.sql.delta.catalog.DeltaCatalog"))

   spark = configure_spark_with_delta_pip(builder).getOrCreate()
   spark.sparkContext.setLogLevel("ERROR")
   ```

3. **Read the Delta Lake table**: To read a Delta Lake table, we need to specify the format as `delta` and provide a path to the Delta Lake table. For PySpark, we use the `spark.read.format("delta")` syntax and load the table using `.load("/opt/workspace/data/delta_lake/netflix_titles")`:

```
# For PySpark:
df = spark.read.format("delta").load("/opt/workspace/data/delta_lake/netflix_titles")
```

4. **Optimize the Delta table**: The optimization step improves query performance by optimizing the organization and layout of data within the table. The subsequent compaction operation merges smaller files into larger ones, reducing storage overhead and further enhancing query performance. By applying these optimizations, the code aims to maximize the efficiency and effectiveness of working with the Delta Lake table. To apply compaction on a Delta table in Python, run the following code:

```
deltaTable = DeltaTable.forPath(spark, "/opt/workspace/data/delta_lake/netflix_titles")
# For Hive metastore-based tables: deltaTable = DeltaTable.forName(spark, tableName)
deltaTable.optimize().executeCompaction()
```

Alternatively, to apply compaction on a Delta table in SQL, run the following code:

```
%%sparksql
-- Optimizes the path-based Delta Lake table
OPTIMIZE "/opt/workspace/data/delta_lake/netflix_titles"
```

This will result in the following output:

Figure 3.10 – The output from the OPTIMIZE command is a list of files affected

> **Note**
>
> If you run `OPTIMIZE` twice on the same dataset, the second run will have no effect. This is also known as an idempotent operation.
>
> When you read data from Delta tables, the process is not affected by the `OPTIMIZE` operation. `OPTIMIZE` removes unnecessary files without making any changes to the actual data in the table. It merges smaller files into bigger ones to enhance storage efficiency and performance. This means that if you read the data before and after running `OPTIMIZE`, you will get the same results.

5. **Apply Z-ordering on the Delta table**: Z-ordering is a technique in Delta Lake that organizes data based on selected columns to improve query performance. It groups related data together physically, reducing the amount of data it takes to read during queries. It works well for high-cardinality columns and benefits filtering, joining, and aggregating operations. Z-ordering is a data write optimization that requires considering query patterns and selected columns for optimal results. To perform Z-ordering on our Netflix title table based on country in Python, we can run the following code:

```
deltaTable = DeltaTable.forPath(spark, "/opt/workspace/data/
delta_lake/netflix_titles")  # path-based table
# For Hive metastore-based tables: deltaTable = DeltaTable.
forName(spark, tableName)
deltaTable.optimize().executeZOrderBy("country")
```

Alternatively, to perform Z-ordering on our Netflix title table based on country in SQL, we can run the following code:

```
%%sparksql
OPTIMIZE "/opt/workspace/data/delta_lake/netflix_titles" ZORDER
BY (country)
```

This will result in the following output:

Figure 3.11 – OPTIMIZE outputs the stats of files it changes

> **Note**
>
> Each time Z-ordering is executed, it will attempt to create a new way of organizing the data in all files within a partition. This means that even existing files that were previously part of a Z-ordering operation will be reorganized along with new files. Z_ORDER is not an idempotent operation.

6. **Stop the Spark session**: Finally, we need to stop the Spark session to release the resources used by Spark:

```
spark.stop()
```

There's more...

When optimizing data in Delta tables, there are additional topics you may have to consider. Here are some of them, along with code examples in Python to illustrate the scenarios.

Partitioning

Partitioning in Delta Lake is a technique that organizes data files in a table or data lake based on specific column values. It helps optimize query performance by reducing the amount of data that needs to be scanned during queries. When a table is partitioned, the data is physically stored in directories that corresponds to the unique values of the partitioned column.

In this example, the data is read from a CSV file (`data.csv`) and then written to a Delta Lake table using the `partitionBy` method. The `date` column is chosen as the partitioning column. This means that the data will be stored in separate directories based on distinct values in the `date` column:

```
# Read the data as a DataFrame
df = (spark.read.format("json")
        .option("multiLine", "true")
        .load("../data/nobel_prizes.json"))

# Write the data to a Delta Lake table with partitioning
(df.write.format("delta")
 .mode("overwrite")
 .partitionBy("year")
 .save("/opt/workspace/data/delta_lake/nobel_prizes"))

# Query the partitioned table
df = spark.read.format("delta").load("/opt/workspace/data/delta_lake/
nobel_prizes")
df.show()
```

Partitioning can significantly improve query performance when you need to filter data based on the partitioning column. It allows the query engine to skip reading unnecessary data from directories that don't match the specified partition values.

Multi-part checkpointing

Multi-part checkpointing in Delta Lake is a feature that allows you to create checkpoints in smaller parts instead of a single large checkpoint file. This improves the performance of Delta Lake operations, especially when dealing with large tables.

To enable multi-part checkpointing in Delta Lake, you need to set the `spark.databricks.delta.checkpointPolicies` configuration property to `multiPart`. This can be done using the `SparkSession` or `SparkContext` object.

In the following code, `spark.conf.set("spark.databricks.delta.checkpointPo-licies", "multiPart")` sets the `spark.databricks.delta.checkpointPolicies` configuration property to `multiPart`, enabling multi-part checkpointing:

```
# Enable multi-part checkpointing
spark.conf.set("spark.databricks.delta.checkpointPolicies",
"multiPart")
```

Once multi-part checkpointing is enabled, Delta Lake will create checkpoints in smaller parts, resulting in improved performance during operations such as optimization, compaction, and recovery.

Common issues with optimize and Z-ordering operations in Delta Lake include the following:

- **Performance impact**: Both optimize and Z-ordering operations can be resource-intensive and time-consuming, especially for large tables. Running these operations frequently or on tables with heavy workloads may impact query performance and overall system performance.

- **Choosing the right columns**: Selecting the right columns for Z-ordering is crucial. If poorly chosen, it may not yield significant query performance improvements or, in some cases, may even worsen performance. Understanding query patterns and data access patterns is important for making informed decisions.

- **Data distribution**: Uneven data distributions within the selected columns can lead to imbalanced data placement after Z-ordering, reducing its effectiveness. If data values are not evenly distributed, some partitions may end up with much larger files than others, impacting query performance.

- **Data evolution**: Z-ordering is not dynamic; it remains static after the data is written. If the data distribution changes significantly over time, the initially well-optimized Z-ordering may become less effective, requiring periodic re-evaluation and re-optimization.

- **Metadata overhead**: Both optimize and Z-ordering operations may introduce additional metadata overhead, as they need to track the changes made to the table's layout. This can impact the storage requirements for the Delta Lake table.

See also

- Delta Lake optimization techniques: `https://docs.delta.io/2.0.2/optimizations-oss.html`

- Delta Lake on GitHub: `https://github.com/delta-io/delta`

- Partitioning data in Delta Lake: `https://docs.delta.io/latest/delta-batch.html#-partition-data`

- Multi-part checkpointing: `https://docs.delta.io/latest/optimizations-oss.html#multi-part-checkpointing`

- Choosing the right partition column: `https://docs.delta.io/latest/best-practices.html#-choose-the-right-partition-column`

Versioning and time travel for Delta Lake tables

To version, time travel, and restore data in Delta Lake, we need to use the Delta Lake library, which provides ACID transactions and other data management capabilities on top of Apache Spark. In this hands-on recipe, we will explore how to accomplish these tasks using Delta Lake in Python.

How to do it...

1. **Import the required libraries**: Start by importing the necessary libraries for working with Delta Lake. In this case, we need the `delta` module and the `SparkSession` class from the `pyspark.sql` module:

```
from delta import configure_spark_with_delta_pip, DeltaTable
from pyspark.sql import SparkSession
```

2. **Create a SparkSession object**: To interact with Spark and Delta Lake, you need to create a `SparkSession` object:

```
builder = (SparkSession.builder
                  .appName("time-travel-delta-table")
                  .master("spark://spark-master:7077")
                  .config("spark.executor.memory", "512m")
                  .config("spark.sql.extensions", "io.delta.sql.
DeltaSparkSessionExtension")
                  .config("spark.sql.catalog.spark_catalog", "org.
apache.spark.sql.delta.catalog.DeltaCatalog"))

spark = configure_spark_with_delta_pip(builder).getOrCreate()
spark.sparkContext.setLogLevel("ERROR")
```

3. **Read old data**: Read older versions of data using time travel:

```
df = spark.read.format("delta").option("versionAsOf", 1).load("/
opt/workspace/data/delta_lake/netflix_titles")
df.show(5)
```

Alternatively, you could do this in SQL:

```
%%sparksql
SELECT * FROM delta.'/opt/workspace/data/delta_lake/netflix_
titles' VERSION AS OF 1 LIMIT 3;
```

4. **Restore table**: Restore a Delta table to an earlier state:

```
deltaTable = DeltaTable.forPath(spark, "/opt/workspace/data/
delta_lake/netflix_titles")  # path-based tables, or
deltaTable.restoreToVersion(3) # restore table to oldest version
```

Alternatively, you could do this in SQL:

```
%%sparksql
RESTORE TABLE delta.'/opt/workspace/data/delta_lake/netflix_
titles' TO VERSION AS OF 3;
```

This will result in the following output:

table_size_after_restore	num_of_files_after_restore	num_removed_files	num_restored_files	removed_files_size	restored_files_size
1968555	1	1	1	1968555	1968555

[13]:
```
%%sparksql
RESTORE TABLE delta.'/opt/workspace/data/delta_lake/netflix_title' TO VERSION AS OF 3;
```

Figure 3.12 – Output from the RESTORE command is stats on files and size

5. **Stop the Spark session**: Finally, we need to stop the Spark session to release the resources used by Spark:

```
spark.stop()
```

There's more...

To time travel and access previous versions of data in a Delta table, you need to keep both the log and data files for that version; these files are stored alongside the data files in the table's folder within a subfolder called _delta_log. The data files in a Delta table are never deleted automatically, except when you run VACUUM. VACUUM doesn't delete log files; they are cleaned up automatically after checkpoints are written.

Two Delta table properties control data retention:

- `delta.logRetentionDuration`: Determines how long the table's history is kept. The default is 30 days.

- `delta.deletedFileRetentionDuration`: Determines how long a file must be deleted before it can be cleaned up by VACUUM. The default is 7 days.

You can update these properties to fit your use cases. For example, to access historical data for up to 60 days, even after running VACUUM on the table, set `delta.deletedFileRetentionDuration` and `delta.logRetentionDuration` to `"interval 30 days"`. Just keep in mind that this might increase your storage costs.

See also

- **Querying an older snapshot of a table (time travel)**: You can read more about this at https://docs.delta.io/latest/delta-batch.html#query-an-older-snapshot-of-a-table-time-travel

- **Delta Lake GitHub repository**: Explore the Delta Lake GitHub repository for additional resources, sample code, and discussions: https://github.com/delta-io/delta

Managing Delta Lake tables

To apply table constraints, clone tables, and alter columns in Delta Lake, you need to have a basic understanding of Delta Lake and its features. In this recipe, we will walk you through the process of applying table constraints, cloning tables, and altering columns using Delta Lake. We will use Python as the programming language for the code examples.

How to do it...

1. **Import the required libraries**: Start by importing the necessary libraries for working with Delta Lake. In this case, we need the `delta` module and the `SparkSession` class from the `pyspark.sql` module:

    ```
    from delta import configure_spark_with_delta_pip, DeltaTable
    from pyspark.sql import SparkSession
    ```

2. **Create a SparkSession**: To interact with Spark and Delta Lake, you need to create a SparkSession object:

    ```
    builder = (SparkSession.builder
                    .appName("optimize-delta-table")
                    .master("spark://spark-master:7077")
                    .config("spark.executor.memory", "512m")
                    .config("spark.sql.extensions", "io.delta.sql.
    DeltaSparkSessionExtension")
                    .config("spark.sql.catalog.spark_catalog", "org.
    apache.spark.sql.delta.catalog.DeltaCatalog"))

    spark = configure_spark_with_delta_pip(builder).getOrCreate()
    spark.sparkContext.setLogLevel("ERROR")
    ```

3. **Read the Delta Lake table**: To read a Delta Lake table, we need to specify the format as `delta` and provide a path to the Delta Lake table. For PySpark, we use the `spark.read.format("delta")` syntax and load the table using `.load("/opt/workspace/data/delta_lake/netflix_titles")`:

    ```
    # For PySpark:
    df = spark.read.format("delta").load("/opt/workspace/data/delta_
    lake/netflix_titles")
    ```

4. **Apply table constraints**: In Delta Lake, you can apply table constraints to enforce data integrity and define rules for your tables. Table constraints help ensure that data adheres to specific conditions or requirements. Delta Lake supports two types of constraints:

   - NOT NULL: This constraint ensures that values in specific columns cannot be null

   - CHECK: This constraint verifies that a specified Boolean expression must evaluate to true for each input row

   The following code will apply a NOT NULL constraint to the "title" column in our "netflix_titles" Delta table:

   ```
   %%sparksql
   ALTER TABLE delta.`/opt/workspace/data/delta_lake/netflix_
   titles` ALTER COLUMN title DROP NOT NULL
   ```

5. You can also apply CHECK constraints on the table with a valid Boolean expression. The following code will add a CHECK constraint named validType to the netflix_title table. The constraint ensures that the values in the type column of the table can only be one of three specified values: 'Movie', 'Show', or 'TV Show':

   ```
   %%sparksql
   ALTER TABLE delta.'/opt/workspace/data/delta_lake/netflix_
   titles' ADD CONSTRAINT validType CHECK (type IN ('Movie',
   'Show','TV Show'));
   ```

6. **Cloning tables in Delta Lake**: Delta Lake allows you to clone tables using the SHALLOW CLONE command. Only the metadata of the original table is cloned, while the data files are not physically copied.

   The following code creates a new table named netflix_titles_shallow_clone using the SHALLOW CLONE command in Delta Lake. The new table will have the same schema and metadata as the original netflix_titles table but will not copy the actual data files. Instead, it will reference the same data files as the original table, resulting in a shallow clone:

   ```
   %%sparksql
   CREATE OR REPLACE TABLE delta.'/opt/workspace/data/delta_lake/
   netflix_titles_shallow_clone' SHALLOW CLONE delta.'/opt/
   workspace/data/delta_lake/netflix_titles'
   ```

7. **Updating the Delta Lake table schema**: There are different ways to update the schema of a Delta Lake table, depending on your needs and preferences. Here are some methods you can use:

   I.   **Adding columns**: The following code alters the netflix_titles table in Delta Lake and adds a new column named ID with the INT data type:

   ```
   %%sparksql
   ALTER TABLE delta.'/opt/workspace/data/delta_lake/netflix_
   titles' ADD COLUMNS (ID INT)
   ```

II. **Reordering columns**: The following code alters the `netflix_titles` table and modifies the position of the `country` column. By using the `ALTER COLUMN` clause with the `AFTER` keyword, you can specify where the column should be positioned in relation to other columns in the table. In this example, the `country` column will be moved to appear after the `show_id` column in the table's column order:

```
%%sparksql
ALTER TABLE delta.'/opt/workspace/data/delta_lake/netflix_
titles' ADD COLUMNS (ID INT)
```

III. **Enabling column mapping**: Column mapping in Delta Lake enables flexibility in naming and evolving table columns without the need to rewrite the underlying Parquet files. It allows Delta tables to have different column names than the corresponding Parquet file columns. This feature is useful for performing operations such as renaming columns or dropping columns in Delta tables without physically modifying the Parquet files.

Additionally, the column mapping feature allows users to name Delta table columns using characters that are not typically allowed in Parquet, such as spaces. This capability eliminates the need to rename columns when ingesting CSV or JSON data directly into Delta, as the column mapping handles the transformation between the column names used in Delta and the corresponding column names expected by Parquet.

To enable `columnMapping` on our `netflix_title` table, we can run the following command:

```
%%sparksql
ALTER TABLE delta.'/opt/workspace/data/delta_lake/netflix_
title' SET TBLPROPERTIES (
    'delta.minReaderVersion' = '2',
    'delta.minWriterVersion' = '5',
    'delta.columnMapping.mode' = 'name'
)
```

IV. **Renaming a column**: We can rename columns without rewriting the data now that `columnMapping` is enabled on the table. To rename the `lister_in` column `genres`, we can run the following command:

```
%%sparksql
ALTER TABLE delta.'/opt/workspace/data/delta_lake/netflix_
title' RENAME COLUMN listed_in TO genres
```

V. **Dropping a column**: We can also drop columns with the `ALTER` command with the `DROP COLUMN` clause. The following code will delete the `ID` column we added previously:

```
%%sparksql
ALTER TABLE delta.'/opt/workspace/data/delta_lake/netflix_
title' DROP COLUMN ID
```

8. **Stop the Spark session**: Finally, we need to stop the Spark session to release the resources used by Spark:

```
spark.stop()
```

See also

- *Delta column mapping* – `https://docs.delta.io/latest/delta-column-mapping.html`

- *Replace table schema*– `https://docs.delta.io/latest/delta-batch.html#replace-table-schema`

- *Constraints* – `https://docs.delta.io/latest/delta-constraints.html#constraints`

4
Ingesting Streaming Data

Using the Spark SQL engine, Apache Spark Structured Streaming provides a stream processing engine that can handle large-scale and reliable data streams. You can write your streaming computation using the same syntax as a batch computation on static data. The Spark SQL engine will run your computation in an incremental and continuous manner and keep the final result updated as new streaming data arrives. The computation is performed on the same efficient Spark SQL engine. The system also ensures that the computation is fault-tolerant from end to end by using checkpointing and write-ahead logs.

Apache Spark Structured Streaming is favored for real-time data processing due to its high-level, unified API that seamlessly integrates both streaming and batch data processing. This unified approach simplifies development, making it accessible to those familiar with Spark SQL. It offers a wide range of benefits, including built-in fault tolerance mechanisms, support for exactly once semantics, scalability for handling large data volumes, and the ability to perform windowed aggregations for time-based analytics. Additionally, it provides compatibility with various data sources, event time processing for out-of-order data, and an extensive ecosystem and community support. Adaptive query optimization and state management through checkpoints enhance its performance and reliability. Structured Streaming is a powerful choice for organizations across industries, allowing them to process and analyze streaming data with efficiency and ease while meeting low-latency requirements.

In this chapter, we will look at recipes for setting up Spark Structured Streaming for real-time data processing. We will cover creating a streaming query, managing checkpoints and stateful queries, reading data from real-time sources, such as Apache Kafka, and handling schema evolution and metadata.

The recipes covered in this chapter are as follows:

- Configuring Spark Structured Streaming for real-time data processing
- Reading data from real-time sources, such as Apache Kafka, with Apache Spark Structured Streaming
- Defining transformations and filters on a Streaming DataFrame
- Configuring checkpoints for Structured Streaming in Apache Spark
- Configuring triggers for Structured Streaming in Apache Spark

- Applying window aggregations to streaming data with Apache Spark Structured Streaming

- Handling out-of-order and late-arriving events with watermarking in Apache Spark Structured Streaming

By the end of this chapter, you will be able to master the efficient processing and analysis of streaming data using Structured Streaming by exploring its features such as fault tolerance, scalability, windowed aggregations, event time processing, and compatibility with diverse data sources. You will also know how to set up and manage streaming queries, covering checkpoints, triggers, schema evolution, and metadata.

Technical requirements

Before starting, make sure that your `docker-compose` images are up and running. Open JupyterLab's server, running on localhost (`http://127.0.0.1:8888/lab`). Additionally, ensure that you have cloned the Git repo for this book and have access to the notebook and data used in this chapter.

Remember to stop all services defined in the `docker-compose` file for this book when you are done running the code examples. You can do this by executing this command:

```
$ docker-compose stop
```

You can find the notebooks and data for this chapter at `https://github.com/PacktPublishing/Data-Engineering-with-Databricks-Cookbook/tree/main/Chapter04`.

Configuring Spark Structured Streaming for real-time data processing

In this recipe, you will learn how to configure **Apache Spark Structured Streaming** using Python for real-time data processing. Spark Structured Streaming is used in a variety of scenarios in which you need to ingest and analyze data as they arrive in real time from sources such as IoT devices, social media streams, sensors, or financial transactions. Structured Streaming provides the means to handle these continuous data streams. This configuration is particularly relevant when low-latency processing is crucial for making timely decisions or taking immediate actions based on incoming data. Structured Streaming also becomes essential when dealing with event time-based processing, enabling you to perform time-based aggregations and windowing operations on data with timestamps.

Getting ready

To run this recipe, we first need to set up incoming streaming data. We will feed data by opening a terminal window in the JupyterLab UI and running the following command that uses the `nc` (netcat) utility to create a socket connection on port `9999` and listen for incoming data:

```
nc -lk 9999
```

Once the previous command is running, you can start typing any text on the command line.

For example, you can enter the following text:

```
Fundamentals of Data Engineering: Plan and Build Robust Data Systems
by Joe Reis and Matt Housley. This book provides a concise overview
of the data engineering landscape and a framework of best practices to
assess and solve data engineering problems. It also helps you choose
the best technologies and architectures for your data needs.

Designing Data-Intensive Applications: The Big Ideas Behind Reliable,
Scalable, and Maintainable Systems** by Martin Kleppmann. This book
explains the fundamental principles and trade-offs behind the design
of distributed data systems. It covers topics such as replication,
partitioning, consistency, fault tolerance, batch and stream
processing, and data model
```

Here is a screenshot of the sample data:

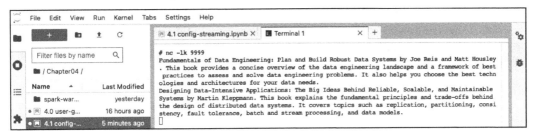

Figure 4.1 – Terminal window with input text to the nc command

How to do it...

1. **Import the required libraries**: Start by importing the necessary libraries for working with Delta Lake. In this case, we need the `delta` module and the `SparkSession` class from the `pyspark.sql` module:

    ```
    from pyspark.sql import SparkSession
    from pyspark.sql.functions import explode, split
    ```

2. **Create a SparkSession object**: To interact with Spark and Delta Lake, you need to create a SparkSession object:

```
spark= (SparkSession.builder
            .appName("config-streaming")
            .master("spark://spark-master:7077")
            .config("spark.executor.memory", "512m")
            .getOrCreate())
spark.sparkContext.setLogLevel("ERROR")
```

3. **Create a streaming DataFrame that represents the input data from the source**: You can use the readStream method of the SparkSession object and specify the source type, options, and schema.

   This code will create a streaming DataFrame that reads data from the socket connection to localhost on port 9999:

```
# Create DataFrame representing the stream of input lines from
connection to localhost:9999
lines = (spark.readStream
            .format("socket")
            .option("host", "localhost")
            .option("port", 9999)
            .load())
```

4. **Apply transformations**: You can use the same methods and functions as you would use on a static DataFrame on the streaming DataFrame. For example, the following code splits the lines into words and creates a new streaming DataFrame with a single column named word using the explode and split PySpark functions:

```
# Split the lines into words
words = lines.select(
    explode(split(lines.value, " ")).alias("word"))
```

5. **Aggregation on streaming DateFrame using groupBy**: This code generates a running word count and creates a new streaming DataFrame with two columns named word and count:

```
# Generate running word count
wordCounts = words.groupBy("word").count()
```

6. **Create a streaming query that writes the output of the streaming DataFrame to the destination**: You can use the writeStream method of the streaming DataFrame and specify the output mode, sink type, options, and trigger. The following code will create a streaming query that writes only the updated results to the console output every five seconds without truncating the output columns:

```
# Start running the query that prints the running counts to the
console
query = (wordCounts.writeStream
```

```
.outputMode("complete")
.format("console")
.start())
```

The output is the following:

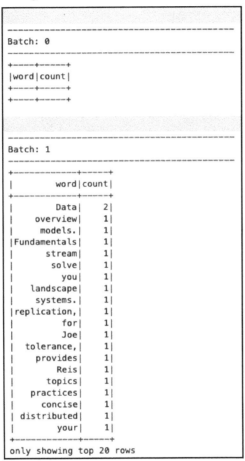

Figure 4.2 – Output for micro-batches

7. **Enter new input data**: Open the terminal and add more data to the netcat listener. See the following example text:

```
Data-Driven Science and Engineering: Machine Learning, Dynamical
Systems, and Control by Steven L. Brunton and J. Nathan Kutz13.
This book teaches you how to apply machine learning and data
analytics techniques to solve complex engineering and scientific
problems. It covers topics such as dimensionality reduction,
sparse sensing, system identification, and control design.
```

A new batch for the stream query is triggered, and the output is updated as shown:

```
-----------------------------------------------------
Batch: 2
-----------------------------------------------------
+--------------+-----+
|          word|count|
+--------------+-----+
|     Dynamical|    1|
|          Data|    2|
|       complex|    1|
|      overview|    1|
|       models.|    1|
|       Science|    1|
|  Fundamentals|    1|
|        stream|    1|
|        Nathan|    1|
|            by|    1|
|         solve|    2|
|           you|    2|
|     landscape|    1|
|            L.|    1|
|      systems.|    1|
|         apply|    1|
|  replication,|    1|
|           for|    1|
|           Joe|    1|
|           how|    1|
+--------------+-----+
only showing top 20 rows
```

Figure 4.3 – Another output triggered after receiving new data

8. **Stop streaming query**: Stop the query by using the stop method of the query:

```
query.stop()
```

9. **Stop the Spark session**: Finally, we need to stop the Spark session to release the resources used by Spark:

```
spark.stop()
```

How it works...

When you create a streaming DataFrame, Spark creates a logical plan that represents your streaming computation as a series of transformations on an unbounded table. This logical plan is similar to a batch query plan, except that it has some additional information, such as a watermark and output mode:

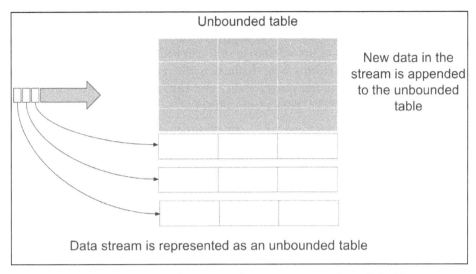

Figure 4.4 – Structured streaming data stream as an unbounded table

When you start a streaming query, Spark creates a physical plan that translates your logical plan into a series of micro-batch jobs or continuous tasks that run on your cluster. The physical plan is optimized by the Spark SQL engine using various techniques, such as predicate pushdown, projection pruning, and join reordering.

Depending on your source type and options, Spark will periodically read new data from the source and append them to an internal buffer. The buffer acts as an input table for your streaming query.

Depending on your trigger type and options, Spark will periodically process new data from the buffer and update an internal state store. The state store acts as an output table for your streaming query. It keeps track of intermediate results such as aggregates, windows, and joins.

Depending on your sink type and options, Spark will periodically write new data from the state store to the destination. The destination acts as a consumer for your streaming query. It receives final results such as counts, averages, and alerts.

There's more...

Let's review some additional information related to Structured Streaming that can help you improve your streaming application and learn more about Spark Structured Streaming:

- **Custom sources and sinks**: You can use various built-in or custom sources and sinks for your streaming application. For example, you can use `file`, `socket`, `rate`, `memory`, and `delta` as sources and `console`, `file`, `memory`, `delta`, `foreach`, and `foreachBatch` as sinks. You can also use Kafka as both source and sink. You can find more details and examples at `https://spark.apache.org/docs/latest/structured-streaming-programming-guide.html#input-sources` and `https://spark.apache.org/docs/latest/structured-streaming-programming-guide.html#output-sinks`.

- **Output mode**: You can use various output modes for your streaming query depending on your application logic and requirements. For example, you can use the append mode to write only new records to the destination, update mode to write only updated records to the destination, or complete mode to write all records to the destination. You can find more details and examples at `https://spark.apache.org/docs/latest/structured-streaming-programming-guide.html#output-modes`.

- **Triggers**: You can use various triggers for your streaming query depending on your latency and throughput needs. For example, you can use a processing time trigger to process data at a fixed interval, an event time trigger to process data based on the event time in the data, or a continuous trigger to process data as soon as it arrives. You can find more details and examples at `https://spark.apache.org/docs/latest/structured-streaming-programming-guide.html#triggers`.

See also

- *Structured Streaming Programming Guide*: `https://spark.apache.org/docs/latest/structured-streaming-programming-guide.html`

- *Spark Structured Streaming API Documentation*: `https://spark.apache.org/docs/latest/api/python/reference/pyspark.streaming.html`

- Spark Structured Streaming examples: `https://github.com/apache/spark/tree/master/examples/src/main/python/sql/streaming`

Reading data from real-time sources, such as Apache Kafka, with Apache Spark Structured Streaming

In this recipe, you will learn how to read data from real-time sources, such as Apache Kafka, with Apache Spark Structured Streaming. You will use the same APIs as when working with batch data. Integrating Apache Spark and Apache Kafka offers a powerful combination of real-time data processing capabilities. It enables real-time data processing as Kafka serves as a highly scalable and fault-tolerant message broker that receives and delivers data streams, which Spark can ingest and analyze as they are generated. Kafka acts as a data buffer, ensuring that data is not lost in cases of processing delays or failures in Spark.

Getting ready

Before we start, we need to make sure that we have a Kafka cluster running and a topic that produces some streaming data. For simplicity, we will use a single-node Kafka cluster and a topic named `users`. Open the `4.0 user-gen-kafka.ipynb` notebook and execute the cell. This notebook produces a user record every few seconds and puts it on a Kafka topic called `users`.

Make sure you have run the previously mentioned notebook and that it is producing records as shown here:

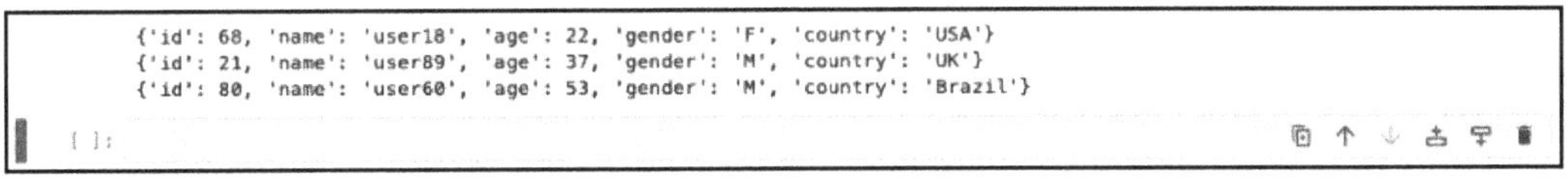

```
{'id': 68, 'name': 'user18', 'age': 22, 'gender': 'F', 'country': 'USA'}
{'id': 21, 'name': 'user89', 'age': 37, 'gender': 'M', 'country': 'UK'}
{'id': 80, 'name': 'user60', 'age': 53, 'gender': 'M', 'country': 'Brazil'}
[ ]:
```

Figure 4.5 – Output from user generation script

How to do it...

1. **Import the required libraries**: Start by importing the necessary libraries for working with Delta Lake. In this case, we need the `delta` module and the `SparkSession` class from the `pyspark.sql` module:

    ```
    from delta import configure_spark_with_delta_pip
    from pyspark.sql import SparkSession
    from pyspark.sql.functions import col, from_json
    from pyspark.sql.types import StructType, StructField,
    IntegerType, StringType
    ```

2. **Create a SparkSession object**: To interact with Spark and Delta Lake, you need to create a `SparkSession` object:

    ```
    builder = (SparkSession.builder
                .appName("connect-kafka-streaming")
                .master("spark://spark-master:7077")
                .config("spark.executor.memory", "512m")
                .config("spark.sql.extensions", "io.delta.sql.
    DeltaSparkSessionExtension")
                .config("spark.sql.catalog.spark_catalog", "org.
    apache.spark.sql.delta.catalog.DeltaCatalog"))

    spark = configure_spark_with_delta_pip(builder,['org.apache.
    spark:spark-sql-kafka-0-10_2.12:3.4.1']).getOrCreate()
    spark.sparkContext.setLogLevel("ERROR")
    ```

3. **Create a DataFrame**: To represent the stream of input data from Kafka, we will need to create a DataFrame using the `spark.readStream.format("kafka")` method. You need to specify some options, such as the bootstrap servers, the topic name or pattern, and the starting and ending offsets:

    ```
    df = (spark.readStream
          .format("kafka")
          .option("kafka.bootstrap.servers", "kafka:9092")
    ```

```
.option("subscribe", "users")
.option("startingOffsets", "earliest")
.load()
```

4. **Parse the JSON messages**: The Kafka messages have a JSON payload in the `value` column. We will use the `from_json` function and provide the schema of the JSON data as an argument to read this payload. We can use the `StructType` and `StructField` classes from the `pyspark.sql.types` module to define the schema:

```
schema = StructType([
    StructField('id', IntegerType(), True),
    StructField('name', StringType(), True),
    StructField('age', IntegerType(), True),
    StructField('gender', StringType(), True),
    StructField('country', StringType(), True)])

df = df.withColumn('value', from_json(col('value').
cast("STRING"), schema))
```

5. **Extract the nested fields**: Since the JSON is nested, we will have to extract the columns from the value column using the `col` function and alias them with meaningful names:

```
df = df.select(
    col('value.id').alias('id'),
    col('value.name').alias('name'),
    col('value.age').alias('age'),
    col('value.gender').alias('gender'),
    col('value.country').alias('country'))
```

6. **Write the transformed DataFrame to a console sink**: We will use the `writeStream` method of the DataFrame to write to a sink. We need to specify the output mode, the format, and the trigger interval as options. We can also use the `option` method to set some additional options, such as showing the watermark and truncating the output:

```
query = (df.writeStream
    .outputMode('append')
    .format('console')
    .start())
```

The output is the following:

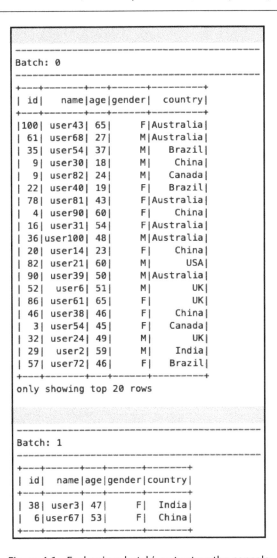

Figure 4.6 – Each micro-batch's output on the console

7. **Stop the streaming query**: Stop the query by using the `stop` method of the query:

    ```
    query.stop()
    ```

8. **Stop the Spark session**: Finally, we need to stop the Spark session to release the resources used by Spark:

    ```
    spark.stop()
    ```

How it works...

- Spark Structured Streaming integrates with Kafka using the `spark-sql-kafka-0-10` library, which provides a source and a sink for Kafka. The source allows you to read data from Kafka topics or partitions as a stream of records, where each record consists of a key, a value, an offset, a partition, a timestamp, and optional headers. The sink allows you to write data to Kafka topics as a stream of records, where each record consists of a key, a value, and optional headers.

- When you create a DataFrame from Kafka using the `readStream` method, you need to specify some options, such as the bootstrap servers, the topic name or pattern, and the starting and ending offsets. The bootstrap servers are the addresses of the Kafka brokers that the source connects to. The topic name or pattern determines which topics or partitions to subscribe to. The start and end offsets determine the range of records to read from each partition. You can use special values such as `earliest`, `latest`, or `none` to indicate the beginning, the end, or no offset, respectively. You can also use JSON strings to specify specific offsets for each partition. You can also enable headers if your Kafka messages have them by setting the `includeHeaders` option to `true`.

- When you write the output of the stream to a destination system using the `writeStream` method, you need to specify some options such as the output mode, the format, the trigger interval, and any additional options for the destination system. The output mode determines how the output table is updated when new data arrives. There are three output modes:

  - **Append mode**: This only adds new rows to the output table and does not modify the existing rows.

  - **Update mode**: This updates the existing rows and adds new rows based on the values in the stream.

  - **Complete mode**: This rewrites the entire output table based on the values in the stream. The output mode that you choose depends on the type of query that you have.

  For example, if you have aggregations without windows, you need to use complete mode because every row in the output table may change when new data arrives. If you have aggregations with windows, you can use update mode because only rows that belong to updated windows may change. If you have no aggregations, you can use append mode because only new rows are added to the output table (compatibility with different types of queries). The format determines what kind of destination system that you want to write to. There are built-in formats such as console, filesystem (`parquet`, `csv`, and `json`), database (`jdbc`), or Kafka. You can also use third-party formats such as S3 (s3a).

- Any additional options for the destination system depend on the format that you choose. For example, if you write to a filesystem, you can specify the path, the partition columns, and the file format. If you write to a database, you can specify the URL, the table name, and the user credentials. If you write to Kafka, you can specify the bootstrap servers, the topic name, and the key and value columns.

- When you start the streaming query using the `start` method, Spark creates a streaming query object that manages the execution of the query. You can use this object to monitor the status of the query, such as the progress, the metrics, and the exceptions. You can also use this object to stop or restart the query if needed.

There's more...

Here are some additional topics that you may want to learn more about:

- **Schema inference**: If your Kafka messages have a structured format, such as JSON or Avro, you can infer their schema automatically using the `from_json` or `from_avro` functions. This way, you can access their nested fields more easily and perform schema validation and evolution.

- **Offset management**: Spark Structured Streaming keeps track of the offsets that it has processed from Kafka using checkpoints and write-ahead logs. This ensures that the query can resume from where it left off in case of failures or restarts. You can also manually manage the offsets using external storage if you want more control over them.

- **Watermarking**: Spark Structured Streaming supports watermarking to handle late and out-of-order data in streams. Watermarking allows you to specify a threshold of how late the data can be and still be considered for processing. For example, if you have a stream of events with timestamps and you want to aggregate them by hour, you can use watermarking to ignore any events that arrive more than an hour late. In this way, you can avoid recomputing previous results and reduce state size.

See also

- *Structured Streaming + Kafka Integration Guide*: `https://spark.apache.org/docs/latest/structured-streaming-kafka-integration.html`

- *Structured Streaming Programming Guide*: `https://spark.apache.org/docs/latest/structured-streaming-programming-guide.html`

- *How to process streams of data with Apache Kafka and Spark*: `https://cloudblogs.microsoft.com/opensource/2018/07/09/how-to-data-processing-apache-kafka-spark/`

- *Real-Time End-to-End Integration with Apache Kafka in Apache Spark's Structured Streaming*: `https://www.databricks.com/blog/2017/04/04/real-time-end-to-end-integration-with-apache-kafka-in-apache-sparks-structured-streaming.html`

- *Apache Spark Structured Streaming — Input Sources*: `https://medium.com/expedia-group-tech/apache-spark-structured-streaming-input-sources-2-of-6-6a72f798838c`

Defining transformations and filters on a Streaming DataFrame

In this recipe, we will show you how to use Spark SQL to define transformations and filters on a streaming DataFrame that reads data from a Kafka topic. Applying transformations and filters on Streaming DataFrames enables the processing and manipulation of streaming data as it traverses the data pipeline. This functionality permits various operations, including data cleaning, enrichment, and reformatting, which are crucial for adapting the data to specific processing requirements. Real-world streaming data often arrive with errors, inconsistencies, or missing values. These operations help clean and sanitize the data by either eliminating problematic records or rectifying issues, ensuring the reliability and accuracy of subsequent analyses.

Getting ready

Before we start, we need to make sure that we have a Kafka cluster running and a topic that produces some streaming data. For simplicity, we will use a single-node Kafka cluster and a topic named `users`. Open the `4.0 user-gen-kafka.ipynb` notebook and execute the cell. This notebook produces a user record every few seconds and puts it on a Kafka topic called `users`.

Make sure you have run this notebook and that it is producing records as shown here:

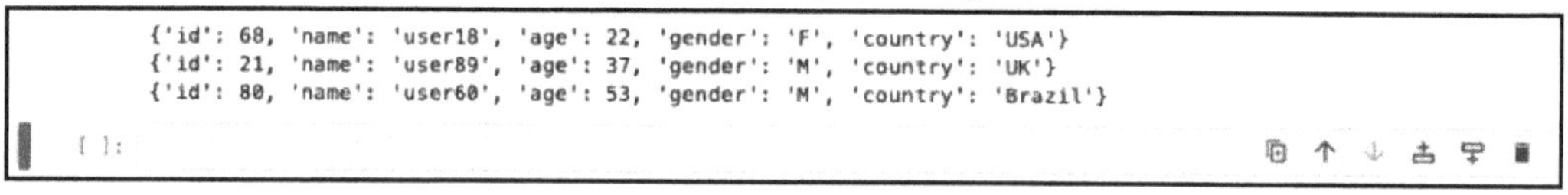

```
{'id': 68, 'name': 'user18', 'age': 22, 'gender': 'F', 'country': 'USA'}
{'id': 21, 'name': 'user89', 'age': 37, 'gender': 'M', 'country': 'UK'}
{'id': 80, 'name': 'user60', 'age': 53, 'gender': 'M', 'country': 'Brazil'}

[ ]:
```

Figure 4.7 – Output from user generation script

How to do it...

1. **Import the required libraries**: Start by importing the necessary libraries for working with Delta Lake. In this case, we need the `delta` module and the `SparkSession` class from the `pyspark.sql` module:

```
from delta import configure_spark_with_delta_pip
from pyspark.sql import SparkSession
from pyspark.sql.functions import col, from_json, avg
from pyspark.sql.types import StructType, StructField,
IntegerType, StringType
```

2. **Create a SparkSession object**: To interact with Spark and Delta Lake, you need to create a `SparkSession` object:

```
builder = (SparkSession.builder
            .appName("transform-filter-streaming")
            .master("spark://spark-master:7077")
```

```
        .config("spark.executor.memory", "512m")
        .config("spark.sql.extensions", "io.delta.sql.
DeltaSparkSessionExtension")
        .config("spark.sql.catalog.spark_catalog", "org.
apache.spark.sql.delta.catalog.DeltaCatalog"))

spark = configure_spark_with_delta_pip(builder,['org.apache.
spark:spark-sql-kafka-0-10_2.12:3.4.1']).getOrCreate()
spark.sparkContext.setLogLevel("ERROR")
```

3. **Create a DataFrame**: To represent the stream of input data from Kafka, we will create a DataFrame using the `spark.readStream.format("kafka")` method. You need to specify some options, such as the bootstrap servers, the topic name or pattern, and the starting and ending offsets. You can also enable headers if your Kafka messages have them:

```
df = (spark.readStream
      .format("kafka")
      .option("kafka.bootstrap.servers", "kafka:9092")
      .option("subscribe", "users")
      .option("startingOffsets", "earliest")
      .load()
```

4. **Parse the JSON messages**: The Kafka messages have a JSON payload in the value column. We will use the `from_json` function and provide the schema of the JSON data as an argument to read this payload. We can use the `StructType` and `StructField` classes from the `pyspark.sql.types` module to define the schema:

```
schema = StructType([
    StructField('id', IntegerType(), True),
    StructField('name', StringType(), True),
    StructField('age', IntegerType(), True),
    StructField('gender', StringType(), True),
    StructField('country', StringType(), True)])

df = df.withColumn('value', from_json(col('value').
cast("STRING"), schema))
```

5. **Extract the nested fields**: Since the JSON is nested, we will have to extract the columns from the value column using the `col` function and alias them with meaningful names:

```
df = df.select(
    col('value.id').alias('id'),
    col('value.name').alias('name'),
    col('value.age').alias('age'),
    col('value.gender').alias('gender'),
    col('value.country').alias('country'))
```

6. **Define transformations and filters**: We can apply the transformations and filters on the DataFrame using Spark SQL expressions. For example, we can group the data by country and gender and calculate the average age for each group using the `groupBy` and `agg` methods. We can also filter out the records where the age is more than 21 using the `filter` method:

```
df = (df.select('age','country', 'gender').filter("age >= 21").
groupBy('country', 'gender').agg(avg('age').alias('avg_age')))
```

7. **Write the transformed and filtered DataFrame to a console sink**: We will use the `writeStream` method of the DataFrame to write to a sink. We need to specify the output mode, the format, and the trigger interval as options.

8. Compare the value of the `avg_age` column between batches. Note how these values change as data is generated on the Kafka source:

```
query = (df.writeStream
    .outputMode('complete')
    .format('console')
    .start())
```

The output is the following:

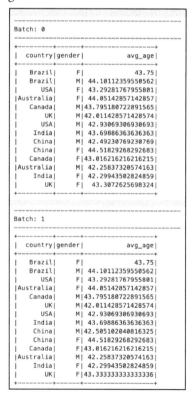

Figure 4.8 – Output of the applied aggregations on every micro-batch

9. **Stop streaming query**: Stop the query by using the `stop` method of the query:

```
query.stop()
```

10. **Stop the Spark session**: Finally, we need to stop the Spark session to release the resources used by Spark:

```
spark.stop()
```

See also

- For more information about Spark Structured Streaming, you can refer to the official documentation: `https://spark.apache.org/docs/latest/structured-streaming-programming-guide.html`

- For more examples of Spark Structured Streaming with Python, you can check out this GitHub repository: `https://github.com/jaceklaskowski/spark-structured-streaming-book`

Configuring checkpoints for Structured Streaming in Apache Spark

In this recipe, we will learn how to configure checkpoints for stateful streaming queries in Apache Spark. Checkpoints are a mechanism to ensure the fault tolerance and reliability of streaming applications by saving the intermediate state of the query to a durable storage system. Checkpoints can also help recover from failures and resume the query from where it left off.

Getting ready

Before we start, we need to make sure that we have a Kafka cluster running and a topic that produces some streaming data. For simplicity, we will use a single-node Kafka cluster and a topic named `users`. Open the `4.0 user-gen-kafka.ipynb` notebook and execute the cell. This notebook produces a user record every few seconds and puts it on a Kafka topic called `users`.

Make sure you have run this notebook and that it is producing records as shown here:

```
{'id': 68, 'name': 'user18', 'age': 22, 'gender': 'F', 'country': 'USA'}
{'id': 21, 'name': 'user89', 'age': 37, 'gender': 'M', 'country': 'UK'}
{'id': 80, 'name': 'user60', 'age': 53, 'gender': 'M', 'country': 'Brazil'}

[ ]:
```

Figure 4.9 – Output from user generation script

How to do it...

1. **Import the required libraries**: Start by importing the necessary libraries for working with Delta Lake. In this case, we need the `delta` module and the `SparkSession` class from the `pyspark.sql` module:

```
from delta import configure_spark_with_delta_pip
from pyspark.sql import SparkSession
from pyspark.sql.functions import col, from_json
from pyspark.sql.types import StructType, StructField,
IntegerType, StringType
```

2. **Create a SparkSession object**: To interact with Spark and Delta Lake, you need to create a `SparkSession` object:

```
builder = (SparkSession.builder
            .appName("transform-filter-streaming")
            .master("spark://spark-master:7077")
            .config("spark.executor.memory", "512m")
            .config("spark.sql.extensions",
                "io.delta.sql.DeltaSparkSessionExtension")
            .config("spark.sql.catalog.spark_catalog",
                "org.apache.spark.sql.delta.catalog.
DeltaCatalog"))

spark = configure_spark_with_delta_pip(builder,['org.apache.
spark:spark-sql-kafka-0-10_2.12:3.4.1']).getOrCreate()
spark.sparkContext.setLogLevel("ERROR")
```

3. **Create a DataFrame**: To represent the stream of input data from Kafka, we will need to create a DataFrame using the `spark.readStream.format("kafka")` method. You need to specify some options, such as the bootstrap servers, the topic name or pattern, and the starting and ending offsets. You can also enable headers if your Kafka messages have them:

```
df = (spark.readStream
        .format("kafka")
        .option("kafka.bootstrap.servers", "kafka:9092")
        .option("subscribe", "users")
        .option("startingOffsets", "earliest")
        .load()
```

4. **Parse the JSON messages**: The Kafka messages have a JSON payload in the value column. We will use the `from_json` function and provide the schema of the JSON data as an argument to read this payload. We can use the `StructType` and `StructField` classes from the `pyspark.sql.types` module to define the schema:

```
schema = StructType([
    StructField('id', IntegerType(), True),
    StructField('name', StringType(), True),
    StructField('age', IntegerType(), True),
    StructField('gender', StringType(), True),
    StructField('country', StringType(), True)])

df = df.withColumn('value', from_json(col('value').
cast("STRING"), schema))
```

5. **Extract the nested fields**: Since the JSON is nested, we will have to extract the columns from the value column using the `col` function and alias them with meaningful names:

```
df = df.select(
    col('value.id').alias('id'),
    col('value.name').alias('name'),
    col('value.age').alias('age'),
    col('value.gender').alias('gender'),
    col('value.country').alias('country'))
```

6. **Writing the streaming data to the Delta Lake table**: We will use the `writeStream` method of the DataFrame to write to a sink. We need to specify the `checkpointLocation` option and specify a directory where we want to store the checkpoint information. We will use the `/opt/workspace/data/checkpoint` directory for this purpose:

```
query = (df.writeStream
    .format("console")
    .outputMode("append")
    .option("checkpointLocation", "/opt/workspace/data/
checkpoint")
    .start())
```

The output is the following:

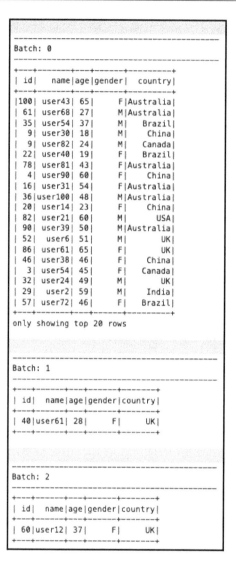

Figure 4.10 – Output of every micro-batch

7. If we run this code, we will see the same output as before in the console, but we will also see a new directory named `checkpoint` in `/opt/workspace/data/checkpoint`. This directory contains the state of our streaming application. For example, the `commits` subdirectory contains information about the batches of data processed so far. If we look inside this subdirectory, we will see something like this:

```
/opt/workspace/data/checkpoint/sources/0/0
/opt/workspace/data/checkpoint/sources/0/1
/opt/workspace/data/checkpoint/sources/0/2
```

8. **Stop the streaming query**: Now, let's simulate a failure by interrupting our streaming application by stopping the streaming query in the notebook:

```
query.stop()
```

9. **Restart the application again with the same code as step 6**: We will see the following output; the application starts with `Batch: 3` since it already processed `Batch 0, 1,` and 2 before the failure/stop:

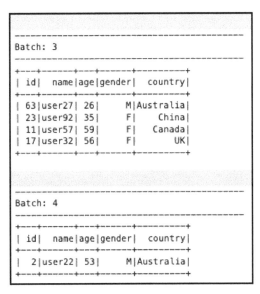

```
-------------------------------------------------
Batch: 3
-------------------------------------------------
+---+------+---+------+---------+
| id|  name|age|gender|  country|
+---+------+---+------+---------+
| 63|user27| 26|     M|Australia|
| 23|user92| 35|     F|    China|
| 11|user57| 59|     F|   Canada|
| 17|user32| 56|     F|       UK|
+---+------+---+------+---------+




-------------------------------------------------
Batch: 4
-------------------------------------------------
+---+------+---+------+---------+
| id|  name|age|gender|  country|
+---+------+---+------+---------+
|  2|user22| 53|     M|Australia|
+---+------+---+------+---------+
```

Figure 4.11 – Output from every micro-batch

10. **Stop the streaming query**: Stop the query by using the `stop` method of the query:

```
query.stop()
```

11. **Stop the Spark session**: Finally, we need to stop the Spark session to release the resources used by Spark:

```
spark.stop()
```

How it works...

Checkpoints store the intermediate state of a streaming query on a HDFS-compatible filesystem. The state includes information such as the following:

- The offsets of the streaming source that have been processed so far
- The current version of the streaming sink that has been written so far

- The configuration and metadata of the streaming query

- The intermediate state of stateful operations such as aggregations and joins

Checkpoints enable fault tolerance and resilience in streaming applications by allowing them to recover from failures or restarts. When a streaming query is restarted, it reads the checkpoint information and resumes from where it left off. This ensures that no data is lost or duplicated in the process.

Checkpoints also enable some advanced features in Spark Structured Streaming, such as the following:

- **Recovery semantics after changes in a streaming query**: Checkpoints allow a streaming query to be modified (such as changing the logic and adding or removing operations) and still resume from where it left off. This is possible because Spark compares the checkpointed query plan with the new query plan and tries to reconcile them.

- **Asynchronous progress tracking**: Checkpoints allow a streaming query to report its progress asynchronously to an external system (such as the Spark UI and SparkListener). This is possible because Spark writes the progress information to the checkpoint directory periodically and allows other systems to read it independently.

There's more...

Here are some additional tips and tricks for configuring checkpoints and triggers in Spark Structured Streaming:

- **Checkpoint location**: The checkpoint location should be a reliable and durable storage system that can handle concurrent writes and reads. It should also be accessible from all nodes in the cluster. It is recommended to use an HDFS-compatible filesystem (such as HDFS, S3, and Azure Blob Storage) for storing checkpoints.

- **Checkpoint cleaning**: Spark Structured Streaming automatically cleans up old checkpoint files that are no longer needed for recovery. However, if you want to manually delete checkpoint files, you should stop the streaming query first and then delete the entire checkpoint directory. You should not delete individual files or subdirectories inside the checkpoint directory, as this may corrupt the state of the streaming query.

- **Checkpoint configuration**: You can configure some aspects of checkpointing using Spark SQL configuration options. For example, you can set `spark.sql.streaming.minBatchesToRetain` to control how many batches of state are retained for recovery. You can also set `spark.sql.streaming.stateStore.providerClass` to choose a different state store implementation (such as RocksDB) for storing the intermediate state.

See also

- *Structured Streaming Programming Guide*: `https://spark.apache.org/docs/latest/structured-streaming-programming-guide.html`

- *Apache Spark Structured Streaming — Checkpoints and Triggers (4 of 6)*: `https://medium.com/expedia-group-tech/apache-spark-structured-streaming-checkpoints-and-triggers-4-of-6-b6f15d5cfd8d`

- *Checkpoint storage in Structured Streaming*: `https://www.waitingforcode.com/apache-spark-structured-streaming/checkpoint-storage-structured-streaming/read`

- *Speed Up Streaming Queries With Asynchronous State Checkpointing* – The Databricks Blog: `https://www.databricks.com/blog/2022/05/02/speed-up-streaming-queries-with-asynchronous-state-checkpointing.html`

Configuring triggers for Structured Streaming in Apache Spark

In this recipe, we will learn how to configure triggers for Structured Streaming in Apache Spark. A trigger is an event that kicks off the processing of a micro-batch in a streaming system. In Spark Structured Streaming, we can configure different types of triggers to control the frequency and timing of micro-batches. To specify the trigger in a streaming query, we use the `trigger` option. Let's see how to do that with some examples.

Getting ready

Before we start, we need to make sure that we have a Kafka cluster running and a topic that produces some streaming data. For simplicity, we will use a single-node Kafka cluster and a topic named `users`. Open the `4.0 user-gen-kafka.ipynb` notebook and execute the cell. This notebook produces a user record every few seconds and puts it on a Kafka topic called `users`.

Make sure you have run this notebook and that it is producing records as shown here:

```
{'id': 68, 'name': 'user18', 'age': 22, 'gender': 'F', 'country': 'USA'}
{'id': 21, 'name': 'user89', 'age': 37, 'gender': 'M', 'country': 'UK'}
{'id': 80, 'name': 'user60', 'age': 53, 'gender': 'M', 'country': 'Brazil'}
```

Figure 4.12 – Output from user generation script

How to do it...

1. **Import the required libraries**: Start by importing the necessary libraries for working with Delta Lake. In this case, we need the `delta` module and the `SparkSession` class from the `pyspark.sql` module:

```
from delta import configure_spark_with_delta_pip
from pyspark.sql import SparkSession
from pyspark.sql.functions import col, from_json
from pyspark.sql.types import StructType, StructField,
IntegerType, StringType
```

2. **Create a SparkSession object**: To interact with Spark and Delta Lake, you need to create a `SparkSession` object:

```
builder = (SparkSession.builder
            .appName("transform-filter-streaming")
            .master("spark://spark-master:7077")
            .config("spark.executor.memory", "512m")
            .config("spark.sql.extensions", "io.delta.sql.
DeltaSparkSessionExtension")
            .config("spark.sql.catalog.spark_catalog", "org.
apache.spark.sql.delta.catalog.DeltaCatalog"))

spark = configure_spark_with_delta_pip(builder,['org.apache.
spark:spark-sql-kafka-0-10_2.12:3.4.1']).getOrCreate()
spark.sparkContext.setLogLevel("ERROR")
```

3. **Create a DataFrame**: To represent the stream of input data from Kafka, we will need to create a DataFrame using the `spark.readStream.format("kafka")` method. You need to specify some options, such as the bootstrap servers, the topic name or pattern, and the starting and ending offsets. You can also enable headers if your Kafka messages have them:

```
df = (spark.readStream
      .format("kafka")
      .option("kafka.bootstrap.servers", "kafka:9092")
      .option("subscribe", "users")
      .option("startingOffsets", "earliest")
      .load())
```

4. **Parse the JSON messages**: The Kafka messages have a JSON payload in the value column. We will use the `from_json` function and provide the schema of the JSON data as an argument to read this payload. We can use the `StructType` and `StructField` classes from the `pyspark.sql.types` module to define the schema:

```
schema = StructType([
    StructField('id', IntegerType(), True),
    StructField('name', StringType(), True),
    StructField('age', IntegerType(), True),
    StructField('gender', StringType(), True),
    StructField('country', StringType(), True)])

df = df.withColumn('value', from_json(col('value').
cast("STRING"), schema))
```

5. **Extract the nested fields**: Since the JSON is nested, we will have to extract the columns from the value column using the `col` function and alias them with meaningful names:

```
df = df.select(
    col('value.id').alias('id'),
    col('value.name').alias('name'),
    col('value.age').alias('age'),
    col('value.gender').alias('gender'),
    col('value.country').alias('country'))
```

6. **Applying default triggers**: The default trigger executes a micro-batch as soon as the previous one finishes. This means that the query will process data as fast as possible, without any delays between micro-batches. This is equivalent to setting the trigger option to `.trigger(processingTime='0 seconds')`. We can use the default trigger by simply omitting the `trigger` option in our streaming query. For example, the following code uses the default trigger:

```
query = (df.writeStream
    .format("console")
    .outputMode("append")
    .start())
```

If we run this code, we will see that the query processes one file in each micro-batch and updates the result table as soon as possible:

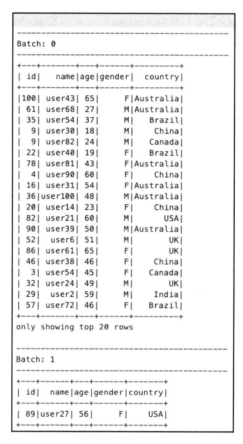

```
--------------------------------------------
Batch: 0
--------------------------------------------

+---+-------+---+------+---------+
| id|   name|age|gender|  country|
+---+-------+---+------+---------+
|100| user43| 65|     F|Australia|
| 61| user68| 27|     M|Australia|
| 35| user54| 37|     M|   Brazil|
|  9| user30| 18|     M|    China|
|  9| user82| 24|     M|   Canada|
| 22| user40| 19|     F|   Brazil|
| 78| user81| 43|     F|Australia|
|  4| user90| 60|     F|    China|
| 16| user31| 54|     F|Australia|
| 36|user100| 48|     M|Australia|
| 20| user14| 23|     F|    China|
| 82| user21| 60|     M|      USA|
| 90| user39| 50|     M|Australia|
| 52|  user6| 51|     M|       UK|
| 86| user61| 65|     F|       UK|
| 46| user38| 46|     F|    China|
|  3| user54| 45|     F|   Canada|
| 32| user24| 49|     M|       UK|
| 29|  user2| 59|     M|    India|
| 57| user72| 46|     F|   Brazil|
+---+-------+---+------+---------+
only showing top 20 rows


--------------------------------------------
Batch: 1
--------------------------------------------
+---+-------+---+------+-------+
| id|   name|age|gender|country|
+---+-------+---+------+-------+
| 89|user27| 56|     F|    USA|
+---+-------+---+------+-------+
```

Figure 4.13 – Every micro-batch triggered as new data arrives

7. **Stop streaming query**: Stop the query by using the `stop` method of the query:

    ```
    query.stop()
    ```

8. **Apply processing time triggers**. The processing time trigger executes a micro-batch at a regular interval based on the processing time (also known as wall-clock time). This means that the query will process data at a fixed rate, with a specified delay between micro-batches. We can set the interval by passing a duration (such as 5 seconds or 1 minute) or a numeric value (such as 5000 or 60000) to the `Trigger.ProcessingTime` function. For example, the following code uses a processing time trigger with an interval of 30 seconds:

    ```
    query = (df.writeStream
        .format("console")
        .outputMode("append")
        .trigger(processingTime='30 seconds')
        .start())
    ```

If we run this code, we will see that the query processes one file in each micro-batch and updates the result table every 10 seconds:

```
------------------------------------------------------
Batch: 0
------------------------------------------------------
+----+-------+---+------+---------+
| id|   name|age|gender|  country|
+----+-------+---+------+---------+
|100| user43| 65|     F|Australia|
| 61| user68| 27|     M|Australia|
| 35| user54| 37|     M|   Brazil|
|  9| user30| 18|     M|    China|
|  9| user82| 24|     M|   Canada|
| 22| user40| 19|     F|   Brazil|
| 78| user81| 43|     F|Australia|
|  4| user90| 60|     F|    China|
| 16| user31| 54|     F|Australia|
| 36|user100| 48|     M|Australia|
| 20| user14| 23|     F|    China|
| 82| user21| 60|     M|      USA|
| 90| user39| 50|     M|Australia|
| 52|  user6| 51|     M|       UK|
| 86| user61| 65|     F|       UK|
| 46| user38| 46|     F|    China|
|  3| user54| 45|     F|   Canada|
| 32| user24| 49|     M|       UK|
| 29|  user2| 59|     M|    India|
| 57| user72| 46|     F|   Brazil|
+----+-------+---+------+---------+
only showing top 20 rows


------------------------------------------------------
Batch: 1
------------------------------------------------------
+----+-------+---+------+---------+
| id|   name|age|gender|  country|
+----+-------+---+------+---------+
| 55|user41| 43|     M|   Canada|
| 28|user38| 21|     F|    China|
| 34|user99| 26|     M|Australia|
+----+-------+---+------+---------+
```

Figure 4.14 – Every micro-batch triggered at a specified time interval

9. **Stop streaming query**: Stop the query by using the `stop` method of the query:

    ```
    query.stop()
    ```

10. **Apply a one-time trigger**: The one-time trigger executes a single micro-batch and then terminates the query. This means that the query will process data only once, without any repetition. We can use the one-time trigger for testing purposes or for running batch queries periodically.

We can use the one-time trigger by setting the `trigger` option to `.trigger(once=True)`. For example, the following code uses a one-time trigger:

```
query = (df.writeStream
    .format("console")
    .outputMode("append")
    .trigger(once=True)
    .start())
```

When we run this code, we will see that the query processes one file in a single micro-batch and updates the result table once, and then stops:

```
-----------------------------------------------------------
Batch: 0
-----------------------------------------------------------
+---+-------+---+------+---------+
| id|   name|age|gender|  country|
+---+-------+---+------+---------+
|100| user43| 65|     F|Australia|
| 61| user68| 27|     M|Australia|
| 35| user54| 37|     M|   Brazil|
|  9| user30| 18|     M|    China|
|  9| user82| 24|     M|   Canada|
| 22| user40| 19|     F|   Brazil|
| 78| user81| 43|     F|Australia|
|  4| user90| 60|     F|    China|
| 16| user31| 54|     F|Australia|
| 36|user100| 48|     M|Australia|
| 20| user14| 23|     F|    China|
| 82| user21| 60|     M|      USA|
| 90| user39| 50|     M|Australia|
| 52|  user6| 51|     M|       UK|
| 86| user61| 65|     F|       UK|
| 46| user38| 46|     F|    China|
|  3| user54| 45|     F|   Canada|
| 32| user24| 49|     M|       UK|
| 29|  user2| 59|     M|    India|
| 57| user72| 46|     F|   Brazil|
+---+-------+---+------+---------+
only showing top 20 rows
```

Figure 4.15 – Only one micro-batch triggered

11. **Stop streaming query**: Stop the query by using the `stop` method of the query:

    ```
    query.stop()
    ```

12. **Stop the Spark session**: Finally, we need to stop the Spark session to release the resources used by Spark:

    ```
    spark.stop()
    ```

How it works...

Triggers control the frequency and timing of micro-batches in a streaming system. Triggers affect the latency and throughput of a streaming application by determining how often and when data is processed.

Spark Structured Streaming supports different types of triggers, such as the following:

- **Default trigger**: This trigger executes a micro-batch as soon as the previous one finishes. It maximizes the throughput of a streaming application by processing data as fast as possible. However, this trigger may also increase the latency of a streaming application by introducing variability in the processing time of each micro-batch.

- **Processing a time trigger**: This trigger executes a micro-batch at a regular interval based on the processing time. It allows a streaming application to control the latency and throughput trade-off by setting a fixed rate of processing data. However, this trigger may also introduce some delay or data loss in a streaming application if the processing time exceeds the interval or if there are failures or restarts.

- **One-time trigger**: This trigger executes a single micro-batch and then terminates the query. It allows a streaming application to process data only once, without any repetition. This trigger can be useful for testing purposes or for running batch queries periodically.

See also

- *Structured Streaming Programming Guide*: https://spark.apache.org/docs/latest/structured-streaming-programming-guide.html

- *Speed Up Streaming Queries With Asynchronous State Checkpointing* – The Databricks Blog: https://www.databricks.com/blog/2022/05/02/speed-up-streaming-queries-with-asynchronous-state-checkpointing.html

- *Apache Spark Structured Streaming — Checkpoints and Triggers (4 of 6)*: https://medium.com/expedia-group-tech/apache-spark-structured-streaming-checkpoints-and-triggers-4-of-6-b6f15d5cfd8d

- *Checkpoint storage in Structured Streaming*: https://www.waitingforcode.com/apache-spark-structured-streaming/checkpoint-storage-structured-streaming/read

Applying window aggregations to streaming data with Apache Spark Structured Streaming

In this recipe, we will learn how to configure window aggregations on streaming queries in Apache Spark. Window aggregations are a common operation in stream processing, where we want to compute some aggregate function (such as count, sum, and average) over a sliding window of time or rows. For example, we might want to know the number of orders per minute, the average revenue per hour, or the top 10 products per day.

Getting ready

Before we start, we need to make sure that we have a Kafka cluster running and a topic that produces some streaming data. For simplicity, we will use a single-node Kafka cluster and a topic named events. Open the 4.0 events-gen-kafka.ipynb notebook and execute the cell. This notebook produces an event record every second and puts it on a Kafka topic called events.

Make sure you have run this notebook and that it is producing records as shown here:

```
{'user_id': 27, 'event_type': 'share', 'event_time': '08/06/2023, 14:19:07', 'processing_time': '08/06/2023, 14:19:14'}
{'user_id': 49, 'event_type': 'purchase', 'event_time': '08/06/2023, 14:19:06', 'processing_time': '08/06/2023, 14:19:15'}
{'user_id': 86, 'event_type': 'click', 'event_time': '08/06/2023, 14:19:06', 'processing_time': '08/06/2023, 14:19:16'}
{'user_id': 54, 'event_type': 'view', 'event_time': '08/06/2023, 14:19:13', 'processing_time': '08/06/2023, 14:19:17'}
{'user_id': 32, 'event_type': 'click', 'event_time': '08/06/2023, 14:19:17', 'processing_time': '08/06/2023, 14:19:18'}
{'user_id': 55, 'event_type': 'view', 'event_time': '08/06/2023, 14:19:06', 'processing_time': '08/06/2023, 14:19:19'}
{'user_id': 12, 'event_type': 'click', 'event_time': '08/06/2023, 14:19:15', 'processing_time': '08/06/2023, 14:19:20'}
{'user_id': 29, 'event_type': 'click', 'event_time': '08/06/2023, 14:19:20', 'processing_time': '08/06/2023, 14:19:21'}
{'user_id': 78, 'event_type': 'click', 'event_time': '08/06/2023, 14:19:03', 'processing_time': '08/06/2023, 14:19:22'}
{'user_id': 74, 'event_type': 'share', 'event_time': '08/06/2023, 14:19:01', 'processing_time': '08/06/2023, 14:19:23'}
{'user_id': 8, 'event_type': 'view', 'event_time': '08/06/2023, 14:19:21', 'processing_time': '08/06/2023, 14:19:24'}
{'user_id': 6, 'event_type': 'like', 'event_time': '08/06/2023, 14:19:13', 'processing_time': '08/06/2023, 14:19:25'}
{'user_id': 30, 'event_type': 'like', 'event_time': '08/06/2023, 14:19:18', 'processing_time': '08/06/2023, 14:19:26'}
{'user_id': 55, 'event_type': 'like', 'event_time': '08/06/2023, 14:19:15', 'processing_time': '08/06/2023, 14:19:27'}
{'user_id': 25, 'event_type': 'purchase', 'event_time': '08/06/2023, 14:19:20', 'processing_time': '08/06/2023, 14:19:28'}
```

Figure 4.16 – Output from events generation script

How to do it...

1. **Import the required libraries**: Start by importing the necessary libraries for working with Delta Lake. In this case, we need the delta module and the SparkSession class from the pyspark.sql module:

```
from delta import configure_spark_with_delta_pip
from pyspark.sql import SparkSession
from pyspark.sql.functions import col, from_json, window, count, to_timestamp
from pyspark.sql.types import StructType, StructField, IntegerType, StringType
```

2. **Create a SparkSession object**: To interact with Spark and Delta Lake, you need to create a `SparkSession` object:

```
builder = (SparkSession.builder
            .appName("apply-window-aggregations")
            .master("spark://spark-master:7077")
            .config("spark.executor.memory", "512m")
            .config("spark.sql.extensions", "io.delta.sql.
DeltaSparkSessionExtension")
            .config("spark.sql.catalog.spark_catalog", "org.
apache.spark.sql.delta.catalog.DeltaCatalog"))

spark = configure_spark_with_delta_pip(builder,['org.apache.
spark:spark-sql-kafka-0-10_2.12:3.4.1']).getOrCreate()
spark.sparkContext.setLogLevel("ERROR")
```

3. **Create a DataFrame**: To represent the stream of input data from Kafka, we will need to create a DataFrame using the `spark.readStream.format("kafka")` method. You need to specify some options, such as the bootstrap servers, the topic name or pattern, and the starting and ending offsets. You can also enable headers if your Kafka messages have them:

```
df = (spark.readStream
        .format("kafka")
        .option("kafka.bootstrap.servers", "kafka:9092")
        .option("subscribe", "events")
        .option("startingOffsets", "earliest")
        .load())
```

4. **Parse the JSON messages**: The Kafka messages have a JSON payload in the value column. We will use the `from_json` function and provide the schema of the JSON data as an argument to read this payload. We can use the `StructType` and `StructField` classes from the `pyspark.sql.types` module to define the schema:

```
schema = StructType([
    StructField('user_id', IntegerType(), True),
    StructField('event_type', StringType(), True),
    StructField('event_time', StringType(), True),
    StructField('processing_time', StringType(), True)])

df = df.withColumn('value', from_json(col('value').
cast("STRING"), schema))
```

5. **Extract the nested fields**: Since the JSON is nested, we will have to extract the columns from the value column using the `col` function and alias them with meaningful names:

```
df = (df
    .select(
        col('value.user_id').alias('user_id'),
        col('value.event_type').alias('event_type'),
        col('value.event_time').alias('event_time'),
        col('value.processing_time')
            .alias('processing_time')
    .withColumn("event_time"
        ,to_timestamp(col("event_time")
        , "MM/dd/yyyy, HH:mm:ss" ))
    .withColumn("processing_time"
        ,to_timestamp(col("processing_time")
        , "MM/dd/yyyy, HH:mm:ss")))
```

6. **Apply a window aggregation function on the DataFrame**: For example, to compute the number of users per hour for each event type, we can use the following code:

```
df = (df.groupBy(
    window(col("event_time"), "60 minute", "60 minute")
    , col("event_type"))
    .agg(count(col("user_id")).alias("NumberOfUsers")))
```

7. **Write the output of the window aggregation to the console**: We will use the `writeStream` method of the DataFrame to write to a sink. We need to specify the output mode, the format, and the trigger interval as options:

```
query = (df.writeStream
    .outputMode('complete')
    .format('console')
    .start())
```

The output is as follows:

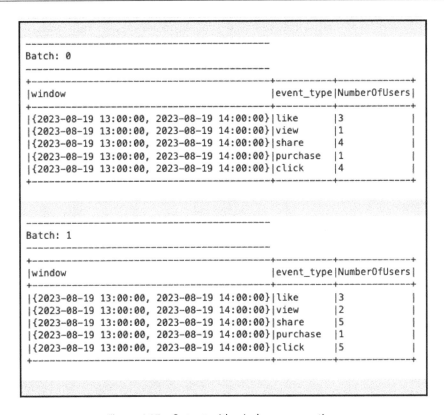

```
-----------------------------------------------
Batch: 0
-----------------------------------------------

+----------------------------------------------+----------+-------------+
|window                                        |event_type|NumberOfUsers|
+----------------------------------------------+----------+-------------+
|{2023-08-19 13:00:00, 2023-08-19 14:00:00}|like    |3            |
|{2023-08-19 13:00:00, 2023-08-19 14:00:00}|view    |1            |
|{2023-08-19 13:00:00, 2023-08-19 14:00:00}|share   |4            |
|{2023-08-19 13:00:00, 2023-08-19 14:00:00}|purchase|1            |
|{2023-08-19 13:00:00, 2023-08-19 14:00:00}|click   |4            |
+----------------------------------------------+----------+-------------+




-----------------------------------------------
Batch: 1
-----------------------------------------------

+----------------------------------------------+----------+-------------+
|window                                        |event_type|NumberOfUsers|
+----------------------------------------------+----------+-------------+
|{2023-08-19 13:00:00, 2023-08-19 14:00:00}|like    |3            |
|{2023-08-19 13:00:00, 2023-08-19 14:00:00}|view    |2            |
|{2023-08-19 13:00:00, 2023-08-19 14:00:00}|share   |5            |
|{2023-08-19 13:00:00, 2023-08-19 14:00:00}|purchase|1            |
|{2023-08-19 13:00:00, 2023-08-19 14:00:00}|click   |5            |
+----------------------------------------------+----------+-------------+
```

Figure 4.17 – Output with window aggregations

8. **Stop streaming query**: Stop the query by using the `stop` method of the query:

```
query.stop()
```

9. **Stop the Spark session**: Finally, we need to stop the Spark session to release the resources used by Spark:

```
spark.stop()
```

There's more...

We can use watermarking to handle late or out-of-order data in our streaming query. Watermarking is a mechanism that allows us to specify how late we expect our data to be and how long we want to wait for late data before dropping the old state. To use watermarking, we need to use the `withWatermark` method on our DataFrame before applying any window operation. For example, if we expect our data to be at most 10 minutes late, we can use the following:

```
df = df.withWatermark("timestamp", "10 minutes")
```

We can use different types of window aggregations depending on our use case. For example, if we want to compute an `aggregate` function over fixed intervals of time (such as every hour or every day), we can use tumbling windows, which have no overlap between consecutive windows. If we want to compute an aggregate function over sliding intervals of time (such as every 15 minutes or every 30 seconds), we can use sliding windows, which have some overlap between consecutive windows. If we want to compute an `aggregate` function over sessions of activity (such as user sessions or shopping sessions), we can use session windows, which have variable durations based on gaps of inactivity. To use different types of window aggregations, we need to specify different window specifications in our `groupBy` method:

```
# Tumbling window
df = (df.groupBy(window(col("event_time"), "1 hour"), col("event_
type"))
    .agg(...))

# Sliding window
df = (df.groupBy(window(col("event_time"), "15 minutes", "5 minutes"),
col("event_type"))
    .agg(...))

# Session
df = (df.groupBy(session_window(col("event_time"), "15 minutes"),
col("event_type"))
.agg(…))
```

We can use different output modes depending on our use case. For example, if we only want to see the new or updated rows in each window, we can use the update output mode, which only shows the rows that have changed since the last trigger. If we only want to see the new rows that have been added to the result table, we can use the append output mode, which only shows the rows that are new in each trigger. To use different output modes, we need to specify them in our `writeStream` method:

```
# Update output mode
query = (df.writeStream.outputMode("update")
    .format("console")
    .option("truncate", False)
    .start())

# Append output mode
query = (df.writeStream.outputMode("append")
    .format("console")
    .option("truncate", False)
    .start())
```

See also

- *Structured Streaming Programming Guide*: `https://spark.apache.org/docs/latest/structured-streaming-programming-guide.html`

- *Window Operations on Event Time*: `https://spark.apache.org/docs/latest/structured-streaming-programming-guide.html#window-operations-on-event-time`

- *Watermarking*: `https://spark.apache.org/docs/latest/structured-streaming-programming-guide.html#handling-late-data-and-watermarking`

- *Output Modes*: `https://spark.apache.org/docs/latest/structured-streaming-programming-guide.html#output-modes`

Handling out-of-order and late-arriving events with watermarking in Apache Spark Structured Streaming

In this recipe, you will learn how to use watermarking to handle out-of-order and late-arriving events in a streaming application that computes the average temperature of different cities over a sliding window of time. You will use Spark SQL to define the streaming query and the watermark logic. You will also learn how to monitor the progress and performance of your streaming application using the Spark UI.

Watermarking is a technique that allows Apache Spark Structured Streaming to handle out-of-order and late-arriving events in streaming applications. It enables the system to specify how late the data can be and handle old data or data that arrives after the expected window accordingly. Watermarking also allows the system to free up states and resources by discarding old data that is no longer relevant.

Getting ready

Before we start, we need to make sure that we have a Kafka cluster running and a topic that produces some streaming data. For simplicity, we will use a single-node Kafka cluster and a topic named `events`. Open the `4.0 events-gen-kafka.ipynb` notebook and execute the cell. This notebook produces an event record every second and puts it on a Kafka topic called `events`.

Make sure you have run this notebook and that it is producing records as shown here:

```
{'user_id': 27, 'event_type': 'share', 'event_time': '08/06/2023, 14:19:07', 'processing_time': '08/06/2023, 14:19:14'}
{'user_id': 49, 'event_type': 'purchase', 'event_time': '08/06/2023, 14:19:06', 'processing_time': '08/06/2023, 14:19:15'}
{'user_id': 86, 'event_type': 'click', 'event_time': '08/06/2023, 14:19:06', 'processing_time': '08/06/2023, 14:19:16'}
{'user_id': 54, 'event_type': 'view', 'event_time': '08/06/2023, 14:19:13', 'processing_time': '08/06/2023, 14:19:17'}
{'user_id': 32, 'event_type': 'click', 'event_time': '08/06/2023, 14:19:17', 'processing_time': '08/06/2023, 14:19:18'}
{'user_id': 55, 'event_type': 'view', 'event_time': '08/06/2023, 14:19:06', 'processing_time': '08/06/2023, 14:19:19'}
{'user_id': 12, 'event_type': 'click', 'event_time': '08/06/2023, 14:19:15', 'processing_time': '08/06/2023, 14:19:20'}
{'user_id': 29, 'event_type': 'click', 'event_time': '08/06/2023, 14:19:20', 'processing_time': '08/06/2023, 14:19:21'}
{'user_id': 78, 'event_type': 'click', 'event_time': '08/06/2023, 14:19:03', 'processing_time': '08/06/2023, 14:19:22'}
{'user_id': 74, 'event_type': 'share', 'event_time': '08/06/2023, 14:19:01', 'processing_time': '08/06/2023, 14:19:23'}
{'user_id': 8, 'event_type': 'view', 'event_time': '08/06/2023, 14:19:21', 'processing_time': '08/06/2023, 14:19:24'}
{'user_id': 6, 'event_type': 'like', 'event_time': '08/06/2023, 14:19:13', 'processing_time': '08/06/2023, 14:19:25'}
{'user_id': 30, 'event_type': 'like', 'event_time': '08/06/2023, 14:19:18', 'processing_time': '08/06/2023, 14:19:26'}
{'user_id': 55, 'event_type': 'like', 'event_time': '08/06/2023, 14:19:15', 'processing_time': '08/06/2023, 14:19:27'}
{'user_id': 25, 'event_type': 'purchase', 'event_time': '08/06/2023, 14:19:20', 'processing_time': '08/06/2023, 14:19:28'}
```

Figure 4.18 – Output from events generation script

How to do it...

1. **Import the required libraries**: Start by importing the necessary libraries for working with Delta Lake. In this case, we need the `delta` module and the `SparkSession` class from the `pyspark.sql` module:

```
from delta import configure_spark_with_delta_pip
from pyspark.sql import SparkSession
from pyspark.sql.functions import col, from_json, window, count, to_timestamp
from pyspark.sql.types import StructType, StructField, IntegerType, StringType
```

2. **Create a SparkSession object**: To interact with Spark and Delta Lake, you need to create a `SparkSession` object:

```
builder = (SparkSession.builder
           .appName("handle-late-and-out-of-order-data")
           .master("spark://spark-master:7077")
           .config("spark.executor.memory", "512m")
           .config("spark.sql.extensions", "io.delta.sql.DeltaSparkSessionExtension")
           .config("spark.sql.catalog.spark_catalog", "org.apache.spark.sql.delta.catalog.DeltaCatalog"))

spark = configure_spark_with_delta_pip(builder,['org.apache.spark:spark-sql-kafka-0-10_2.12:3.4.1']).getOrCreate()
spark.sparkContext.setLogLevel("ERROR")
```

3. **Create a DataFrame**: To represent the stream of input data from Kafka, we will need to create a DataFrame using the `spark.readStream.format("kafka")` method. You need to specify some options, such as the bootstrap servers, the topic name or pattern, and the starting and ending offsets. You can also enable headers if your Kafka messages have them:

```
df = (spark.readStream
      .format("kafka")
      .option("kafka.bootstrap.servers", "kafka:9092")
      .option("subscribe", "events")
      .option("startingOffsets", "latest")
      .load())
```

4. **Parse the JSON messages:** The Kafka messages have a JSON payload in the value column. We will use the `from_json` function and provide the schema of the JSON data as an argument to read this payload. We can use the `StructType` and `StructField` classes from the `pyspark.sql.types` module to define the schema:

```
schema = StructType([
    StructField('user_id', IntegerType(), True),
    StructField('event_type', StringType(), True),
    StructField('event_time', StringType(), True),
    StructField('processing_time', StringType(), True)])

df = df.withColumn('value', from_json(col('value').
cast("STRING"), schema))
```

5. **Extract the nested fields**: Since the JSON is nested, we will have to extract the columns from the value column using the `col` function and alias them with meaningful names:

```
df = (df.select(
        col('value.user_id').alias('user_id'),
        col('value.event_type').alias('event_type'),
        col('value.event_time').alias('event_time'),
        col('value.processing_time')
          .alias('processing_time'))
      .withColumn("event_time"
        , to_timestamp(col("event_time")
        , "MM/dd/yyyy, HH:mm:ss" ))
      .withColumn("processing_time"
        , to_timestamp(col("processing_time")
        , "MM/dd/yyyy, HH:mm:ss")))
```

6. **Define the watermark logic for your streaming DataFrame**: We will be using the `withWatermark` method. You need to specify two parameters: the event time column and the threshold. The event time column is the column that contains the timestamp of each event, which is `event_time` in this example. The threshold is the maximum amount of time you want to wait for late or out-of-order events, which is 10 seconds in this example. This means that any event that arrives more than 10 seconds after its expected window will be dropped by the system:

```
# Define the watermark logic for the streaming DataFrame
df = df.withWatermark("event_time", "10 seconds")
```

7. **Apply a window aggregation function on the DataFrame**: For example, to compute the number of users per hour for each event type, we can use the following code:

```
df = (df.groupBy(
            window(col("event_time")
              , "1 minute", "1 minute")
              , col("user_id"))
        .count().alias("NumberOfEvents"))
```

8. **Start the streaming query and write the output to the console**: We will use the `writeStream` method. You need to specify the output mode, the format, and the option for truncating the output. In this example, we use `update`, which means that only the rows that were updated by the new data will be written to the output:

```
query = (df.writeStream
    .outputMode('update')
    .format('console')
    .option("truncate", False)
    .start())
```

The output is the following:

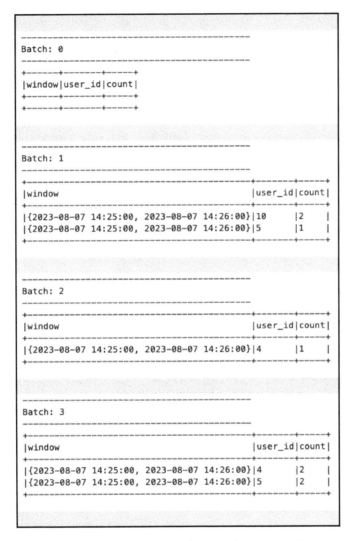

Figure 4.19 – Console output of the window aggregations

9. **Stop streaming query**: Stop the query by using the `stop` method of the query:

    ```
    query.stop()
    ```

10. **Stop the Spark session**: Finally, we need to stop the Spark session to release the resources used by Spark:

    ```
    spark.stop()
    ```

There's more...

Some additional information about watermarking is the following:

- Watermarking is a trade-off between latency and completeness. If you set a lower threshold, you will get faster results, but you may miss some late or out-of-order events. If you set a higher threshold, you will get more complete results, but you may have to wait longer for them.

- Watermarking works well for streaming applications that have a bounded delay in the data source, such as network or sensor data. However, it may not work well for streaming applications that have an unbounded delay in the data source, such as user-generated data or historical data.

- Watermarking also enables some stateful operations on streaming data, such as windowed aggregations, joins, and deduplication. These operations require the system to maintain some state for each key or window, which can grow indefinitely over time. Watermarking allows the system to evict old states that are no longer relevant based on the event time and the threshold.

See also

- *Structured Streaming Programming Guide*: `https://spark.apache.org/docs/latest/structured-streaming-programming-guide.html`

- *Feature Deep Dive: Watermarking in Apache Spark Structured Streaming*: `https://www.databricks.com/blog/2022/08/22/feature-deep-dive-watermarking-apache-spark-structured-streaming.html`

- *Event-time Aggregation and Watermarking in Apache Spark's Structured Streaming*: `https://www.databricks.com/blog/2017/05/08/event-time-aggregation-watermarking-apache-sparks-structured-streaming.html`

- *Spark Structured Streaming (Part 4) – Handling Late Data*: `https://blog.knoldus.com/spark-structured-streaming-part-4-handling-late-data/`

5

Processing Streaming Data

Streaming data is data that is continuously generated and updated in real time, such as sensor readings, weblogs, social media posts, online transactions, and more. Streaming data can provide valuable insights into the current state and trends of various domains, such as e-commerce, finance, health care, gaming, and the **Internet of Things** (**IoT**). However, streaming data also poses many challenges for data ingestion and processing, such as scalability, reliability, fault tolerance, latency, and consistency.

Apache Spark is a popular open source framework for large-scale distributed data processing. Apache Spark Structured Streaming is an extension of Spark SQL that enables scalable and fault-tolerant processing of streaming data using a declarative API based on DataFrames and datasets. Apache Spark Structured Streaming supports various sources and sinks for streaming data, such as Kafka, Flume, **Hadoop Distributed File System** (**HDFS**), **Amazon Simple Storage Service** (**Amazon S3**), Azure Blob storage, and more.

Delta Lake is an open source storage layer that brings ACID transactions and schema enforcement to Apache Spark and big data workloads. Delta Lake enables reliable and efficient management of both batch and streaming data using a common format called Delta files. Delta Lake also provides features such as time travel, data versioning, audit history, and upserts.

In this chapter, you will learn how to use Apache Spark Structured Streaming and Delta Lake to ingest and process streaming data in various scenarios.

We will cover the following recipes in this chapter:

- Writing the output of Apache Spark Structured Streaming to a sink such as Delta Lake
- Idempotent stream writing with Delta Lake and Apache Spark Structured Streaming
- Merging or applying Change Data Capture on Apache Spark Structured Streaming and Delta Lake
- Joining streaming data with static data in Apache Spark Structured Streaming and Delta Lake
- Joining streaming data with streaming data in Apache Spark Structured Streaming and Delta Lake
- Monitoring real-time data processing with Apache Spark Structured Streaming

By the end of this chapter, you will be able to build scalable and reliable streaming pipelines using Apache Spark Structured Streaming and Delta Lake. You will have learned how to ensure data consistency and quality through idempotent and upsert operations, enrich and analyze streaming data with joins, and optimize applications using monitoring tools.

Technical requirements

Before starting, make sure that your `docker-compose` images are up and running, and open Jupyter Lab's server running on the localhost (`http://127.0.0.1:8888/lab`). Also, ensure that you have cloned the Git repo for this book and have access to the notebook and data used in this chapter.

Remember to stop all services defined in the `docker-compose` file for this book when you are done running the code examples. You can do this by executing this command:

```
$ docker-compose stop
```

You can find the notebooks and data for this chapter at `https://github.com/PacktPublishing/Data-Engineering-with-Databricks-Cookbook/tree/main/Chapter05`.

Writing the output of Apache Spark Structured Streaming to a sink such as Delta Lake

In this recipe, you will learn how to write the output of Apache Spark Structured Streaming to a sink such as Delta Lake. Delta Lake can serve as a unified storage layer for various data types, reducing data silos within organizations. By using Delta Lake as a sink for streaming data, you can simplify data pipelines, reducing complexity and streamlining data architecture. Delta Lake enables unified analytics, allowing you to leverage a wide range of analytics tools and frameworks within a single environment, including Apache Spark, Databricks, SQL, and **machine learning** (**ML**) libraries. This versatility makes Delta Lake a valuable choice for real-time data processing and analytics pipelines, enhancing data reliability, durability, and consistency while simplifying data management and supporting compliance requirements.

Getting ready

Before we start, we need to make sure that we have a Kafka cluster running and a topic that produces some streaming data. For simplicity, we will use a single-node Kafka cluster and a topic named `users`. Open the `5.0 user-gen-kafka.ipynb` notebook and execute the cell. This notebook produces a user record every few seconds and puts it on a Kafka topic called `users`.

Make sure you have run this notebook and that it is producing records as shown:

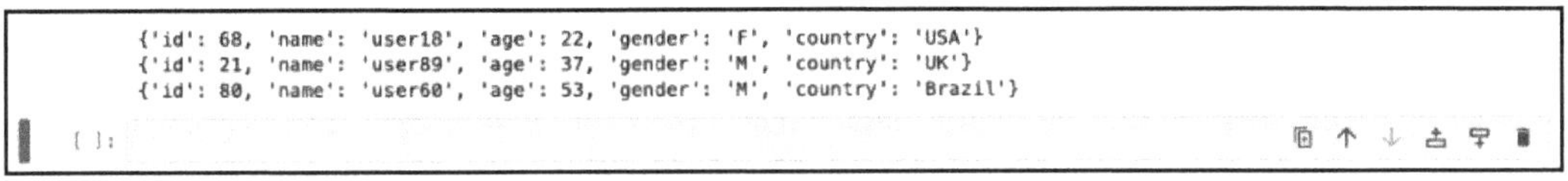

```
{'id': 68, 'name': 'user18', 'age': 22, 'gender': 'F', 'country': 'USA'}
{'id': 21, 'name': 'user89', 'age': 37, 'gender': 'M', 'country': 'UK'}
{'id': 80, 'name': 'user60', 'age': 53, 'gender': 'M', 'country': 'Brazil'}
```

Figure 5.1 – Kafka user data generation output

How to do it...

1. **Import the required libraries**: Start by importing the necessary libraries for working with Delta Lake. In this case, we need the `delta` module and the `SparkSession` class from the `pyspark.sql` module:

```
from delta import configure_spark_with_delta_pip, DeltaTable
from pyspark.sql import SparkSession
from pyspark.sql.functions import col, from_json
from pyspark.sql.types import StructType, StructField,
IntegerType, StringType
```

2. **Create a SparkSession object**: To interact with Spark and Delta Lake, you need to create a `SparkSession` object. We will also be using a `configure_spark_with_delta_pip` function that takes the `SparkSession` builder and a list of Spark packages as parameters to install the `org.apache.spark:spark-sql-kafka-0-10_2.12:3.4.1` package. This package provides integration between Apache Spark's Structured Streaming module and Apache Kafka. Specifically, it allows Spark Structured Streaming to consume and process data from Kafka topics:

```
builder = (SparkSession.builder
           .appName("delta-write-streaming")
           .master("spark://spark-master:7077")
           .config("spark.executor.memory", "512m")
           .config("spark.sql.extensions", "io.delta.sql.
DeltaSparkSessionExtension")
           .config("spark.sql.catalog.spark_catalog", "org.
apache.spark.sql.delta.catalog.DeltaCatalog"))

spark = configure_spark_with_delta_pip(builder,['org.apache.
spark:spark-sql-kafka-0-10_2.12:3.4.1']).getOrCreate()
spark.sparkContext.setLogLevel("ERROR")
```

3. **Create a Delta table**: Before we load data into the table, we need to create a table in the Delta catalog. This can be done with the CREATE OR REPLACE TABLE command. You can specify the table name, which is default.netflix_titles, and also specify the data format as DELTA and the location of the table data, which in our case is /opt/workspace/data/delta_lake/delta-write-streaming/users:

```
%%sparksql
CREATE OR REPLACE TABLE default.users (
    id INT,
    name STRING,
    age INT,
    gender STRING,
    country STRING
) USING DELTA LOCATION '/opt/workspace/data/delta_lake/delta-write-streaming/users';
```

4. **Create a DataFrame**: To represent the stream of input data from Kafka, we need to create a DataFrame using the spark.readStream.format("kafka") method. You need to specify some options, such as the bootstrap servers, the topic name or pattern, and the starting and ending offsets. You can also enable headers if your Kafka messages have them:

```
df = (spark.readStream
    .format("kafka")
    .option("kafka.bootstrap.servers", "kafka:9092")
    .option("subscribe", "users")
    .option("startingOffsets", "earliest")
    .load())
```

5. **Parse the JSON messages**: Parse the data in the value column using the from_json function and provide the schema of the JSON data as an argument. We can use the StructType and StructField classes from the pyspark.sql.types module to define the schema:

```
schema = StructType([
    StructField('id', IntegerType(), True),
    StructField('name', StringType(), True),
    StructField('age', IntegerType(), True),
    StructField('gender', StringType(), True),
    StructField('country', StringType(), True)])

df = df.withColumn('value', from_json(col('value').cast("STRING"), schema))
```

6. **Extract the nested fields**: From the `value` column, extract the fields using the `col` function and alias them with meaningful names:

```
df = df.select(
    col('value.id').alias('id'),
    col('value.name').alias('name'),
    col('value.age').alias('age'),
    col('value.gender').alias('gender'),
    col('value.country').alias('country'))
```

7. **Write the streaming data to the Delta Lake table**: You can use the `writeStream` method to write the streaming data to the Delta Lake table. You need to specify the format as `delta` and the location as the path of the Delta table. You also need to specify the output mode as `append`, `complete`, or `update`. For example, you can use the following Python code to write the streaming data to the Delta table in `append` mode:

```
query = (df.writeStream
    .format("delta")
    .outputMode("append")
    .option("checkpointLocation", "/opt/workspace/data/delta_
lake/delta-write-streaming/users/_checkpoints/")
    .start("/opt/workspace/data/delta_lake/delta-write-
streaming/users"))
```

8. **Query the Delta table**: Run the SQL `SELECT` command to query the Delta table:

```
%%sparksql
SELECT COUNT(*) FROM delta.`/opt/workspace/data/delta_lake/
delta-write-streaming/users`;
```

This will result in the following output:

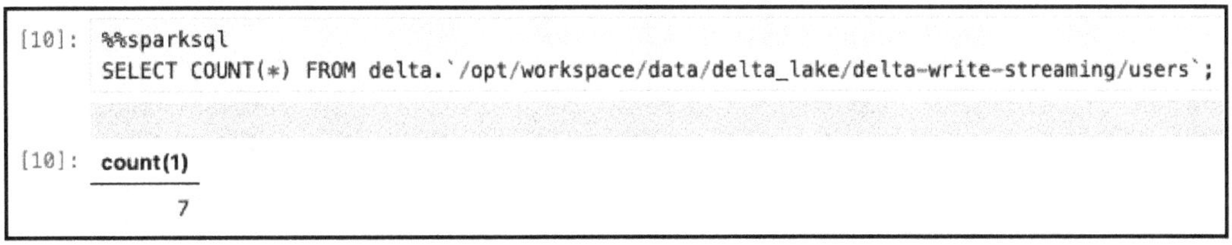

```
[10]: %%sparksql
      SELECT COUNT(*) FROM delta.`/opt/workspace/data/delta_lake/delta-write-streaming/users`;


[10]: count(1)

          7
```

Figure 5.2 – Output from user count query

9. **Query the Delta table again**: Running the same query after a few seconds have passed will yield more records:

```
%%sparksql
SELECT COUNT(*) FROM delta.`/opt/workspace/data/delta_lake/
delta-write-streaming/users`;
```

This will result in the following output:

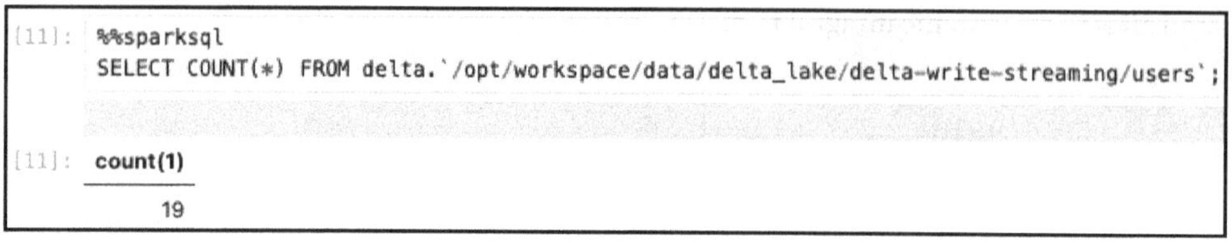

```
[11]:  %%sparksql
       SELECT COUNT(*) FROM delta.'/opt/workspace/data/delta_lake/delta-write-streaming/users';


[11]:  count(1)
            19
```

Figure 5.3 – Count of user records ingested

10. **Stop the streaming query**: Stop the query using the `stop` method of the query:

```
query.stop()
```

11. **Stop the Spark session**: Finally, we need to stop the Spark session to release the resources used by Spark.

```
spark.stop()
```

How it works...

When you write the output of Apache Spark Structured Streaming to a sink such as Delta Lake, you are leveraging the deep integration of Delta Lake with Structured Streaming. Delta Lake addresses numerous challenges commonly associated with streaming systems, such as the following:

- Maintaining "exactly-once" processing with more than one stream (or concurrent batch jobs)
- Efficiently discovering which files are new when using files as the source for a stream
- Handling schema evolution and enforcing schema on write
- Supporting updates and deletes on streaming data
- Enabling time travel and audit history on streaming data

Delta Lake uses a transaction log to keep track of all changes made to a table. The transaction log is an ordered list of atomic, deterministic, and serializable commits that have occurred on a table. Each commit records metadata such as the following:

- The timestamp of the commit
- The user who performed the commit
- The operation that was performed (such as write, update, delete, merge, and so on)
- A list of files that were added or removed in each partition
- The statistics of each file (such as number of rows, size, and so on)
- The schema of each file

When you write a streaming query to a Delta table, Delta Lake ensures that only one writer can commit at a time by using **optimistic concurrency control** (**OCC**). If concurrent writers attempt to write to the same partition, Delta Lake will throw a `ConcurrentAppendException` error. Delta Lake will also validate that the schema of the incoming data matches the schema of the table and rejects any incompatible records. It also allows you to query any version or timestamp of the table using time travel syntax.

See also

- *Table streaming reads and writes* – Delta Lake documentation: `https://docs.delta.io/latest/delta-streaming.html`

- *Simplifying Streaming Data Ingestion into Delta Lake* – Databricks blog: `https://www.databricks.com/blog/2022/09/12/simplifying-streaming-data-ingestion-delta-lake.html`

- *From Kafka to Delta Lake using Apache Spark Structured Streaming*: `https://blogit.michelin.io/kafka-to-delta-lake-using-apache-spark-streaming-avro/`

- *What is Delta Lake?* – Azure Databricks | *Microsoft Learn*: `https://learn.microsoft.com/en-us/azure/databricks/delta/`

- *Best practices: Structured Streaming with Kinesis* – Databricks: `https://docs.databricks.com/structured-streaming/kinesis-best-practices.html`

- *Delta table as a sink*: `https://docs.databricks.com/structured-streaming/delta-lake.html#delta-table-as-a-sink`

Idempotent stream writing with Delta Lake and Apache Spark Structured Streaming

In this recipe, you will learn how to perform idempotent stream writing with Delta Lake and Apache Spark Structured Streaming. Idempotent stream writing means that the same data can be written to a Delta table multiple times without changing the final result. This is useful for scenarios where you need to ensure exactly-once processing of streaming data, such as deduplicating records, upserting data, or handling failures and retries.

Getting ready

Before we start, we need to make sure that we have a Kafka cluster running and a topic that produces some streaming data. For simplicity, we will use a single-node Kafka cluster and a topic named `users`. Open the `5.0 user-gen-kafka.ipynb` notebook and execute the cell. This notebook produces a user record every few seconds and puts it on a Kafka topic called `users`.

Make sure you have run this notebook and that it is producing records as shown:

```
{'id': 68, 'name': 'user18', 'age': 22, 'gender': 'F', 'country': 'USA'}
{'id': 21, 'name': 'user89', 'age': 37, 'gender': 'M', 'country': 'UK'}
{'id': 80, 'name': 'user60', 'age': 53, 'gender': 'M', 'country': 'Brazil'}
```

Figure 5.4 – Kafka user data generation output

How to do it...

1. **Import the required libraries**: Start by importing the necessary libraries for working with Delta Lake. In this case, we need the `delta` module and the `SparkSession` class from the `pyspark.sql` module:

    ```
    from delta import configure_spark_with_delta_pip, DeltaTable
    from pyspark.sql import SparkSession
    from pyspark.sql.functions import col, from_json
    from pyspark.sql.types import StructType, StructField,
    IntegerType, StringType
    ```

2. **Create a SparkSession object**: To interact with Spark and Delta Lake, you need to create a `SparkSession` object:

    ```
    builder = (SparkSession.builder
                .appName("idempotent-stream-write-delta")
                .master("spark://spark-master:7077")
                .config("spark.executor.memory", "512m")
                .config("spark.sql.extensions", "io.delta.sql.
    DeltaSparkSessionExtension")
                .config("spark.sql.catalog.spark_catalog", "org.
    apache.spark.sql.delta.catalog.DeltaCatalog"))

    spark = configure_spark_with_delta_pip(builder,['org.apache.
    spark:spark-sql-kafka-0-10_2.12:3.4.1']).getOrCreate()
    spark.sparkContext.setLogLevel("ERROR")
    ```

3. **Create a Delta table**: Before we load data into the table, we will need to create a table in the Delta catalog. This can be done with the `CREATE OR RELACE TABLE` command. You can specify the table name, which is `default.users`, and also specify the data format as `DELTA` and the location of the table data, which in our case is `/opt/workspace/data/delta_lake/idempotent-stream-write-delta/users`:

    ```
    %%sparksql
    CREATE OR REPLACE TABLE default.users (
        id INT,
    ```

```
    name STRING,
    age INT,
    gender STRING,
    country STRING
) USING DELTA LOCATION '/opt/workspace/data/delta_lake/
idempotent-stream-write-delta/users';
```

4. **Create a DataFrame**: To represent the stream of input data from Kafka as a Spark DataFrame, we will be using the `spark.readStream.format("kafka")` method. You need to specify some options such as the bootstrap servers, the topic name or pattern, and the starting and ending offsets. You can also enable headers if your Kafka messages have them:

```
df = (spark.readStream
    .format("kafka")
    .option("kafka.bootstrap.servers", "kafka:9092")
    .option("subscribe", "users")
    .option("startingOffsets", "earliest")
    .load())
```

5. **Parse the JSON messages**: Parse the data in the `value` column using the `from_json` function and provide the schema of the JSON data as an argument. We can use the `StructType` and `StructField` classes from the `pyspark.sql.types` module to define the schema:

```
schema = StructType([
    StructField('id', IntegerType(), True),
    StructField('name', StringType(), True),
    StructField('age', IntegerType(), True),
    StructField('gender', StringType(), True),
    StructField('country', StringType(), True)])

df = df.withColumn('value', from_json(col('value').
cast("STRING"), schema))
```

6. **Extract the nested fields**: From the `value` column, extract the fields using the `col` function and alias them with meaningful names:

```
df = df.select(
    col('value.id').alias('id'),
    col('value.name').alias('name'),
    col('value.age').alias('age'),
    col('value.gender').alias('gender'),
    col('value.country').alias('country'))
```

7. **Write the streaming data to the Delta Lake table**: You can use the `writeStream` method to write the streaming data to the Delta Lake table. You need to specify the format as `delta` and the location as the path of the Delta table. You also need to specify the output mode as `append`, `complete`, or `update`. For example, you can use the following Python code to write the streaming data to the Delta table in `append` mode:

```python
query = (df.writeStream
    .format("delta")
    .outputMode("append")
    .option("checkpointLocation", "/opt/workspace/data/delta_
lake/idempotent-stream-write-delta/users/_checkpoints/")
    .start("/opt/workspace/data/delta_lake/idempotent-stream-
write-delta/users"))
```

8. **Write a function to batch write to two Delta tables**: When writing the function, you can pass two options: `txnVersion` and `txnAppId`. The `txnVersion` option specifies a unique identifier for each batch of data, and the `txnAppId` option specifies a unique identifier for the application that performs the write operation. These options are used by Delta Lake to check if the same batch of data has already been written by the same application, and if so, it will skip the write operation to avoid duplication. When using these options, the same write operation will produce the same result even if it is executed multiple times:

```python
# Define a function writing to two destinations
app_id = 'idempotent-stream-write-delta'
def writeToDeltaLakeTableIdempotent(batch_df, batch_id):
    # location 1
    (batch_df.filter("country IN ('India','China')")
    .write
    .format("delta")
    .mode("append")
    .option("txnVersion", batch_id)
    .option("txnAppId", app_id)
    .save("/opt/workspace/data/delta_lake/idempotent-stream-
write-delta/user_asia"))
    # location 2
    (batch_df.filter("country IN ('USA','Canada','Brazil')")
    .write
    .format("delta")
    .mode("append")
    .option("txnVersion", batch_id)
    .option("txnAppId", app_id)
    .save("/opt/workspace/data/delta_lake/idempotent-stream-
write-delta/user_americas"))
```

9. **Write with foreach**: You use the `foreachBatch` method to apply a custom function to each batch of data in your streaming DataFrame:

```
# Apply the function against the micro-batches using
'foreachBatch'
write_query = (df
 .writeStream
 .format("delta")
 .queryName("Users By Region")
 .foreachBatch(writeToDeltaLakeTableIdempotent)
 .start())
```

> **Note**
>
> When you use `foreachBatch` with Delta Lake in Apache Spark Structured Streaming, you have the opportunity to implement custom logic within the provided function. This function can include writing the streaming data to a Delta Lake table, and if you design the logic to be idempotent, it ensures that even if the write operation is retried or executed multiple times, the final state of the data remains consistent. This is particularly important in streaming scenarios where data may be processed and written continuously, and ensuring the integrity of the data is crucial. To explain what we built in this recipe, please reference the following diagram:
>
>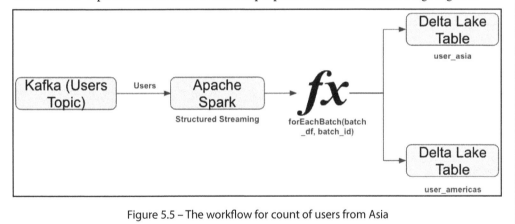
>
> Figure 5.5 – The workflow for count of users from Asia

10. **Query the Delta table**: Query the Delta table to get the count of rows inserted into the two tables, first for Asia:

```
%%sparksql
SELECT COUNT(*) FROM delta.`/opt/workspace/data/delta_lake/
idempotent-stream-write-delta/user_asia`;
```

The output for this query is as shown:

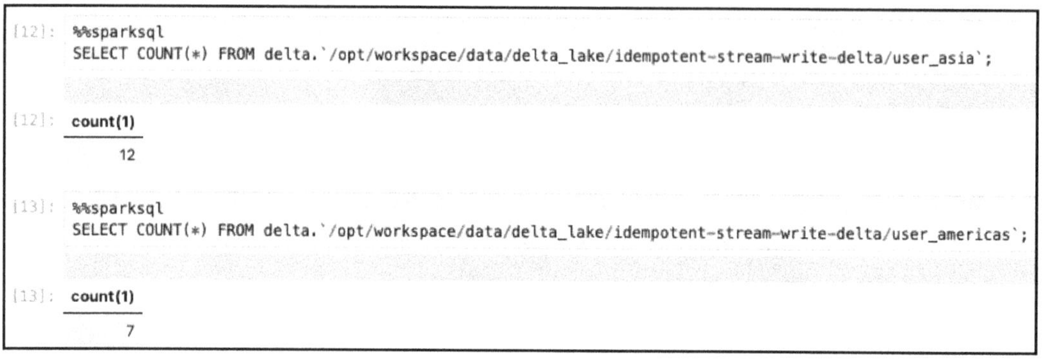

```
[12]: %%sparksql
      SELECT COUNT(*) FROM delta.`/opt/workspace/data/delta_lake/idempotent-stream-write-delta/user_asia`;


[12]: count(1)

          12

[13]: %%sparksql
      SELECT COUNT(*) FROM delta.`/opt/workspace/data/delta_lake/idempotent-stream-write-delta/user_americas`;


[13]: count(1)

           7
```

Figure 5.6 – Count of users from Asia

And then we run the following query to get the count of rows for Americas:

```
%%sparksql
SELECT COUNT(*) FROM delta.`/opt/workspace/data/delta_lake/
idempotent-stream-write-delta/user_americas`;
```

This query will produce the following output:

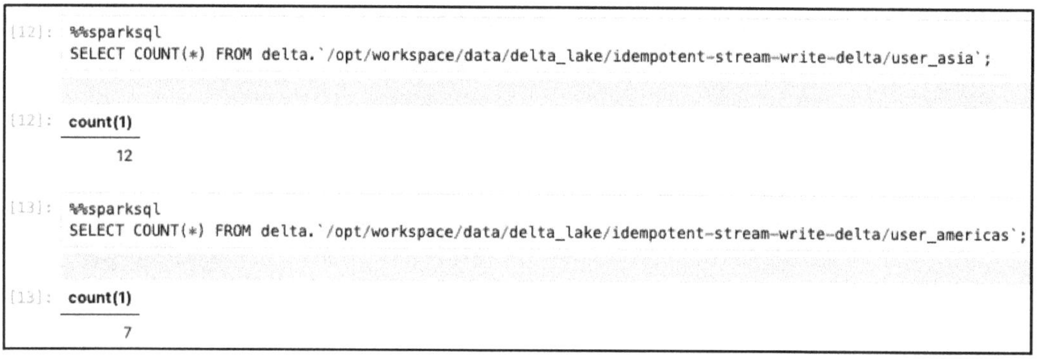

```
[12]: %%sparksql
      SELECT COUNT(*) FROM delta.`/opt/workspace/data/delta_lake/idempotent-stream-write-delta/user_asia`;


[12]: count(1)

          12

[13]: %%sparksql
      SELECT COUNT(*) FROM delta.`/opt/workspace/data/delta_lake/idempotent-stream-write-delta/user_americas`;


[13]: count(1)

           7
```

Figure 5.7 – Count of users from the Americas

11. **Query the Delta table again**: Running the same query after a few seconds have passed will yield more records, first for Asia:

```
%%sparksql
SELECT COUNT(*) FROM delta.`/opt/workspace/data/delta_lake/
idempotent-stream-write-delta/user_asia`;
```

The updated output is as shown:

```
[16]:  %%sparksql
       SELECT COUNT(*) FROM delta.`/opt/workspace/data/delta_lake/idempotent-stream-write-delta/user_asia`;


[16]:  count(1)

          13


[17]:  %%sparksql
       SELECT COUNT(*) FROM delta.`/opt/workspace/data/delta_lake/idempotent-stream-write-delta/user_americas`;


[17]:  count(1)

          17
```

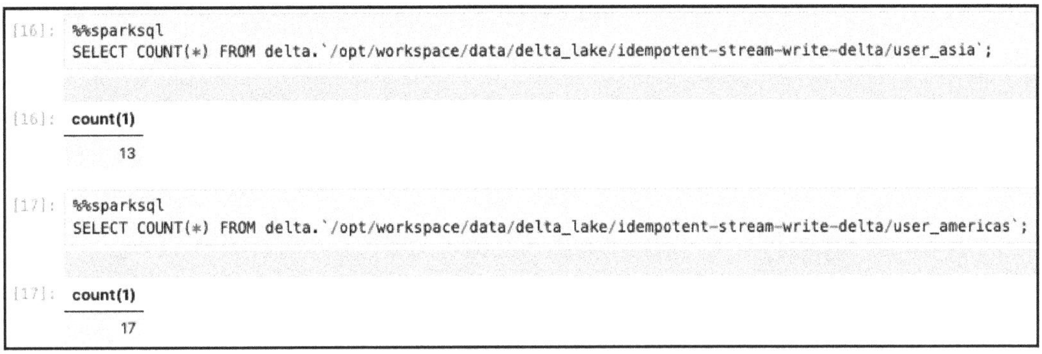

Figure 5.8 – Count of users from Asia

And then for Americas:

```
%%sparksql
SELECT COUNT(*) FROM delta.`/opt/workspace/data/delta_lake/
idempotent-stream-write-delta/user_americas`;
```

The updated output is as shown:

```
[16]:  %%sparksql
       SELECT COUNT(*) FROM delta.`/opt/workspace/data/delta_lake/idempotent-stream-write-delta/user_asia`;


[16]:  count(1)

          13


[17]:  %%sparksql
       SELECT COUNT(*) FROM delta.`/opt/workspace/data/delta_lake/idempotent-stream-write-delta/user_americas`;


[17]:  count(1)

          17
```

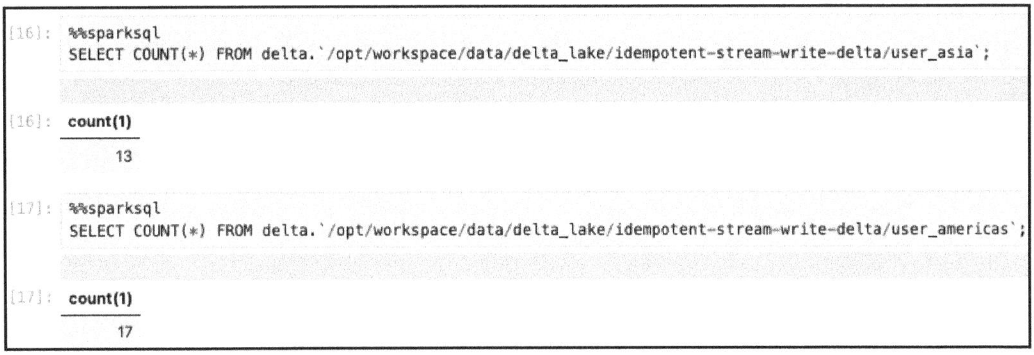

Figure 5.9 – Count of users from the Americas

12. **Stop the streaming query**: Stop the query by using the `stop` method of the query:

```
query.stop()
write_query.stop()
```

13. **Stop the Spark session**: Finally, we need to stop the Spark session to release the resources used by Spark:

```
spark.stop()
```

See also

- *Table Streaming Reads and Writes* – Delta Lake documentation: `https://docs.delta.io/0.4.0/delta-streaming.html`

- *Structured streaming writing to multiple streams* – Stack Overflow: `https://stackoverflow.com/questions/57367226/structured-streaming-writing-to-multiple-streams`

Merging or applying Change Data Capture on Apache Spark Structured Streaming and Delta Lake

In this recipe, we will use Spark Structured Streaming to read data from a Kafka topic that contains change events and merge them into a Delta Lake table that maintains the latest state of the users.

CDC is a technique to capture changes made to a data source and apply them to a target system. CDC can be used for data replication, data integration, data warehousing, and stream processing.

Getting ready

Before we start, we need to make sure that we have a Kafka cluster running and a topic that produces some streaming data. For simplicity, we will use a single-node Kafka cluster and a topic named `users`. Open the `5.0 user-gen-kafka.ipynb` notebook and execute the cell. This notebook produces a user record every few seconds and puts it on a Kafka topic called `users`.

Make sure you have run this notebook and that it is producing records as shown:

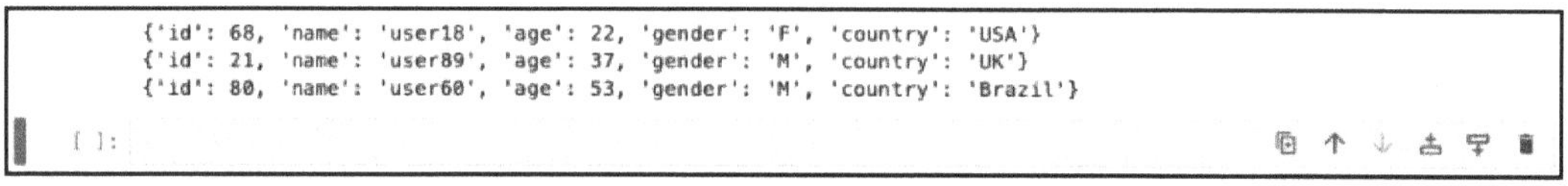

```
{'id': 68, 'name': 'user18', 'age': 22, 'gender': 'F', 'country': 'USA'}
{'id': 21, 'name': 'user89', 'age': 37, 'gender': 'M', 'country': 'UK'}
{'id': 80, 'name': 'user60', 'age': 53, 'gender': 'M', 'country': 'Brazil'}
```

Figure 5.10 – Kafka user data generation output

How to do it...

1. **Import the required libraries**: Start by importing the necessary libraries for working with Delta Lake. In this case, we need the `delta` module and the `SparkSession` class from the `pyspark.sql` module:

```
from delta import configure_spark_with_delta_pip, DeltaTable
from pyspark.sql import SparkSession
from pyspark.sql.functions import col, from_json
from pyspark.sql.types import StructType, StructField,
IntegerType, StringType
```

2. **Create a SparkSession object**: To interact with Spark and Delta Lake, you need to create a `SparkSession` object:

```
builder = (SparkSession.builder
    .appName("merge-cdc-streaming")
    .master("spark://spark-master:7077")
    .config("spark.executor.memory", "512m")
    .config("spark.sql.extensions",
        "io.delta.sql.DeltaSparkSessionExtension")
    .config("spark.sql.catalog.spark_catalog",
        "org.apache.spark.sql.delta.catalog.DeltaCatalog"))

spark = configure_spark_with_delta_pip(builder,
    ['org.apache.spark:spark-sql-kafka-0-10_2.12:3.4.1']).
getOrCreate()
spark.sparkContext.setLogLevel("ERROR")
```

3. **Create a Delta table**: Before we load data, we will need to create a table in the Delta catalog. This can be done with the `CREATE OR REPLACE TABLE` command. You can specify the table name, which is `default.users`, and also specify the data format as `DELTA` and the location of the table data, which in our case is `/opt/workspace/data/delta_lake/merge-cdc-streaming/users`:

```
%%sparksql
CREATE OR REPLACE TABLE default.users (
    id INT,
    name STRING,
    age INT,
    gender STRING,
    country STRING
) USING DELTA LOCATION '/opt/workspace/data/delta_lake/merge-cdc-streaming/users';
```

4. **Create a DataFrame**: Create a DataFrame that represents the stream of input data from Kafka using the `spark.readStream.format("kafka")` method. You need to specify some options such as the bootstrap servers, the topic name or pattern, and the start and end offsets. You can also enable headers if your Kafka messages have them:

```
df = (spark.readStream
    .format("kafka")
    .option("kafka.bootstrap.servers", "kafka:9092")
    .option("subscribe", "users")
    .option("startingOffsets", "earliest")
    .load()
```

5. **Parse the JSON messages**: Parse the data in the value column using the `from_json` function and provide the schema of the JSON data as an argument. We can use the `StructType` and `StructField` classes from the `pyspark.sql.types` module to define the schema:

```
schema = StructType([
    StructField('id', IntegerType(), True),
    StructField('name', StringType(), True),
    StructField('age', IntegerType(), True),
    StructField('gender', StringType(), True),
    StructField('country', StringType(), True)])

df = df.withColumn('value', from_json(col('value').
cast("STRING"), schema))
```

6. **Extract the nested fields**: From the value column, extract the fields using the `col` function and alias them with meaningful names:

```
df = df.select(
    col('value.id').alias('id'),
    col('value.name').alias('name'),
    col('value.age').alias('age'),
    col('value.gender').alias('gender'),
    col('value.country').alias('country'))
```

7. **Define the merge logic**: We will use Delta Lake's `mergeInto` method to merge the streaming DataFrame into a Delta Lake table that maintains the latest state of the customers. The `mergeInto` method takes two arguments:

   - `target`: The Delta Lake table to merge into, which can be specified by its path or name

   - `source`: The streaming DataFrame to merge from, which must have a column named `key` that matches the primary key of the target table

The `mergeInto` method also supports a SQL-like syntax to define the merge logic using `whenMatched` and `whenNotMatched` clauses. The `whenMatched` clause specifies what to do when a row in the source matches a row in the target based on the `key` column. The `whenNotMatched` clause specifies what to do when a row in the source does not match any row in the target based on the `key` column. We will use the following logic for our merge operation:

   - When a row in the source matches a row in the target, and the source value is not `null`, update the target row with the source value

   - When a row in the source matches a row in the target, and the source value is `null`, delete the target row

- When a row in the source does not match any row in the target, and the source value is not `null`, insert the source value as a new row in the target

We will use PySpark's `expr` method to write the merge logic as follows:

```
def upsertToDelta(microBatchDf, batchId):
    deltaTable = DeltaTable.forPath(spark, "/opt/workspace/data/
delta_lake/merge-cdc-streaming/users" )
    (deltaTable.alias("dt")
    .merge(source=microBatchDf.alias("sdf"),
        condition="sdf.id = dt.id")
    .whenMatchedUpdate(set={
        "id": "sdf.id",
        "name": "sdf.name",
        "age": "sdf.gender",
        "country": "sdf.country" })
    .whenNotMatchedInsert(values={
        "id": "sdf.id",
        "name": "sdf.name",
        "age": "sdf.gender",
        "country": "sdf.country" })
    .execute())
```

8. **Write the merged data to a Delta Lake table**: We will use PySpark's `writeStream` method to write the streaming DataFrame to a Delta Lake table. We will specify the output mode, checkpoint location, and trigger interval as options. We will also use the `foreachBatch` method to apply the merge logic to each micro-batch of the streaming DataFrame:

```
query = (df.writeStream
    .format("delta")
    .foreachBatch(upsertToDelta)
    .outputMode("update")
    .option("checkpointLocation", "/opt/workspace/data/delta_
lake/merge-cdc-streaming/users/_checkpoints/")
    .start("/opt/workspace/data/delta_lake/merge-cdc-streaming/
users"))
```

9. **Query the Delta table history**: Query the history of the Delta table to see various versions being merged:

```
%%sparksql
DESCRIBE HISTORY delta.`/opt/workspace/data/delta_lake/merge-
cdc-streaming/users`;
```

This will result in the following output:

version	timestamp	userId	userName	operation	operationParameters	job	notebook	clusterId	readVersion	isolationLevel	isBlindAppend	operationMetrics	userMetadata	engineInfo
2	2023-08-26 11:51:11.377000	null	null	MERGE	{'matchedPredicates': '[{"actionType":"update"}]', 'predicate': '["(id#357 = id#1172)"]', 'notMatchedBySourcePredicates': '[]', 'notMatchedPredicates': '[{"actionType":"insert"}]'}	null	null	null	1	Serializable	False	{'numOutputRows': '263', 'numTargetBytesAdded': '2822', 'numTargetRowsInserted': '0', 'numTargetFilesAdded': '1', 'numTargetRowsMatchedDeleted': '0', 'numTargetFilesRemoved': '1', 'numTargetRowsMatchedUpdated': '1', 'executionTimeMs': '8758', 'numTargetRowsCopied': '262', 'rewriteTimeMs': '4588', 'numTargetRowsUpdated': '1', 'numTargetRowsDeleted': '0', 'scanTimeMs': '4126', 'numSourceRows': '1', 'numTargetChangeFilesAdded': '0', 'numTargetRowsNotMatchedBySourceUpdated': '0', 'numTargetRowsNotMatchedBySourceDeleted': '0', 'numTargetBytesRemoved': '2817'}	null	Apache-Spark/3.4.1 Delta-Lake/2.4.0
1	2023-08-26 11:50:56.577000	null	null	MERGE	{'matchedPredicates': '[{"actionType":"update"}]', 'predicate': '["(id#357 = id#387)"]', 'notMatchedBySourcePredicates': '[]', 'notMatchedPredicates': '[{"actionType":"insert"}]'}	null	null	null	0	Serializable	False	{'numOutputRows': '263', 'numTargetBytesAdded': '2817', 'numTargetRowsInserted': '263', 'numTargetFilesAdded': '1', 'numTargetRowsMatchedDeleted': '0', 'numTargetFilesRemoved': '0', 'numTargetRowsMatchedUpdated': '0', 'executionTimeMs': '7877', 'numTargetRowsCopied': '0', 'rewriteTimeMs': '2561', 'numTargetRowsUpdated': '0', 'numTargetRowsDeleted': '0', 'scanTimeMs': '5247', 'numSourceRows': '263', 'numTargetChangeFilesAdded': '0', 'numTargetRowsNotMatchedBySourceUpdated': '0', 'numTargetRowsNotMatchedBySourceDeleted': '0', 'numTargetBytesRemoved': '0'}	null	Apache-Spark/3.4.1 Delta-Lake/2.4.0

Figure 5.11 – Merge operation metrics from Delta table history

10. **Stop the streaming query**: Wait for the query to terminate using the `stop` method of the query:

    ```
    query.stop()
    ```

11. **Stop the Spark session**: Finally, we need to stop the Spark session to release the resources used by Spark:

    ```
    spark.stop()
    ```

There's more…

We can use Spark's built-in functions and expressions to manipulate the data in the streaming DataFrame, such as `coalesce`, `struct`, `from_json`, `to_json`, and so on. For example, we can simplify the code for extracting the `key` and `value` columns as follows:

```
# Extract the key and value columns using Spark's built-in functions
df = df.withColumn("key", coalesce(col("value.before.id"),
    col("value.after.id"))).withColumn("value", when(col("value.op")
== "d",
    lit(None)).otherwise(col("value.after")))
```

We can use Spark's watermark feature to handle late or out-of-order data in the streaming DataFrame. Watermarking allows us to specify a threshold of how late we expect the data to be and how long we want to wait for it. For example, we can add a watermark on the `ts_ms` column as follows:

```
# Add a watermark on the ts_ms column
df = df.withWatermark("ts_ms", "10 seconds")
```

This means that we expect the data to arrive within 10 seconds of its timestamp and we will drop any data that is older than the watermark.

We can use Spark's trigger feature to control the frequency of updates in the streaming query. Triggering allows us to specify how often we want to process new data and generate new results. For example, we can use a processing time trigger of 5 seconds as follows:

```
# Use a processing time trigger of 5 seconds
query = df.writeStream.trigger(processingTime="5 seconds")
.start()
```

This means that we will process new data every 5 seconds and update the result accordingly.

Joining streaming data with static data in Apache Spark Structured Streaming and Delta Lake

In this recipe, you will learn how to join streaming data with static data in Apache Spark Structured Streaming and Delta Lake. This is a common use case for many applications that need to enrich streaming data with additional information from a historical or reference dataset. For example, you may want to join a stream of user events with a static table of user profiles, or a stream of orders with a static table of product details.

Getting ready

Before we start, we need to make sure that we have a Kafka cluster running and a topic that produces some streaming data. For simplicity, we will use a single-node Kafka cluster and a topic named `orders`. Open the `5.0 orders-gen-kafka.ipynb` notebook and execute the cell. This notebook simulates streaming data of online orders, which contains the order ID, the product ID, the quantity, and the timestamp.

Make sure you have run this notebook and that it is producing records as shown:

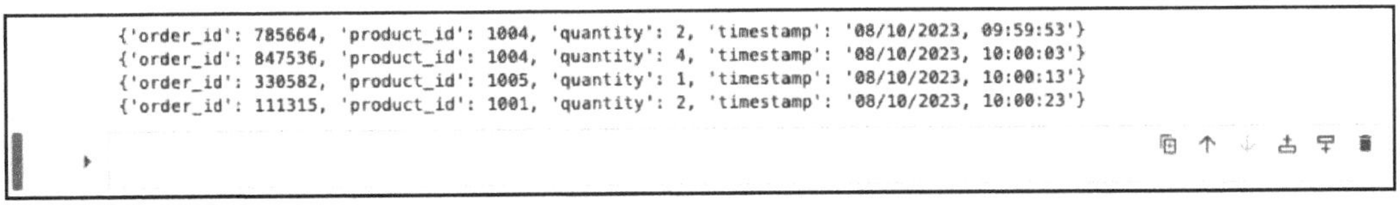

```
{'order_id': 785664, 'product_id': 1004, 'quantity': 2, 'timestamp': '08/10/2023, 09:59:53'}
{'order_id': 847536, 'product_id': 1004, 'quantity': 4, 'timestamp': '08/10/2023, 10:00:03'}
{'order_id': 330582, 'product_id': 1005, 'quantity': 1, 'timestamp': '08/10/2023, 10:00:13'}
{'order_id': 111315, 'product_id': 1001, 'quantity': 2, 'timestamp': '08/10/2023, 10:00:23'}
```

Figure 5.12 – Kafka order data generation output

How to do it...

1. **Import the required libraries**: Start by importing the necessary libraries for working with Delta Lake. In this case, we need the `delta` module and the `SparkSession` class from the `pyspark.sql` module:

```
from delta import configure_spark_with_delta_pip, DeltaTable
from pyspark.sql import SparkSession
from pyspark.sql.functions import col, from_json, to_timestamp
from pyspark.sql.types import StructType, StructField,
IntegerType, StringType
```

2. **Create a SparkSession object**: To interact with Spark and Delta Lake, you need to create a `SparkSession` object:

```
builder = (SparkSession.builder
    .appName("joining-stream-static-data")
    .master("spark://spark-master:7077")
    .config("spark.executor.memory", "512m")
    .config("spark.sql.extensions",
        "io.delta.sql.DeltaSparkSessionExtension")
    .config("spark.sql.catalog.spark_catalog",
        "org.apache.spark.sql.delta.catalog.DeltaCatalog"))

spark = configure_spark_with_delta_pip(builder,
    ['org.apache.spark:spark-sql-kafka-0-10_2.12:3.4.1']).
getOrCreate()
spark.sparkContext.setLogLevel("ERROR")
```

3. **Create a DataFrame**: To represent the stream of input data from Kafka using the `spark.readStream.format("kafka")` method, we need to create a DataFrame. You need to specify some options such as the bootstrap servers, the topic name or pattern, and the starting and ending offsets.

   Read the streaming data from the `streaming_orders` folder using the `readStream` method of the `SparkSession` object. Specify the format as `json` and the schema as a `StructType` object with the fields of the order data. You can also set the option for `maxFilesPerTrigger` to 1, which means that only one file will be processed in each micro-batch:

```
# Define the schema of the streaming data
streaming_schema = StructType([
    StructField("order_id", IntegerType()),
    StructField("product_id", IntegerType()),
    StructField("quantity", IntegerType()),
    StructField("timestamp", IntegerType())])

streaming_df = (spark.readStream
```

```
        .format("kafka")
        .option("kafka.bootstrap.servers", "kafka:9092")
        .option("subscribe", "orders")
        .option("startingOffsets", "earliest")
        .option("failOnDataLoss", "false")
        .load()
        .withColumn('value', from_json(col('value').cast("STRING"),
    streaming_schema)))

    streaming_df = (streaming_df
        .select(
            col('value.order_id').alias('order_id'),
            col('value.product_id').alias('product_id'),
            col('value.quantity').alias('quantity'),
            to_timestamp(col("timestamp"), "MM/dd/yyyy,
                HH:mm:ss" ).alias('timestamp')))
```

4. **Read the static data**: The static data can be read into a DataFrame from a list of tuples named product_details:

```
# Define a list of tuples
product_details = [
    (1001, "Laptop", 999.99),
    (1002, "Mouse", 19.99),
    (1003, "Keyboard", 29.99),
    (1004, "Monitor", 199.99),
    (1005, "Speaker", 49.99)]

# Define a list of column names
columns = ["product_id", "name", "price"]

# Create a DataFrame from the list of tuples
static_df = spark.createDataFrame(product_details, columns)
static_df.show()
```

5. **Join the streaming data with the static data**: Using the join method of the DataFrame, we can join two DataFrames. Specify the join condition as the equality of the product_id column in both DataFrames. You can also specify the join type as inner, which means that only matching rows will be included in the result:

```
# Join the streaming data with the static data
joined_df = (streaming_df
    .join(static_df,streaming_df.product_id == static_
df.product_id,"inner")
    .drop(static_df.product_id)
```

```
                .withColumn('invoice_amount',
                    streaming_df.quantity*static_df.price))
```

6. **Write the streaming data to the Delta Lake table**: You can use the `writeStream` method to write the streaming data to the Delta Lake table. You need to specify the format as `delta` and the location as the path of the Delta table. You also need to specify the output mode as `append`, `complete`, or `update`. For example, you can use the following Python code to write the streaming data to the Delta table in `append` mode:

```
query = (joined_df.writeStream
    .format("delta")
    .outputMode("append")
        .option("failOnDataLoss", "true")
    .option("checkpointLocation", "/opt/workspace/data/delta_
lake/joining-stream-static/orders/_checkpoints/")
        .start("/opt/workspace/data/delta_lake/joining-stream-static/
orders"))
```

7. **Query the Delta table**: Query streaming data joined with static data from the Delta table with the following statement:

```
%%sparksql
SELECT * FROM delta.`/opt/workspace/data/delta_lake/joining-
stream-static/orders`;
```

This will result in the following output:

```
[10]:  %%sparksql
       SELECT * FROM delta.`/opt/workspace/data/delta_lake/joining-stream-static/orders`;
```

	order_id	product_id	quantity	timestamp	name	price	invoice_amount
[10]:	360173	1002	1	2023-08-26 12:03:59.301000	Mouse	19.99	19.99
	519915	1002	4	2023-08-26 12:04:09.303000	Mouse	19.99	79.96
	685163	1005	3	2023-08-26 12:04:39.335000	Speaker	49.99	149.97
	106261	1004	2	2023-08-26 12:04:19.314000	Monitor	199.99	399.98
	410064	1004	4	2023-08-26 12:03:49.295000	Monitor	199.99	799.96
	108816	1005	1	2023-08-26 12:04:49.339000	Speaker	49.99	49.99
	944946	1001	2	2023-08-26 12:04:29.324000	Laptop	999.99	1999.98

Figure 5.13 – Output of the table with streaming orders enriched with static data

8. **Stop the streaming query**: Wait for the query to terminate using the `stop` method of the query:

```
query.stop()
```

9. **Stop the Spark session**: Finally, we need to stop the Spark session to release the resources used by Spark:

```
spark.stop()
```

There's more...

Here are some additional tips and tricks for joining streaming data with static data in Apache Spark Structured Streaming and Delta Lake:

- You can use different types of joins for your streaming query, such as left outer join, right outer join, full outer join, or left semi join. However, you need to be aware of the implications of each join type on the latency and completeness of your results. For example, a left outer join may delay the output of some rows until the right-side data arrives, while a full outer join may never output some rows if the data from both sides never arrives.

- You can also use watermarking and event-time constraints to handle late or out-of-order data in your streaming query. As mentioned in the previous recipes, watermarking is a technique that allows you to specify a threshold of how late the data can be and discard any data that is older than the threshold. Event-time constraints are conditions that you can apply on the event-time columns of your join, such as range or equality, to filter out irrelevant or duplicate data. You can read more about watermarking and event-time constraints in the Spark documentation.

- You can monitor and troubleshoot your streaming query using the Spark UI, which provides various metrics and statistics about the query's progress, performance, and state. You can access the Spark UI by opening the `http://<driver-node>:4040` URL in your browser, where `<driver-node>` is the hostname or IP address of the node where your Spark driver is running. You can read more about the Spark UI in the Spark documentation.

See also

- Join operations with Apache Spark: `https://spark.apache.org/docs/latest/structured-streaming-programming-guide.html#join-operations`

Joining streaming data with streaming data in Apache Spark Structured Streaming and Delta Lake

In Apache Spark Structured Streaming, stream-to-stream joins refer to the capability of combining two or more streaming DataFrames or Datasets based on a common key. This operation enables the merging of ongoing, real-time data streams, allowing for dynamic and continuous analysis of correlated information. The result is a new streaming DataFrame that evolves over time as the input streams are updated, facilitating real-time processing and analytics on streaming data.

In this recipe, you will learn how to join two streams of data using Apache Spark Structured Streaming and Delta Lake. You will also learn how to handle late-arriving and out-of-order data, and how to update the join results as new data arrives. Here is a diagram that shows how the two streams are joined in this recipe:

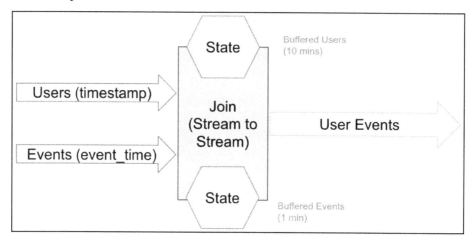

Figure 5.14 – Stream-to-stream joins in structured streaming

Getting ready

Before we start, we need to make sure that we have a Kafka cluster running and a topic that produces some streaming data. For simplicity, we will use a single-node Kafka cluster and a topic named events. Open the 5.0 events-gen-kafka.ipynb notebook and execute the cell. This notebook produces events record every few seconds and puts it on a Kafka topic called events.

Make sure you have run this notebook and that it is producing records as shown:

```
{'user_id': 27, 'event_type': 'share', 'event_time': '08/06/2023, 14:19:07', 'processing_time': '08/06/2023, 14:19:14'}
{'user_id': 49, 'event_type': 'purchase', 'event_time': '08/06/2023, 14:19:06', 'processing_time': '08/06/2023, 14:19:15'}
{'user_id': 86, 'event_type': 'click', 'event_time': '08/06/2023, 14:19:06', 'processing_time': '08/06/2023, 14:19:16'}
{'user_id': 54, 'event_type': 'view', 'event_time': '08/06/2023, 14:19:13', 'processing_time': '08/06/2023, 14:19:17'}
{'user_id': 32, 'event_type': 'click', 'event_time': '08/06/2023, 14:19:17', 'processing_time': '08/06/2023, 14:19:18'}
{'user_id': 55, 'event_type': 'view', 'event_time': '08/06/2023, 14:19:06', 'processing_time': '08/06/2023, 14:19:19'}
{'user_id': 12, 'event_type': 'click', 'event_time': '08/06/2023, 14:19:15', 'processing_time': '08/06/2023, 14:19:20'}
{'user_id': 29, 'event_type': 'click', 'event_time': '08/06/2023, 14:19:20', 'processing_time': '08/06/2023, 14:19:21'}
{'user_id': 78, 'event_type': 'click', 'event_time': '08/06/2023, 14:19:03', 'processing_time': '08/06/2023, 14:19:22'}
{'user_id': 74, 'event_type': 'share', 'event_time': '08/06/2023, 14:19:01', 'processing_time': '08/06/2023, 14:19:23'}
{'user_id': 8, 'event_type': 'view', 'event_time': '08/06/2023, 14:19:21', 'processing_time': '08/06/2023, 14:19:24'}
{'user_id': 6, 'event_type': 'like', 'event_time': '08/06/2023, 14:19:13', 'processing_time': '08/06/2023, 14:19:25'}
{'user_id': 30, 'event_type': 'like', 'event_time': '08/06/2023, 14:19:18', 'processing_time': '08/06/2023, 14:19:26'}
{'user_id': 55, 'event_type': 'like', 'event_time': '08/06/2023, 14:19:15', 'processing_time': '08/06/2023, 14:19:27'}
{'user_id': 25, 'event_type': 'purchase', 'event_time': '08/06/2023, 14:19:20', 'processing_time': '08/06/2023, 14:19:28'}
```

Figure 5.15 – Kafka event data generation output

Also, open the 5.0 user-gen-kafka.ipynb notebook and execute the cell. This notebook produces a user record every few seconds and puts it on a Kafka topic called events.

Make sure you have run this notebook and that it is producing records as shown:

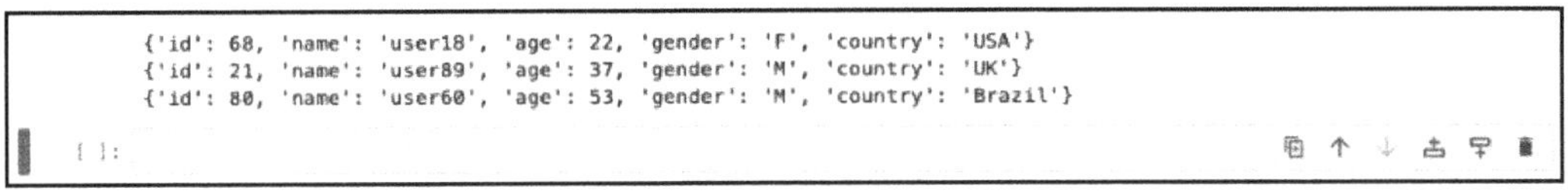

```
{'id': 68, 'name': 'user18', 'age': 22, 'gender': 'F', 'country': 'USA'}
{'id': 21, 'name': 'user89', 'age': 37, 'gender': 'M', 'country': 'UK'}
{'id': 80, 'name': 'user60', 'age': 53, 'gender': 'M', 'country': 'Brazil'}

[ ]:
```

Figure 5.16 – Kafka user data generation output

How to do it...

1. **Import the required libraries**: Start by importing the necessary libraries for working with Delta Lake. In this case, we need the `delta` module and the `SparkSession` class from the `pyspark.sql` module:

    ```
    from delta import configure_spark_with_delta_pip, DeltaTable
    from pyspark.sql import SparkSession
    from pyspark.sql.functions import col, from_json, to_timestamp
    from pyspark.sql.types import StructType, StructField,
    IntegerType, StringType
    ```

2. **Create a SparkSession object**: To interact with Spark and Delta Lake, you need to create a `SparkSession` object:

    ```
    builder = (SparkSession.builder
        .appName("joining-stream-stream-data")
        .master("spark://spark-master:7077")
        .config("spark.executor.memory", "512m")
        .config("spark.sql.extensions",
            "io.delta.sql.DeltaSparkSessionExtension")
        .config("spark.sql.catalog.spark_catalog",
            "org.apache.spark.sql.delta.catalog.DeltaCatalog"))

    spark = configure_spark_with_delta_pip(builder,
        ['org.apache.spark:spark-sql-kafka-0-10_2.12:3.4.1']).
    getOrCreate()
    spark.sparkContext.setLogLevel("ERROR")
    ```

3. **Create a streaming DataFrame for users**: To represent the stream of input data from Kafka, we will use the `spark.readStream.format("kafka")` method. You need to specify some options such as the bootstrap servers, the topic name or pattern, and the starting and ending offsets. Read the streaming data from the user Kafka topic to the `user_df` DataFrame:

    ```
    schema = StructType([
        StructField('id', IntegerType(), True),
        StructField('name', StringType(), True),
        StructField('age', IntegerType(), True),
    ```

```
        StructField('gender', StringType(), True),
        StructField('country', StringType(), True)])

users_df = (spark.readStream
    .format("kafka")
    .option("kafka.bootstrap.servers", "kafka:9092")
    .option("subscribe", "users")
    .option("startingOffsets", "earliest")
    .load()
    .withColumn('value', from_json(col('value').cast("STRING"),
schema)))

users_df = users_df.select(
    col('value.id').alias('id'),
    col('value.name').alias('name'),
    col('value.age').alias('age'),
    col('value.gender').alias('gender'),
    col('value.country').alias('country'))
```

4. **Create a streaming DataFrame for events**: To represent the stream of input data from Kafka using the `spark.readStream.format("kafka")` method, we need to create a DataFrame. You need to specify some options such as the bootstrap servers, the topic name or pattern, and the starting and ending offsets. You can also enable headers if your Kafka messages have them.

    Read the streaming data from the `events` Kafka topic into the `events_df` DataFrame:

```
schema = StructType([
    StructField('user_id', IntegerType(), True),
    StructField('event_type', StringType(), True),
    StructField('event_time', StringType(), True),
    StructField('processing_time', StringType(), True)])

events_df = (spark.readStream
    .format("kafka")
    .option("kafka.bootstrap.servers", "kafka:9092")
    .option("subscribe", "events")
    .option("startingOffsets", "latest")
    .load()
    .withColumn('value', from_json(col('value').cast("STRING"),
schema)))

events_df = (events_df
    .select(
        col('value.user_id').alias('user_id'),
        col('value.event_type').alias('event_type'),
```

```
           col('value.event_time').alias('event_time'),
           col('value.processing_time').alias('processing_time')
       .withColumn("event_time", to_timestamp(col("event_time"),
           "MM/dd/yyyy, HH:mm:ss" ))
       .withColumn("processing_time", to_timestamp(col("processing_
   time"),
           "MM/dd/yyyy, HH:mm:ss")))
```

5. **Join the two streaming DataFrames**: We will be using the `join` method to join the two streaming DataFrames. You need to specify the join condition, the join type, and the watermark policy. The watermark policy defines how to handle late-arriving and out-of-order data. For example, you can use event-time watermarking to specify a threshold for how late the data can be and how long to wait for late data. You can use the `withWatermark` method to define a `watermark` column and delay threshold for each stream; for example:

```
# Join the two streaming DataFrames on user
join_df = (events_df
    .join(
        users_df.withWatermark("timestamp", "10 minutes"), #
Define watermark for users stream
        events_df.user_id == users_df.id, # Join condition
        "inner")# Join type
        .withWatermark("event_time", "1 minutes") # Define
watermark for orders stream
        .drop(users_df.id))
```

6. **Write the join result to the Delta table**: We will be using the `writeStream.` `format("delta")` method. You need to specify the output mode, the checkpoint location, and the table location. The output mode defines how to update the table with new data. For example, you can use `append` mode to append new rows to the table or `update` mode to update existing rows and append new rows. The checkpoint location is a directory where Spark will store the state of the stream. The table location is the HDFS directory where the Delta table is stored; for example:

```
query = (join_df.writeStream
    .format("delta")
    .outputMode("append")
        .option("failOnDataLoss", "true")
    .option("checkpointLocation", "/opt/workspace/data/delta_
lake/joining-stream-stream/user_events/_checkpoints/")
    .start("/opt/workspace/data/delta_lake/joining-stream-stream/
user_events"))
```

7. **Query the Delta table**: To query the stream-stream join data, we can query the Delta table with the following statement:

> **Note**
>
> You may have to wait for several minutes to get matching records on both streams and subsequently to have the following query produce results.

```
%%sparksql
SELECT event_type, gender, country,count(user_id)
FROM delta.`/opt/workspace/data/delta_lake/joining-stream-
stream/user_events`
GROUP BY ALL;
```

This will result in the following output:

```
[123]: %%sparksql
       SELECT event_type, gender, country,count(user_id)
       FROM delta.`/opt/workspace/data/delta_lake/joining-stream-stream/user_events`
       GROUP BY ALL;


       only showing top 20 row(s)
```

event_type	gender	country	count(user_id)
purchase	M	India	1
click	F	USA	2
view	M	India	1
click	M	Canada	2
like	M	China	2
purchase	F	Australia	3
share	M	Brazil	1
purchase	M	China	2
purchase	F	Brazil	4
like	M	Australia	3
share	M	USA	2
share	F	UK	3
purchase	M	Australia	2
like	F	USA	1
click	M	China	1
click	F	China	3
click	M	India	4
like	F	Canada	4
purchase	M	Canada	2
share	F	Canada	1

Figure 5.17 – Stream-to-stream-joined data output

8. **Stop the streaming query**: Wait for the query to terminate using the `stop` method of the query:

```
query.stop()
```

9. **Stop the Spark session**: Finally, we need to stop the Spark session to release the resources used by Spark:

```
spark.stop()
```

There's more...

Here are some additional tips and tricks for joining streaming data with streaming data in Apache Spark Structured Streaming and Delta Lake:

- You can use different join types for stream-stream joins, such as `inner`, `leftOuter`, `rightOuter`, or `fullOuter`. However, some join types may require more state management than others, which may affect the performance and scalability of your stream processing. The state data of a streaming query, which includes aggregations, joins, and any other stateful operations, is managed by a state store provider. You can use the `spark.sql.streaming.stateStore.providerClass` configuration option to choose a different state store implementation for your stream-stream joins.

- You can use different watermark policies for stream-stream joins, such as event-time watermarking or processing-time watermarking. Event-time watermarking is based on timestamps of the events in the stream, while processing-time watermarking is based on the system clock of the processing engine. Event-time watermarking is more robust to handle late-arriving and out-of-order data, but it requires you to have reliable timestamps in your data. Processing-time watermarking is simpler to implement, but it may not handle late-arriving and out-of-order data correctly.

- You can use different output modes for stream-stream joins, such as `append`, `update`, or `complete`. However, some output modes may not be supported for some join types or watermark policies. For example, `complete` mode is not supported for stream-stream joins because it requires all the previous data to be stored in memory. You can use the `spark.sql.streaming.outputMode` configuration option to choose a different output mode for your stream-stream joins.

See also

- Stream-stream joins in Apache Spark: `https://spark.apache.org/docs/latest/structured-streaming-programming-guide.html#stream-stream-joins`

- Join operations with Apache Spark: `https://spark.apache.org/docs/latest/structured-streaming-programming-guide.html#join-operations`

Monitoring real-time data processing with Apache Spark Structured Streaming

In this recipe, you will learn how to do the following:

- Use the `status` and `recentProgress` attributes of a streaming query to get information about the input rate, processing rate, latency, state size, and more

- Use the `StreamingQueryListener` API to register a custom listener that can handle events related to the start, progress, and termination of a streaming query

To monitor the performance and progress of your streaming queries, Structured Streaming provides various metrics and APIs that you can use to access them.

Getting ready

Before we start, we need to make sure that we have a Kafka cluster running and a topic that produces some streaming data. For simplicity, we will use a single-node Kafka cluster and a topic named `users`. Open the `5.0 user-gen-kafka.ipynb` notebook and execute the cell. This notebook produces a user record every few seconds and puts it on a Kafka topic called `users`.

Make sure you have run this notebook and that it is producing records as shown:

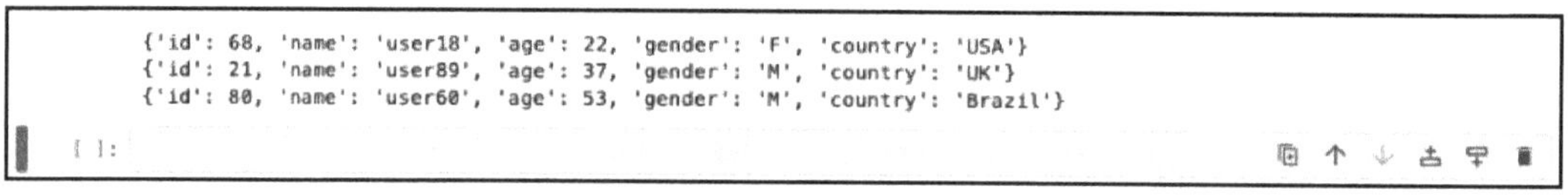

```
{'id': 68, 'name': 'user18', 'age': 22, 'gender': 'F', 'country': 'USA'}
{'id': 21, 'name': 'user89', 'age': 37, 'gender': 'M', 'country': 'UK'}
{'id': 80, 'name': 'user60', 'age': 53, 'gender': 'M', 'country': 'Brazil'}

[ ]:
```

Figure 5.18 – Kafka user data generation output

How to do it...

1. **Import the required libraries**: Start by importing the necessary libraries for working with Delta Lake. In this case, we need the `delta` module and the `SparkSession` class from the `pyspark.sql` module:

```
from delta import configure_spark_with_delta_pip, DeltaTable
from pyspark.sql import SparkSession
from pyspark.sql.functions import col, from_json,to_timestamp
from pyspark.sql.types import StructType, StructField,
IntegerType, StringType
```

2. **Create a SparkSession object**: To interact with Spark and Delta Lake, you need to create a
    SparkSession object:

```
builder = (SparkSession.builder
    .appName("monitor-stream")
    .master("spark://spark-master:7077")
    .config("spark.executor.memory", "512m")
    .config("spark.sql.extensions",
        "io.delta.sql.DeltaSparkSessionExtension")
    .config("spark.sql.catalog.spark_catalog",
        "org.apache.spark.sql.delta.catalog.DeltaCatalog"))

spark = configure_spark_with_delta_pip(builder,
    ['org.apache.spark:spark-sql-kafka-0-10_2.12:3.4.1']).
getOrCreate()
spark.sparkContext.setLogLevel("ERROR")
```

3. **Create a DataFrame**: To represent the stream of input data from Kafka as a Spark DataFrame,
    we will be using the `spark.readStream.format("kafka")` method. You need to
    specify some options such as the bootstrap servers, the topic name or pattern, and the starting
    and ending offsets. You can also enable headers if your Kafka messages have them.

    Parse the JSON messages in the `value` column using the `from_json` function and provide the
    schema of the JSON data as an argument. We can use the `StructType` and `StructField`
    classes from the `pyspark.sql.types` module to define the schema.

    Extract the nested fields from the `value` column using the `col` function and alias them with
    meaningful names:

```
schema = StructType([
    StructField('id', IntegerType(), True),
    StructField('name', StringType(), True),
    StructField('age', IntegerType(), True),
    StructField('gender', StringType(), True),
    StructField('country', StringType(), True),
    StructField('timestamp', StringType(), True)])

users_df = (spark.readStream
    .format("kafka")
    .option("kafka.bootstrap.servers", "kafka:9092")
    .option("subscribe", "users")
    .option("startingOffsets", "latest")
    .load()
    .withColumn('value', from_json(col('value').cast("STRING"),
schema)))
```

```
users_df = users_df.select(
    col('value.id').alias('id'),
    col('value.name').alias('name'),
    col('value.age').alias('age'),
    col('value.gender').alias('gender'),
    col('value.country').alias('country'),
    to_timestamp(col('value.timestamp'), "MM/dd/yyyy,
HH:mm:ss").alias('timestamp'))
```

4. **Write the streaming data to the Delta Lake table**: You can use the `writeStream` method to write the streaming data to the Delta Lake table. You need to specify the format as `delta` and the location as the path of the Delta table. You also need to specify the output mode as `append`, `complete`, or `update`. For example, you can use the following Python code to write the streaming data to the Delta table in `append` mode:

```
query = (users_df.writeStream
    .format("delta")
    .queryName("user-kafka-stream")
    .outputMode("append")
    .option("checkpointLocation", "/opt/workspace/data/delta_
lake/monitor-streams/users/_checkpoints/")
    .start("/opt/workspace/data/delta_lake/monitor-streams/
users"))
```

5. **Use the status and recentProgress attributes**: One of the simplest ways to monitor your streaming query is to use the `status` and `recentProgress` attributes of the `StreamingQuery` object. The `status` attribute returns a dictionary containing information about the query's current state, such as the ID, name, timestamp, input rate, processing rate, latency, state size, and so on. The `recentProgress` attribute returns a list of dictionaries that contain information about the progress of each batch in the query, such as the batch ID, timestamp, duration, input rows, processed rows, state operators' metrics, and so on.

You can access these attributes in your PySpark script by adding some `print` statements after starting the query; for example:

```
# Create a streaming query that writes the output to console
sink
query = agg_df.writeStream.outputMode("complete").
format("console").option("truncate", False).start()

# Print the current status of the query
print(query.status)

# Print the recent progress of the query
print(query.recentProgress)
```

You should see some output similar to the following:

```
[6]:  query.status

[6]:  {'message': 'Processing new data',
       'isDataAvailable': True,
       'isTriggerActive': True}
       [Stage 8:===================================>            (30 + 2) / 50]

[7]:  query.recentProgress

       [Stage 8:===================================>            (34 + 2) / 50]
[7]:  [{'id': '44996f2d-15c6-49d2-9185-72fd08ab7132',
        'runId': '8b493789-ebd9-4e40-afc8-26094de17eb3',
        'name': 'user-kafka-stream',
        'timestamp': '2023-08-26T12:32:12.028Z',
        'batchId': 0,
        'numInputRows': 0,
        'inputRowsPerSecond': 0.0,
        'processedRowsPerSecond': 0.0,
        'durationMs': {'addBatch': 20871,
         'commitOffsets': 162,
         'getBatch': 28,
         'latestOffset': 997,
         'queryPlanning': 506,
         'triggerExecution': 22715,
         'walCommit': 116},
        'stateOperators': [],
        'sources': [{'description': 'KafkaV2[Subscribe[users]]',
         'startOffset': None,
         'endOffset': {'users': {'0': 310}},
         'latestOffset': {'users': {'0': 310}},
         'numInputRows': 0,
         'inputRowsPerSecond': 0.0,
         'processedRowsPerSecond': 0.0,
         'metrics': {'avgOffsetsBehindLatest': '0.0',
          'maxOffsetsBehindLatest': '0',
          'minOffsetsBehindLatest': '0'}}],
        'sink': {'description': 'DeltaSink[/opt/workspace/data/delta_lake/monitor-streams/users]',
         'numOutputRows': -1}}]
```

Figure 5.19 – recentProgress output

You can use these attributes to monitor the performance and progress of your streaming query in real time. However, these attributes only provide textual information and do not support any visualization or alerting features.

6. **Use the StreamingQueryListener API**: Another way to monitor your streaming query is to use the `StreamingQueryListener` API to register a custom listener that can handle events related to the start, progress, and termination of a streaming query. The listener is a class that implements the `StreamingQueryListener` interface and overrides three methods: `onQueryStarted`, `onQueryProgress`, and `onQueryTerminated`. Each method receives an `event` object that contains information about the query and the event.

    You can use the listener to perform various actions based on events, such as logging, debugging, notifying, and so on. For example, you can use the listener to print the query name and ID when it starts, print the input rate and processing rate when it progresses, and print an exception message when it terminates.

To use the listener, you need to import it from the `pyspark.sql.streaming` module and register it with the `spark.streams.addListener` method; for example:

```
from pyspark.sql.streaming import StreamingQueryListener

# Define a custom listener class
class MyListener(StreamingQueryListener):

    # Override the onQueryStarted method
    def onQueryStarted(self, event):
        # Print the query name and id when it starts
        print(f"Query {event.name} with id {event.id} started")

    # Override the onQueryProgress method
    def onQueryProgress(self, event):
        # Print the input rate and processing rate when it
progresses
        print(f"Query {event.name} with id {event.id} has input
rate {event.progress['inputRate']} and processing rate {event.
progress['processingRate']}")

    # Override the onQueryTerminated method
    def onQueryTerminated(self, event):
        # Print the exception message when it terminates
        if event.exception:
            print(f"Query {event.name} with id {event.id}
terminated with exception: {event.exception}")
        else:
            print(f"Query {event.name} with id {event.id}
terminated normally")

# Create an instance of the listener class
listener = MyListener()

# Register the listener with spark.streams
spark.streams.addListener(listener)
```

You should see some output similar to the following:

```
Query made progress: {
  "id" : "44996f2d-15c6-49d2-9185-72fd08ab7132",
  "runId" : "8b493789-ebd9-4e40-afc8-26094de17eb3",
  "name" : "user-kafka-stream",
  "timestamp" : "2023-08-26T12:33:42.109Z",
  "batchId" : 8,
  "numInputRows" : 1,
  "inputRowsPerSecond" : 71.42857142857143,
  "processedRowsPerSecond" : 0.33377837116154874,
  "durationMs" : {
    "addBatch" : 2774,
    "commitOffsets" : 110,
    "getBatch" : 0,
    "latestOffset" : 3,
    "queryPlanning" : 9,
    "triggerExecution" : 2996,
    "walCommit" : 99
  },
  "stateOperators" : [ ],
  "sources" : [ {
    "description" : "KafkaV2[Subscribe[users]]",
    "startOffset" : {
      "users" : {
        "0" : 318
      }
    },
    "endOffset" : {
      "users" : {
        "0" : 319
      }
    },
    "latestOffset" : {
      "users" : {
        "0" : 319
      }
    },
    "numInputRows" : 1,
    "inputRowsPerSecond" : 71.42857142857143,
    "processedRowsPerSecond" : 0.33377837116154874,
    "metrics" : {
      "avgOffsetsBehindLatest" : "0.0",
      "maxOffsetsBehindLatest" : "0",
      "minOffsetsBehindLatest" : "0"
    }
  } ],
  "sink" : {
    "description" : "DeltaSink[/opt/workspace/data/delta_lake/monitor-streams/users]",
    "numOutputRows" : -1
  }
}
```

Figure 5.20 – StreamingQueryListener output

You can use the listener to monitor your streaming query in a customized way. However, the listener only provides programmatic access to events and does not support any visualization or alerting features.

7. **Stop the streaming query**: Stop the query using the `stop` method of the query:

    ```
    query.stop()
    ```

8. **Stop the Spark session**: Finally, we need to stop the Spark session to release the resources used by Spark:

    ```
    spark.stop()
    ```

There's more...

In this section, we will provide some additional information about how to make the most of your streaming query monitoring.

Using the Spark UI

One of the most useful tools for monitoring your streaming query is the Spark UI, which is a web interface that shows various information about your Spark application, such as stages, tasks, executors, storage, environment, and so on. The Spark UI also has a dedicated **Structured Streaming** tab, which shows detailed information about your streaming queries, such as input rate, processing rate, latency, state size, event timeline, and so on.

To access the Spark UI, you need to open a web browser and navigate to `http://<driver-node>:4040/`, where `<driver-node>` is the hostname or IP address of the machine where your Spark driver is running. If you have multiple Spark applications running on the same machine, you may need to use a different port number than `4040`.

The Spark UI allows you to monitor your streaming query in a comprehensive and visual way. You can also use the Spark UI to debug and optimize your streaming query by inspecting the execution plan, metrics, logs, and so on.

Using the checkpoint location

Another useful feature for monitoring your streaming query is the checkpoint location, which is a directory where Spark writes the metadata and state of your streaming query. The checkpoint location allows you to resume your streaming query from the last processed batch in case of a failure or a restart. The checkpoint location also contains some useful information about your streaming query, such as offsets, metrics, configuration, and so on.

To use the checkpoint location, you need to specify it when you create your streaming query using the `option` method; for example:

```
# Create a streaming query that writes the output to console sink
query = agg_df.writeStream.outputMode("complete").format("console").
option("truncate", False).option("checkpointLocation", "/tmp/
checkpoint").start()
```

The checkpoint location can be a local directory or a distributed filesystem, such as HDFS or S3. You can use any filesystem tool to browse and inspect the checkpoint location. You can also use the `spark.sql.streaming.checkpointLocation` configuration property to set a default checkpoint location for all your streaming queries.

The checkpoint location allows you to monitor your streaming query in a persistent and reliable way. You can also use the checkpoint location to recover and resume your streaming query in case of a failure or a restart.

See also

- *Structured Streaming Programming Guide*: `https://spark.apache.org/docs/latest/structured-streaming-programming-guide.html`

- *Monitoring Streaming Queries*: `https://spark.apache.org/docs/latest/structured-streaming-programming-guide.html#monitoring-streaming-queries`

6

Performance Tuning with Apache Spark

Apache Spark is a powerful and versatile framework for large-scale data processing. It offers high-level APIs in Scala, Java, Python, and R, as well as low-level access to the Spark core engine. Spark supports a variety of workloads, such as batch processing, streaming, machine learning, graph analytics, and SQL queries. However, to get the most out of Spark, you need to know how to optimize its performance and avoid common pitfalls.

In this chapter, you will learn how to performance-tune Apache Spark applications.

We will cover the following recipes in this chapter:

- Monitoring Spark jobs in the Spark UI
- Using broadcast variables
- Optimizing Spark jobs by minimizing data shuffling
- Avoiding data skew
- Caching and persistence
- Partitioning and repartitioning
- Optimizing join strategies

By the end of this chapter, you will have a solid understanding of how to tune Apache Spark for optimal performance and how to avoid or solve performance problems. You will also learn some useful tips and tricks to make your Spark code more efficient and elegant.

Technical requirements

Before starting, make sure that your `docker-compose` images are up and running, and open JupyterLab's server, running on localhost (`http://127.0.0.1:8888/lab`). Additionally, ensure that you have cloned the Git repo for this book and have access to the notebook and data used in this chapter.

Remember to stop all services defined in the `docker-compose` file for this book when you are done running the code examples. You can do this by executing the following command:

```
$ docker-compose stop
```

You can find the notebooks and data for this chapter at `https://github.com/PacktPublishing/Data-Engineering-with-Databricks-Cookbook/tree/main/Chapter06`.

Monitoring Spark jobs in the Spark UI

The Spark UI can be used to track the progress and performance of your Spark cluster and its applications. The Spark UI web-based interfaces show you the status and resource usage of your cluster, as well as the details of your Spark jobs, stages, tasks, and SQL queries. The Spark UI is a helpful tool for debugging and optimizing your Spark applications.

In this recipe, we will see how to monitor your Spark jobs in the Spark UI using an example application that reads a CSV file, infers its schema, filters some rows, groups by a column, and counts the number of groups:

How to do it...

1. **Run the Spark application**: Execute the following code to run a sample Spark application that will read a CSV file into a Spark DataFrame with a specific schema, then filter using `release_year` and group by `country`, and finally, display the DataFrame:

    ```
    from pyspark.sql import SparkSession

    # Create a new SparkSession
    spark = (SparkSession
        .builder
        .appName("monitor-spark-ui")
    ```

```python
    .master("spark://spark-master:7077")
    .config("spark.executor.memory", "512m")
    .getOrCreate())

# Set log level to ERROR
spark.sparkContext.setLogLevel("ERROR")

from pyspark.sql.types import StructType, StructField,
StringType, IntegerType, DateType

# Define a Schema
schema = StructType([
    StructField("show_id", StringType(), True),
    StructField("type", StringType(), True),
    StructField("title", StringType(), True),
    StructField("director", StringType(), True),
    StructField("cast", StringType(), True),
    StructField("country", StringType(), True),
    StructField("date_added", DateType(), True),
    StructField("release_year", IntegerType(), True),
    StructField("rating", StringType(), True),
    StructField("duration", StringType(), True),
    StructField("listed_in", StringType(), True),
    StructField("description", StringType(), True)])

# Read CSV file into a DataFrame
df = (spark.read.format("csv")
    .option("header", "true")
    .schema(schema)
    .load("../data/netflix_titles.csv"))

# Filter rows where release_year is greater than 2020
df = df.filter(df.release_year > 2020)

# Group by country and count
df = df.groupBy("country").count()

# Show the result
df.show()
```

2. **Monitor the Spark application**: To monitor this application in the Spark UI, you will need to go to the URL `http://127.0.0.1:4040/`. The UI will take you to the **Jobs** tab by default. On this screen, you can see the status, duration, progress, and event timeline of each job:

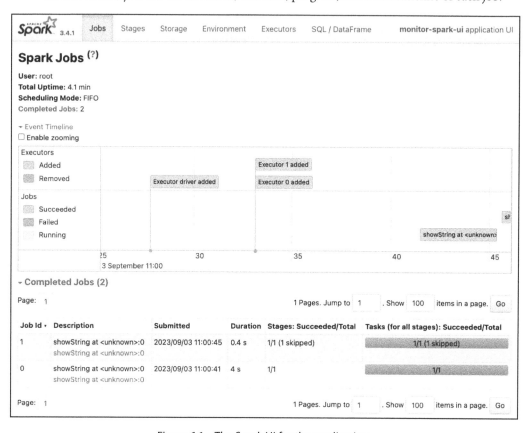

Figure 6.1 – The Spark UI for the application

3. **See jobs details**: Click on a specific job to see its details and see the status, number of stages, associated SQL query, event timeline, and **directed acyclic graph** (**DAG**) visualization of the specific job:

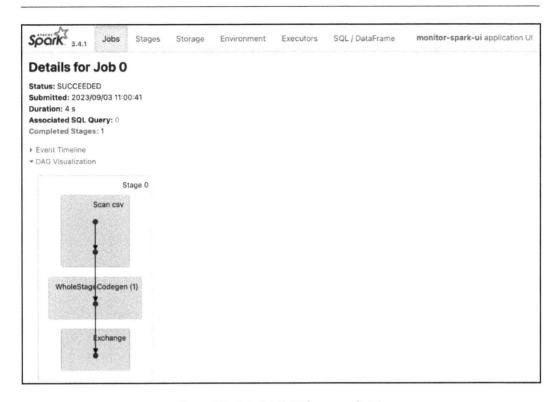

Figure 6.2 – Job details UI for a specific job

You can also see the list of stages grouped by state (active, pending, completed, skipped, or failed):

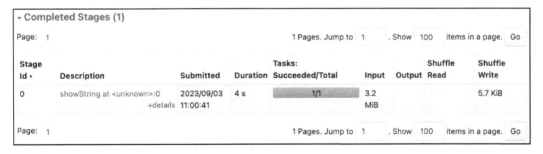

Figure 6.3 – List of stages grouped by status within a Spark job

4.   **See stage details**: Click on a specific stage to see the details and status, number of tasks, input/output/shuffle metrics, event timeline, and DAG visualization of the stage:

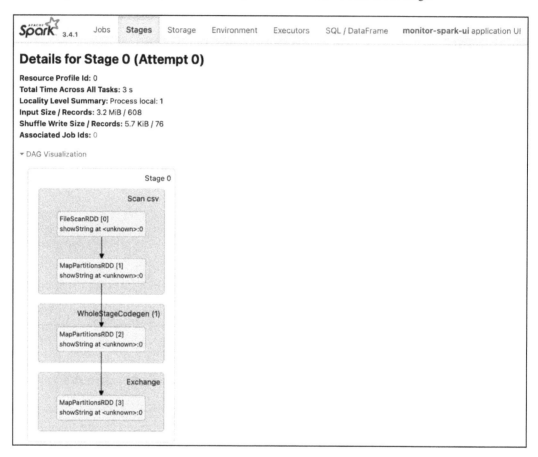

Figure 6.4 – DAG visualization for a specific stage

Figure 6.5 – Event timelines and aggregated metrics for a stage by executor

You can also see the status, executor ID, host, index, attempt, launch time, duration, input/output/shuffle metrics, and error message (if any) of the task:

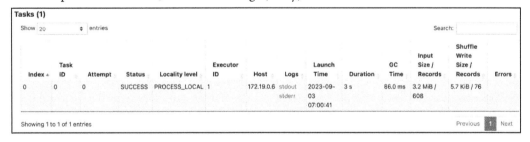

Figure 6.6 – List of tasks within a stage

5. **See environment details**: Go to the **Environment** tab and see the information about the Spark configuration and system properties. You can see the Spark version, master URL, application, and `ID/name/directory/user/jar/options/driver log URLs/default parallelism/executor instances/memory/cores/driver memory/cores/options/environment variables/log4j properties/hadoop configuration/system properties/classpath` entries:

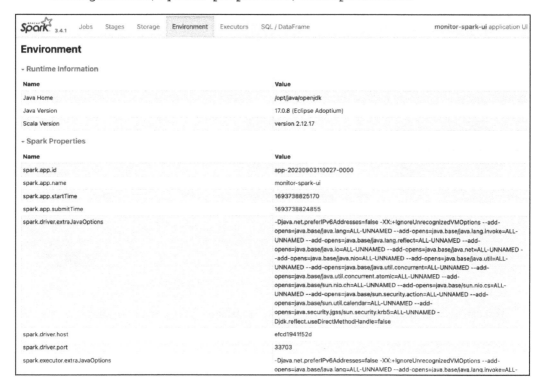

Figure 6.7 – Environment details for the Spark application

6. **See executor details**: Go to the **Executors** tab and see the summary of all executors in the cluster. You can see the executor `ID/host/port/status/threads/memory/disk/log URLs/active/completed/failed tasks/input/output/shuffle metrics/gc time/blacklisted` until the time of each executor. You can also kill an executor by clicking on the kill link:

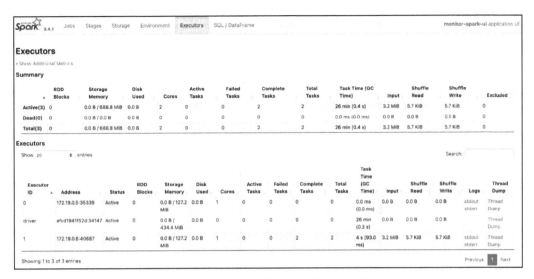

Figure 6.8 – Executor details for the Spark application

7. **See list of queries**: Go to the **SQL/DataFrame** tab and see the summary of all SQL queries executed by the application. You can see the query ID/name/submission time/ duration/job IDs/status/physical plan/graphical plan/metrics/ details link of each query:

Figure 6.9 – Details of all queries for a Spark application

8. **See query details**: Click on a specific query to see its details, such as the query `ID/name/ submission time/duration/job IDs/status/physical plan/graphical plan/metrics/details link` of each query. You can also see the execution plan, the SQL tab metrics, and the SQL performance tab of the query:

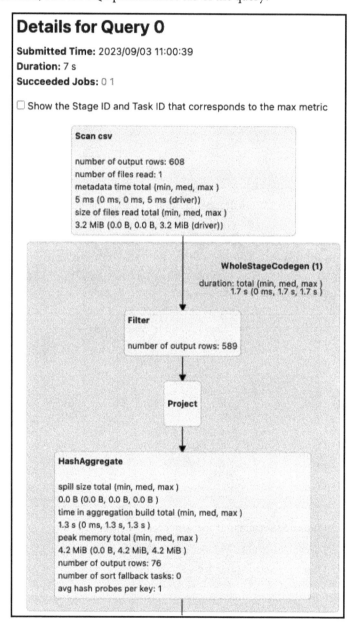

Figure 6.10 – SQL query DAG

```
== Physical Plan ==
AdaptiveSparkPlan (16)
+- == Final Plan ==
   CollectLimit (9)
   +- * HashAggregate (8)
      +- AQEShuffleRead (7)
         +- ShuffleQueryStage (6), Statistics(sizeInBytes=3.2 KiB, rowCount=76)
            +- Exchange (5)
               +- * HashAggregate (4)
                  +- * Project (3)
                     +- * Filter (2)
                        +- Scan csv  (1)
+- == Initial Plan ==
   CollectLimit (15)
   +- HashAggregate (14)
      +- Exchange (13)
         +- HashAggregate (12)
            +- Project (11)
               +- Filter (10)
                  +- Scan csv  (1)


(1) Scan csv
Output [2]: [country#5, release_year#7]
Batched: false
Location: InMemoryFileIndex [file:/opt/workspace/data/netflix_titles.csv]
PushedFilters: [IsNotNull(release_year), GreaterThan(release_year,2020)]
ReadSchema: struct<country:string,release_year:int>

(2) Filter [codegen id : 1]
Input [2]: [country#5, release_year#7]
Condition : (isnotnull(release_year#7) AND (release_year#7 > 2020))

(3) Project [codegen id : 1]
Output [1]: [country#5]
Input [2]: [country#5, release_year#7]

(4) HashAggregate [codegen id : 1]
Input [1]: [country#5]
Keys [1]: [country#5]
Functions [1]: [partial_count(1)]
Aggregate Attributes [1]: [count#47L]
Results [2]: [country#5, count#48L]
```

Figure 6.11 – SQL query plan

9. **Stop the Spark session**: Finally, we need to stop the Spark session to release the resources used by Spark:

    ```
    spark.stop()
    ```

See also

- *Web UI – Spark 3.4.1 Documentation*: `https://spark.apache.org/docs/latest/web-ui.html`

- *Spark Web UI – Understanding Spark Execution*: `https://sparkbyexamples.com/spark/spark-web-ui-understanding/`

- *Beginner's Guide to Spark UI: How to Monitor and Analyze Spark Jobs*: `https://medium.com/@suffyan.asad1/beginners-guide-to-spark-ui-how-to-monitor-and-analyze-spark-jobs-b2ada58a85f7`

- *How to Debug Queries by Just Using Spark UI*: `https://medium.com/swlh/spark-ui-to-debug-queries-3ba43279efee`

- *Debugging with the Apache Spark UI*: `https://learn.microsoft.com/en-us/azure/databricks/clusters/debugging-spark-ui`

Using broadcast variables

Broadcast variables are a feature of Apache Spark that allows you to send large, read-only data to all the executors in a cluster efficiently. This can be useful when you have a large dataset that needs to be used for multiple tasks, but you don't want to send the data over the network for each task. For example, if you have a lookup table that maps country codes to country names and you want to use it in a transformation on a large DataFrame, you can broadcast the lookup table to avoid sending it with every task.

In this recipe, you will learn how to create and use broadcast variables in Apache Spark using Python. You will also learn how broadcast variables work under the hood and what some of their benefits and limitations are.

How to do it...

1. **Import the required libraries**: Start by importing the necessary libraries for working with Delta Lake. In this case, we need the `delta` module and the `SparkSession` class from the `pyspark.sql` module:

```
from pyspark.sql import SparkSession
from pyspark.sql.functions import rand, when, pandas_udf,
PandasUDFType
from pyspark.sql.types import BooleanType
```

2. **Create a SparkSession object**: To interact with Spark and Delta Lake, you need to create a `SparkSession` object:

```
# Create a new SparkSession
spark = (SparkSession
```

```
.builder
.appName("broadcast-variables")
.master("spark://spark-master:7077")
.config("spark.executor.memory", "512m")
.getOrCreate())

# Set log level to ERROR
spark.sparkContext.setLogLevel("ERROR")
```

3. **Generate a large DataFrame**: We can generate a randomized 1-million-record DataFrame with salary, gender, and country code:

```
# Create some sample data frames
# A large data frame with 1 million rows
large_df = (spark.range(0, 1000000)
    .withColumn("salary", 100*(rand() * 100).cast("int"))
    .withColumn("gender", when((rand() * 2).cast("int") == 0,
"M").otherwise("F"))
    .withColumn("country_code",
    .when((rand() * 4).cast("int") == 0, "US")
    .when((rand() * 4).cast("int") == 1, "CN")
    .when((rand() * 4).cast("int") == 2, "IN")
    .when((rand() * 4).cast("int") == 3, "BR")))
large_df.show(5)
```

4. **Create a broadcast variable**: Define a lookup table as a Python dictionary that maps country codes to country names. For example, "US" maps to "United States" and "CN" maps to "China". Then, create a broadcast variable from the lookup table by using the `sparkContext.broadcast()` method. This will return a `pyspark.Broadcast` object that can be accessed through its `value` attribute:

```
# Define lookup table
lookup = {"US": "United States", "CN": "China", "IN": "India",
"BR": "Brazil", "RU": "Russia"}

# Create broadcast variable
broadcast_lookup = spark.sparkContext.broadcast(lookup)
```

5. **Define a pandas UDF for transformations**: Now that you have created a broadcast variable, you can use it in a UDF that looks up the country code and returns the country name:

```
@pandas_udf('string', PandasUDFType.SCALAR)
def country_convert(s):
    return s.map(broadcast_lookup.value)
```

6. **Apply the UDF to a DataFrame**: Apply the scalar pandas `country_convert` UDF to the `country_code` column to generate the `country_name` column:

    ```
    large_df.withColumn("country_name", country_convert(large_
    df.country_code)).show(5)
    ```

    The output is as following:

    ```
    +---+------+------+------------+-------------+
    | id|salary|gender|country_code| country_name|
    +---+------+------+------------+-------------+
    |  0|  4600|     M|          CN|        China|
    |  1|  9000|     F|          IN|        India|
    |  2|  2700|     M|          US|United States|
    |  3|  6200|     M|        null|         null|
    |  4|  5100|     F|          IN|        India|
    +---+------+------+------------+-------------+
    only showing top 5 rows
    ```

7. **Define a pandas UDF for filtering**: You can also use the broadcast variable in a filter UDF and leverage the `isin` function to return a Boolean value:

    ```
    @pandas_udf(BooleanType(), PandasUDFType.SCALAR)
    def filter_unknown_country(s):
        return s.isin(broadcast_lookup.value)
    ```

8. **Apply the filter UDF to the DataFrame**: Apply the filtering scalar pandas `filter_unknown_country` UDF to filter the DataFrame:

    ```
    large_df.filter(filter_unknown_country(large_df.country_code)).
    show(5)
    ```

    The output is as following:

    ```
    +---+------+------+------------+
    | id|salary|gender|country_code|
    +---+------+------+------------+
    |  0|  4600|     M|          CN|
    |  1|  9000|     F|          IN|
    |  2|  2700|     M|          US|
    |  4|  5100|     F|          IN|
    |  5|  8800|     M|          US|
    +---+------+------+------------+
    only showing top 5 rows
    ```

9. **Stop the Spark session**: Finally, we need to stop the Spark session to release the resources used by Spark:

```
spark.stop()
```

How it works...

Broadcast variables work by caching the data on each executor node rather than sending it with every task. This reduces the network traffic and improves the performance of the Spark job. Broadcast variables are read-only and cannot be modified once created. When you create a broadcast variable using the broadcast method of the Spark context, Spark does not immediately send the data to the executors. Instead, it waits until the broadcast variable is used for the first time in a transformation or an action. Once the data is distributed, it is cached on each executor node in a serialized form. When a task needs to access the data, it deserializes it and uses it in the computation. The data remains cached until the broadcast variable is destroyed or the Spark context is stopped.

There's more...

Broadcast variables are a powerful feature of Apache Spark, but they also have some limitations and best practices that you should be aware of:

- Broadcast variables are not automatically garbage collected. You need to explicitly destroy them using the `destroy` method of the `Broadcast` class when you are done with them. This will remove the data from the memory and disk of the executors.

- Broadcast variables are not checkpointed. If an executor node fails and restarts, it will need to fetch the data again from another node or from the driver.

- Broadcast variables should be used sparingly and carefully. Creating too many broadcast variables can consume a lot of memory and disk space on the executors and affect the performance of your Spark job.

Optimizing Spark jobs by minimizing data shuffling

In this recipe, you will learn how to optimize Spark jobs by minimizing data shuffling. Data shuffling is the process of transferring data across different partitions or nodes. Data shuffling can be expensive and time-consuming, as it involves network I/O, disk I/O, and serialization/deserialization of data. Therefore, minimizing data shuffling is one of the key techniques for optimizing Spark performance.

Some of the most common scenarios when data shuffling occurs are the following:

- **When you perform a join operation on two or more DataFrames**: Joining requires shuffling data across partitions or nodes based on the join keys

- **When you perform a global aggregation operation on a DataFrame**: Global aggregation requires shuffling data across partitions or nodes to compute some statistics for the whole DataFrame

- **When you perform a repartition or coalesce operation on a DataFrame**: Repartitioning or coalescing requires shuffling data across partitions or nodes to change the number of partitions or the partitioning criteria of a DataFrame

How to do it...

1. **Import the required libraries**: Start by importing the necessary libraries for working with Delta Lake. In this case, we need the `delta` module and the `SparkSession` class from the `pyspark.sql` module:

```
from pyspark.sql import SparkSession
from pyspark.sql.functions import col, avg, date_sub, current_
date, rand, when, broadcast
```

2. **Create a SparkSession object**: To interact with Spark and Delta Lake, you need to create a `SparkSession` object:

```
# Create a new SparkSession
spark = (SparkSession
    .builder
    .appName("optimize-data-shuffles")
    .master("spark://spark-master:7077")
    .config("spark.executor.memory", "512m")
    .getOrCreate())


# Set log level to ERROR
spark.sparkContext.setLogLevel("ERROR")
```

3. **Generate a large DataFrame**: We can generate a randomized 1-million-record DataFrame using the following:

```
# Create some sample data frames
# A large data frame with 1 million rows
large_df = (spark.range(0, 1000000)
    .withColumn("salary", 100*(rand() * 100).cast("int"))
    .withColumn("gender", when((rand() * 2).cast("int") == 0,
"M").otherwise("F"))
    .withColumn("country_code",
```

```
        .when((rand() * 4).cast("int") == 0, "US")
        .when((rand() * 4).cast("int") == 1, "CN")
        .when((rand() * 4).cast("int") == 2, "IN")
        .when((rand() * 4).cast("int") == 3, "BR")))
large_df.show(5)
```

4. **Reducing shuffle size**: Minimizing data shuffling means reducing the size of the shuffled data. For example, you can reduce the shuffle size in the following instances:

    *   **When you perform a join operation on two DataFrames**: Joining requires shuffling data across partitions or nodes based on the join keys. You can reduce the shuffle size by using a broadcast join or coalesce join instead of a sort-merge join or shuffled `hash` join.

    *   **When you perform a global aggregation on a DataFrame**: Global aggregation requires shuffling data across partitions or nodes to compute some statistics for the whole DataFrame. You can reduce the shuffle size by using partial aggregation or approximate aggregation instead of exact aggregation.

    For example, you can use the following code to join and globally aggregate the sample DataFrame with another DataFrame using a broadcast join and partial aggregation:

    ```
    # Filter the DataFrame by age
    df_filtered = large_df.filter(col("age") >= 55)
    # Map the DataFrame by adding 10% bonus to salary
    df_mapped = df_filtered.withColumn("bonus", col("salary") * 1.1)
    # Locally aggregate the DataFrame by computing the average bonus
    by age
    df_aggregated = df_mapped.groupBy("age").agg(avg("bonus"))
    # Print the result
    df_aggregated.show(5)
    ```

    The output of this code should look something like this:

    ```
    +---+------------------+
    |age|        avg(bonus)|
    +---+------------------+
    | 85|5462.653508771929|
    | 65|5478.981668009669|
    | 78|5457.682827459767|
    | 81| 5482.96132596685|
    | 76| 5465.81568744408|
    +---+------------------+
    only showing top 5 rows
    ```

    The previous code reduces the shuffle size by using a broadcast join and partial aggregation. You can verify this by checking the physical plan of the DataFrame using the `explain` method:

    ```
    df_aggregated.explain()
    ```

The output of the previous code should look something like this:

```
== Physical Plan ==
AdaptiveSparkPlan isFinalPlan=false
+- HashAggregate(keys=[age#542], functions=[avg(bonus#589)])
   +- Exchange hashpartitioning(age#542, 200), ENSURE_REQUIREMENTS, [plan_id=1003]
      +- HashAggregate(keys=[age#542], functions=[partial_avg(bonus#589)])
         +- Project [age#542, (cast(salary#546 as double) * 1.1) AS bonus#589]
            +- Filter (isnotnull(age#542) AND (age#542 >= 55))
               +- Project [age#542, (cast((rand(-2852223682338606353) * 100.0) as int) * 100) AS salary#546]
                  +- Project [cast((rand(-24633094416071200) * 100.0) as int) AS age#542]
                     +- Range (0, 10000000, step=1, splits=2)
```

Figure 6.12 – Aggregated DF explain plan

You can see that there is only one exchange operation in the physical plan, which indicates that only one DataFrame is shuffled across partitions or nodes. The other DataFrame is broadcasted to each node using the `Exchange` operation, which reduces the network I/O and disk I/O.

You can also see that there are two `HashAggregate` operations in the physical plan, which indicates that partial aggregation is performed within each partition before global aggregation is performed across partitions.

5. **Choosing optimal partitioning**: Another way to minimize data shuffling is to choose an optimal partitioning scheme for your DataFrames. Partitioning is the process of dividing your data into smaller and logical units that can be processed in parallel by different executors or cores. Partitioning affects how data is shuffled across partitions or nodes when you perform operations that require shuffling. For example, you can choose an optimal partitioning scheme in the following instances:

    *   **When you create a DataFrame from an external source**: You can specify the number of partitions or the size of each partition when you read data from a file, a database, or a stream. You can also use partition discovery or bucketing to partition your data based on some columns or expressions.

    *   **When you repartition or coalesce a DataFrame**: You can use the repartition or coalesce methods to change the number of partitions or the partitioning criteria of a DataFrame. You can also use the `repartitionByRange` method to partition your data based on a range of values.

    For example, you can use the following code to create and repartition the sample DataFrame using optimal partitioning:

    ```
    # Repartition the DataFrame by gender with 2 partitions
    df_repartitioned = large_df.repartition(col("gender"))

    # Repartition the DataFrame by age range with 5 partitions
    df_repartitioned_by_range = large_df.repartitionByRange(5,
    col("age"))
    ```

The previous code creates and repartitions the sample DataFrame using optimal partitioning. You can verify this by checking the physical plan of each DataFrame using the `explain` method:

```
large_df.explain()
df_repartitioned.explain()
df_repartitioned_by_range.explain()
```

You can see from the following screenshot that the first DataFrame is partitioned by hashing all the columns with 10 partitions using the `Exchange` operation:

```
== Physical Plan ==
*(1) Project [id#4L, date#6, age#9, salary#13, gender#18, CASE WHEN (cast((rand(-5258595656362
598529) * 5.0) as int) = 0) THEN IC WHEN (cast((rand(6677297146942895454) * 5.0) as int) = 1)
THEN IC-2 WHEN (cast((rand(6084707916817199194) * 5.0) as int) = 2) THEN M1 WHEN (cast((rand(7
628756694394173931) * 5.0) as int) = 3) THEN M2 WHEN (cast((rand(-1822131519618029291) * 5.0)
as int) = 4) THEN IC-3 ELSE M3 END AS grade#24]
+- *(1) Project [id#4L, date#6, age#9, salary#13, CASE WHEN (cast((rand(-1512255023260467776)
* 2.0) as int) = 0) THEN M ELSE F END AS gender#18]
   +- *(1) Project [id#4L, date#6, age#9, (cast((rand(900612033348343497) * 100.0) as int) * 1
00) AS salary#13]
      +- *(1) Project [id#4L, date#6, cast((rand(3280734957678084291) * 100.0) as int) AS age#
9]
         +- *(1) Project [id#4L, date_sub(2024-02-21, cast((rand(-5065184338059177050) * 365.
0) as int)) AS date#6]
            +- *(1) Range (0, 1000000, step=1, splits=2)
```

Figure 6.13 – Explain plan for default partition DataFrame

The second DataFrame is partitioned by hashing the `gender` column with two partitions using the `Exchange` operation:

```
== Physical Plan ==
AdaptiveSparkPlan isFinalPlan=false
+- Exchange hashpartitioning(gender#18, 200), REPARTITION_BY_COL, [plan_id=363]
   +- Project [id#4L, date#6, age#9, salary#13, gender#18, CASE WHEN (cast((rand(-525859565636
2598529) * 5.0) as int) = 0) THEN IC WHEN (cast((rand(6677297146942895454) * 5.0) as int) = 1)
THEN IC-2 WHEN (cast((rand(6084707916817199194) * 5.0) as int) = 2) THEN M1 WHEN (cast((rand(7
628756694394173931) * 5.0) as int) = 3) THEN M2 WHEN (cast((rand(-1822131519618029291) * 5.0)
as int) = 4) THEN IC-3 ELSE M3 END AS grade#24]
      +- Project [id#4L, date#6, age#9, salary#13, CASE WHEN (cast((rand(-1512255023260467776)
* 2.0) as int) = 0) THEN M ELSE F END AS gender#18]
         +- Project [id#4L, date#6, age#9, (cast((rand(900612033348343497) * 100.0) as int) *
100) AS salary#13]
            +- Project [id#4L, date#6, cast((rand(3280734957678084291) * 100.0) as int) AS age
#9]
               +- Project [id#4L, date_sub(2024-02-21, cast((rand(-5065184338059177050) * 365.
0) as int)) AS date#6]
                  +- Range (0, 1000000, step=1, splits=2)
```

Figure 6.14 – Explain plan for a partitioned-by-column DataFrame

The third DataFrame is partitioned by sorting the `age` column with five partitions using the Exchange operation:

```
== Physical Plan ==
AdaptiveSparkPlan isFinalPlan=false
+- Exchange rangepartitioning(age#9 ASC NULLS FIRST, 5), REPARTITION_BY_NUM, [plan_id=391]
   +- Project [id#4L, date#6, age#9, salary#13, gender#18, CASE WHEN (cast((rand(-525859565636
2598529) * 5.0) as int) = 0) THEN IC WHEN (cast((rand(66772971469428954454) * 5.0) as int) = 1)
THEN IC-2 WHEN (cast((rand(6084707916817199194) * 5.0) as int) = 2) THEN M1 WHEN (cast((rand(7
628756694394173931) * 5.0) as int) = 3) THEN M2 WHEN (cast((rand(-1822131519618029291) * 5.0)
as int) = 4) THEN IC-3 ELSE M3 END AS grade#24]
      +- Project [id#4L, date#6, age#9, salary#13, CASE WHEN (cast((rand(-15122550232604677776)
* 2.0) as int) = 0) THEN M ELSE F END AS gender#18]
         +- Project [id#4L, date#6, age#9, (cast((rand(900612033348343497) * 100.0) as int) *
100) AS salary#13]
            +- Project [id#4L, date#6, cast((rand(3280734957678084291) * 100.0) as int) AS age
#9]
               +- Project [id#4L, date_sub(2024-02-21, cast((rand(-5065184338059177050) * 365.
0) as int)) AS date#6]
                  +- Range (0, 1000000, step=1, splits=2)
```

Figure 6.15 – Explain plan for a repartitioned-by-range DataFrame

Choosing an optimal partitioning scheme can help you minimize data shuffling by achieving the following:

- Reducing the number of partitions to avoid creating too many small files or tasks that can increase the overhead of shuffling

- Increasing the number of partitions to avoid creating too few large files or tasks that can cause data skew or memory issues

- Aligning the partitioning criteria with the join keys or aggregation keys to avoid shuffling data across partitions or nodes when performing join or aggregation operations

6. **Stop the Spark session**: Finally, we need to stop the Spark session to release the resources used by Spark:

    ```
    spark.stop()
    ```

See also

- *Performance Tuning*: https://spark.apache.org/docs/latest/sql-performance-tuning.html

- *Tuning Spark*: https://spark.apache.org/docs/latest/tuning.html

Avoiding data skew

Data skew is a common problem that can affect the performance and scalability of Apache Spark applications. Data skew occurs when the data being processed is not evenly distributed across partitions, resulting in some tasks taking much longer than others and wasting cluster resources. Data skew can be caused by operations that require shuffling or repartitioning the data, such as `join`, `groupBy`, or `orderBy`.

In this recipe, we will learn how to detect and handle data skew in Apache Spark using various techniques and tips.

How to do it...

1. **Import the required libraries**: Start by importing the necessary libraries for working with Delta Lake. In this case, we need the `delta` module and the `SparkSession` class from the `pyspark.sql` module:

```
from pyspark.sql import SparkSession
from pyspark.sql.functions import rand, col, when, broadcast,
concat, lit
```

2. **Create a SparkSession object**: To interact with Spark and Delta Lake, we need to create a `SparkSession` object:

```
# Create a new SparkSession
spark = (SparkSession
    .builder
    .appName("avoid-data-skew")
    .master("spark://spark-master:7077")
    .config("spark.executor.memory", "512m")
    .getOrCreate())

# Set log level to ERROR
spark.sparkContext.setLogLevel("ERROR")
```

3. **Create a measure time helper function**: We will use a helper function to measure the execution time for each query we run:

```
# Define a function to measure the execution time of a query
import time
def measure_time(query):
    start = time.time()
    query.collect() # Force the query execution by calling an
action
    end = time.time()
    print(f"Execution time: {end - start} seconds")
```

4. **Generate a large DataFrame and a skewed DataFrame**: We can generate a randomized 10-million-record DataFrame and a skewed 1-million-record DataFrame:

```
# Create some sample data frames
# A large data frame with 10 million rows and two columns: id
and value
large_df = spark.range(0, 10000000).withColumn("value",
rand(seed=42))


# A skewed data frame with 1 million rows and two columns: id
and value
skewed_df = spark.range(0, 1000000).withColumn("value",
    rand(seed=42)).withColumn("id", when(col("id")%4 == 0,
0).otherwise(col("id")))
```

5. **Repartition the DataFrames and verify the skew**: By using the `repartition()` method, we can change the number of partitions of both DataFrames to have five partitions based on the `id` column. The output of this code will show the effect of repartitioning on the partition sizes. For `large_df`, which has a uniform distribution of values in the `id` column, the partition sizes should be roughly equal. For `skewed_df`, which has a skewed distribution of values in the `id` column, the partition sizes should be very uneven, with some partitions having many more rows than others:

```
large_df_repartitioned = large_df.repartition(5, "id")
num_partitions = large_df_repartitioned.rdd.getNumPartitions()
print(f"Number of partitions: {num_partitions}")


partition_sizes = large_df_repartitioned.rdd.glom().map(len).
collect()
print(f"Partition sizes: {partition_sizes}")


skewed_df_repartitioned = skewed_df.repartition(5, "id")
num_partitions = skewed_df_repartitioned.rdd.getNumPartitions()
print(f"Number of partitions: {num_partitions}")


partition_sizes = skewed_df_repartitioned.rdd.glom().map(len).
collect()
print(f"Partition sizes: {partition_sizes}")
```

The output is as following:

```
Number of partitions: 5
Partition sizes: [1998962, 2000902, 1999898, 2000588, 1999650]


Number of partitions: 5
Partition sizes: [400054, 150144, 149846, 149903, 150053]
```

6. **Execute a join on skewed data**: Now that we have the skewed partitions, we can get a baseline of skewed performance by joining the two DataFrames, `large_df_repartitioned` and `skewed_df_repartitioned`, using the default join strategy in Spark, which is the `sort-merge` join:

```
# Join the non-skewed DataFrames using the default join strategy
(sort-merge join)
inner_join_df = large_df_repartitioned.join(skewed_df_
repartitioned, "id")
measure_time(inner_join_df)
```

The output is as following:

```
Execution time: 31.603528261184692 seconds
```

We can see that there is a huge data skew in the `inner_join_df` DataFrame, where the `id` value is the same. This means that if we join the two DataFrames on the `id` column, most of the data will be shuffled to one partition, creating a bottleneck and slowing down the join operation.

7. **Isolate skewed data**: One way to handle data skew is to isolate the skewed data from the rest of the data and process it separately. This way, we can avoid shuffling the skewed data and reduce the load on the cluster. To implement this approach, we need to perform the following steps:

I.   Identify the skewed value(s) in the join key column.

II.   Filter out the rows with the skewed value(s) from both DataFrames and save them as separate DataFrames.

III. Join the remaining rows from both DataFrames using the default join strategy (the `sort-merge` join)

IV.   Join the skewed DataFrames using a broadcast `hash` join

V.   Join the results from both joins.

The code is as follows:

```
# Identify the skewed value in the invoice_id column
skewed_value = 0

# Filter out the rows with the skewed value from both DataFrames
large_skewed_df = large_df_repartitioned.filter(large_df_
repartitioned.id == skewed_value)
small_skewed_df = skewed_df_repartitioned.filter(skewed_df_
repartitioned.id == skewed_value)

# Filter out the rows without the skewed value from both
DataFrames
large_non_skewed_df = large_df_repartitioned.filter(large_df_
repartitioned.id != skewed_value)
small_non_skewed_df = skewed_df_repartitioned.filter(skewed_df_
```

```
repartitioned.id != skewed_value)

# Join the non-skewed DataFrames using the default join strategy
(sort-merge join)
non_skewed_join_df = large_non_skewed_df.join(small_non_skewed_
df, "id")

# Join the skewed DataFrames using a broadcast hash join
skewed_join_df = large_skewed_df.join(broadcast(small_skewed_
df), "id")

# Union the results from both joins
final_join_df = non_skewed_join_df.union(skewed_join_df)

measure_time(final_join_df)
```

The output should look like this:

```
Execution time: 15.715256214141846 seconds
```

We can see that by isolating the skewed data, we have avoided shuffling it, reduced the size of the join, and are able to cut the execution time in half. However, this approach has some drawbacks:

- It requires the manual identification of the skewed value(s), which may not be feasible or accurate for large or dynamic datasets

- It requires extra filtering and union operations, which may introduce additional overhead and complexity

- It may not work well if there are multiple or unknown skewed values in the join key column

8. **Broadcast hash join**: Another way to handle data skew is to use a broadcast hash join instead of a sort-merge join. A broadcast hash join is a type of join that broadcasts one of the DataFrames to each executor and builds a hash table in memory. Then, it scans the other DataFrame and probes the hash table for matches. A broadcast hash join can be faster and more efficient than a sort-merge join, especially when one of the DataFrames is small enough to fit in memory.

To implement this approach, we need to perform the following steps:

I.   Identify the smaller DataFrame in terms of size.

II.   Use the broadcast function to mark the smaller DataFrame for broadcasting.

III. Join the two DataFrames using the broadcast function as an argument:

```
smaller_df = skewed_df_repartitioned

# Use the broadcast function to mark the smaller DataFrame for
broadcasting
```

```
from pyspark.sql.functions import broadcast
broadcast_df = broadcast(smaller_df)

# Join the two DataFrames using the broadcast function as an
argument

broadcast_join_df = large_df_repartitioned.join(broadcast_df,
"id")

measure_time(broadcast_join_df)
```

The output should look like this:

```
Execution time: 10.276329278945923 seconds
```

We can see that by using a broadcast hash join, we have avoided sorting and shuffling the data and reduced network traffic. However, this approach has some drawbacks:

- It requires the manual identification of the smaller DataFrame, which may not be feasible or accurate for large or dynamic datasets

- It requires enough memory on each executor to store the broadcasted DataFrame, which may not be available or optimal for resource utilization

- It may not work well if both DataFrames are large or have similar sizes

9. **Key salting**: A third way to handle data skew is to use key salting. Key salting is a technique that modifies the join key column by adding a random suffix (salt) to each value. This way, we can create more partitions and distribute the data more evenly across them. Key salting can improve the parallelism and load balancing of the join operation.

To implement this approach, we need to perform the following steps:

I.   Identify the skewed value(s) in the join key column.

II.   Create a list of salt values to append to the skewed value(s).

III.   Create a new column in both DataFrames that contains the original join key value plus a salt value if it is skewed or just the original join key value otherwise.

IV.   Join the two DataFrames on the new column using the default join strategy (the sort-merge join).

V.   Drop the new column and keep only the original join key column.

The code is as follows:

```
# Import random module
import random

# Identify the skewed value in the id column
```

```
skewed_value = 0

# Create a list of salt values to append to the skewed value
salt_list = ["_A", "_B", "_C", "_D", "_E"]

# Create a new column in both DataFrames that contains the
original invoice_id value plus a salt value if it is skewed, or
just the original invoice_id value otherwise
large_df = (large_df_repartitioned
    .withColumn("salted_id",
        when(large_df_repartitioned.id == skewed_value,
        concat(large_df_repartitioned.id,
            lit(random.choice(salt_list))))
    .otherwise(large_df_repartitioned.id)))
skewed_df = (skewed_df_repartitioned
    .withColumn("salted_id",
        when(skewed_df_repartitioned.id == skewed_value,
        concat(skewed_df_repartitioned.id,
            lit(random.choice(salt_list))))
    .otherwise(skewed_df_repartitioned.id)))

# Join the two DataFrames on the new column using the default
join strategy (sort-merge join)
salted_join_df = large_df.join(skewed_df, "salted_id")

# Drop the new column and keep only the original invoice_id
column
final_join_df = salted_join_df.drop("salted_id")

measure_time(final_join_df)
```

The output should look like this:

```
Execution time: 18.217721939086914 seconds
```

We can see that by using key salting, we have created more partitions and distributed the skewed data more evenly across them. However, this approach has some drawbacks:

- It requires the manual identification of the skewed value(s), which may not be feasible or accurate for large or dynamic datasets

- It requires extra column manipulation and randomization operations, which may introduce additional overhead and complexity

- It may not work well if there are multiple or unknown skewed values in the join key column

10. **Stop the Spark session**: Finally, we need to stop the Spark session to release the resources used by Spark:

```
spark.stop()
```

There's more...

In this section, we will detail some additional considerations while trying to eliminate data skew in your Spark applications:

- Data skew can also occur in other operations that require shuffling or repartitioning data, such as `groupBy` or `orderBy`. The same techniques we used for handling data skew in joins can also be applied to these operations.

- Data skew can be caused by various factors, such as data quality issues, improper partitioning schemes, or business logic. It is important to understand your data and identify potential sources of skew before applying any techniques.

- Data skew can be prevented or minimized by using appropriate partitioning strategies, such as hash partitioning or range partitioning, that can distribute the data more evenly across partitions. You can also use the repartition or coalesce methods to change the number of partitions or the partitioning criteria of your DataFrames.

Caching and persistence

To make Spark applications run faster, developers can use two important techniques: caching and persistence. These techniques allow Spark to store some or all of the data in memory or on disk so that it can be reused without recomputing it. By caching or persisting DataFrames, you can store some intermediate results in the memory (default) or other more durable storage, such as disk space, and/or replicate them. This way, you can avoid recomputing these results when they are needed again in later stages. DataFrames can be cached using the `cache()` or `persist()` methods on them.

In this recipe, we will learn how to cache and persist Spark DataFrames.

How to do it...

1. **Import the required libraries**: Start by importing the necessary libraries for working with Delta Lake. In this case, we need the `delta` module and the `SparkSession` class from the `pyspark.sql` module:

```
from pyspark.sql import SparkSession
from pyspark import StorageLevel
from pyspark.sql.functions import rand, current_date, date_sub
```

2. **Create a SparkSession object**: To interact with Spark and Delta Lake, you need to create a `SparkSession` object:

```
spark = (SparkSession.builder
    .appName("cache-and-persist")
    .master("spark://spark-master:7077")
    .config("spark.executor.memory", "512m")
    .getOrCreate())

spark.sparkContext.setLogLevel("ERROR")
```

3. **Create a measure time helper function**: We will use a helper function to measure the execution time for each query we run:

```
# Define a function to measure the execution time of a query
import time
def measure_time(query):
    start = time.time()
    query.collect() # Force the query execution by calling an action
    end = time.time()
    print(f"Execution time: {end - start} seconds")
```

4. **Generate a large DataFrame**: We can generate a randomized 1-million-record DataFrame with salary, gender, and country code:

```
# Create some sample data frames
# A large data frame with 10 million rows and two columns: id and value
large_df = (spark.range(0, 10000000)
    .withColumn("date", date_sub(current_date(), (rand() * 365).cast("int")))
    .withColumn("ProductId", (rand() * 100).cast("int")))
large_df.show(5)
```

5. **Cache the DataFrame and check its storage level**: We will be using the `cache()` method to let Spark mark it for caching and assign it a default storage level, which is MEMORY_AND_DISK for DataFrames. This means that Spark will try to store the DataFrame in memory as deserialized objects, and if it runs out of memory, it will spill some partitions to the disk:

```
# Cache the DataFrame using cache() method
large_df.cache()
# Check the storage level of the cached DataFrame
print(large_df.storageLevel)
```

The output is as following:

```
Disk Memory Deserialized 1x Replicated
```

6. **Make the DataFrame persistent and check its storage level**: When we call the `persist()` method on a DataFrame, Spark marks it for caching and assigns it a user-specified storage level. We can choose from different storage levels depending on our needs. For example, we can use `MEMORY_AND_DISK_DESER` to store the DataFrame in memory as serialized objects and spill to disk if necessary. This can save some memory space but incur some CPU overhead for serialization and deserialization:

```
# Persist the DataFrame using persist() method with a different
storage level
large_df.persist(MEMORY_AND_DISK_DESER)
# Check the storage level of the persisted DataFrame
print(large_df.storageLevel)
```

The output is as following:

```
Disk Memory Deserialized 1x Replicated
```

7. **Perform some transformations and actions on the cached and persistent DataFrame and measure the execution time**: When we perform some transformations and actions on the cached and persisted DataFrame, Spark will first check whether the DataFrame is already cached in memory or disk. If yes, it will reuse the cached data and avoid recomputing it from scratch. This can save a lot of time and resources, especially if the DataFrame is large or complex to compute. If not, Spark will compute the DataFrame and cache it according to its storage level:

```
results_df = large_df.groupBy("ProductId").agg({"Id": "count"})
measure_time(results_df)
# Show the result
results_df.show(5)
```

The output is as following:

```
Execution time: 9.418130159378052 seconds
+---------+---------+
|ProductId|count(Id)|
+---------+---------+
|       31|   100328|
|       85|    99911|
|       65|    99596|
|       53|   100085|
|       78|    99849|
+---------+---------+
only showing top 5 rows
```

8. **Perform the same transformations again**: When we run the same function on the cached and persisted DataFrame for the second time, we can see that the execution time is much faster than the first time. This is because Spark reuses the cached data and does not need to recompute it again:

```
results_df = large_df.groupBy("ProductId").agg({"Id": "count"})
measure_time(results_df)
# Show the result
results_df.show(5)
```

The output is as following:

```
Execution time: 0.5906562805175781 seconds
+---------+---------+
|ProductId|count(Id)|
+---------+---------+
|       31|   100328|
|       85|    99911|
|       65|    99596|
|       53|   100085|
|       78|    99849|
+---------+---------+
only showing top 5 rows
```

9. **Unpersist the DataFrame using the unpersist() method and check its storage level**: When we call the `unpersist()` method to use on a DataFrame, Spark removes it from the cache and frees up the memory or disk space that was occupied by it. This can help us avoid memory or disk overflow and improve the performance of other DataFrames that need to be cached:

```
# Unpersist the DataFrame using unpersist() method
large_df.unpersist()
# Check the storage level of the unpersisted DataFrame
print(large_df.storageLevel)
```

The output is as following:

```
Serialized 1x Replicated
```

10. **Stop the Spark session**: Finally, we need to stop the Spark session to release the resources used by Spark:

```
spark.stop()
```

There's more...

In this section, we will detail some additional considerations while working with caching and persistence in Apache Spark:

- Caching and persistence are lazy operations in Spark, which means that they do not take effect until an action is performed on them. This allows Spark to optimize the execution plan and avoid unnecessary caching or persistence.

- Caching and persistence are not fault-tolerant in Spark, which means that if a node fails or a partition is lost, Spark will not automatically recover the cached or persistent data. Instead, Spark will recompute the data from the original source or transformations that created it. To achieve fault-tolerance for cached or persistent data, we can use the checkpoint() method on a DataFrame, which will save it to a reliable storage system such as HDFS or S3. However, checkpointing is an expensive operation that involves writing data to disk, so it should be used sparingly and only when necessary.

- Caching and persistence are not the only ways to improve performance of Spark applications. We can also use other techniques, such as partitioning, broadcasting, and tuning. Partitioning helps distribute the data across multiple nodes and reduces data shuffling. Broadcasting helps send a small dataset to all the nodes and avoid network traffic. Tuning helps adjust the configuration parameters and optimize resource allocation.

Partitioning and repartitioning

Partitioning is a way to split the data into multiple chunks that can be processed in parallel by different nodes in a cluster. Repartitioning is a way to change the number or the distribution of partitions in an existing dataset. Both partitioning and repartitioning are important techniques to optimize the performance and scalability of Spark applications.

In this recipe, you will learn how to partition and repartition data using Spark DataFrames in Python. You will also learn how to choose the appropriate partitioning method and number of partitions for your use case and how to deal with some common issues and challenges related to partitioning.

How to do it...

1. **Import the required libraries**: Start by importing the necessary libraries for working with Delta Lake. In this case, we need the delta module and the SparkSession class from the pyspark.sql module:

```
from pyspark.sql import SparkSession
from pyspark.sql.functions import rand, when
```

2. **Create a SparkSession object**: To interact with Spark and Delta Lake, you need to create a `SparkSession` object:

```
# Create a new SparkSession
spark = (SparkSession
    .builder
    .appName("broadcast-variables")
    .master("spark://spark-master:7077")
    .config("spark.executor.memory", "512m")
    .getOrCreate())


# Set log level to ERROR
spark.sparkContext.setLogLevel("ERROR")
```

3. **Generate a large DataFrame**: We can generate a randomized 1-million-record DataFrame with salary, gender, and country code:

```
# Create some sample data frames
# A large data frame with 1 million rows
large_df = (spark.range(0, 1000000)
    .withColumn("salary", 100*(rand() * 100).cast("int"))
    .withColumn("gender", when((rand() * 2).cast("int") == 0,
"M").otherwise("F"))
    .withColumn("country_code",
        .when((rand() * 4).cast("int") == 0, "US")
        .when((rand() * 4).cast("int") == 1, "CN")
        .when((rand() * 4).cast("int") == 2, "IN")
        .when((rand() * 4).cast("int") == 3, "BR")))
large_df.show(5)
```

The output should look something like this:

```
+---+------+------+------------+
| id|salary|gender|country_code|
+---+------+------+------------+
|  0|  3500|     F|          US|
|  1|  1300|     F|        null|
|  2|  3600|     F|          US|
|  3|  5200|     M|          IN|
|  4|  3100|     M|        null|
+---+------+------+------------+
only showing top 5 rows
```

4. **Check the number and size of partitions in a DataFrame**: To check how many partitions a DataFrame has, you can use the `rdd.getNumPartitions()` method. This method returns an integer that represents the number of partitions in the underlying DataFrame.

To check how many rows each partition has, you can use the `rdd.glom().map(len).collect()` method. This method returns a list of integers that represent the number of rows in each partition:

```
num_partitions = df_hash.rdd.getNumPartitions()
print(f"Number of partitions: {num_partitions}")

partition_sizes = df.rdd.glom().map(len).collect()
print(f"Partition sizes: {partition_sizes}")
```

The output should look something like this:

```
Number of partitions: 2
Partition sizes: [500000, 500000]
```

As you can see, the sample DataFrame has two partitions by default, and each partition has 500,000 rows. This is because Spark creates partitions based on the number of cores available on the cluster. In this case, there are two worker nodes with four cores each, so Spark creates eight partitions.

5. **Partition a DataFrame by a column using hash partitioning**: One way to partition a DataFrame by a column is to use hash partitioning. Hash partitioning assigns rows to partitions based on the hash value of the column value. This ensures that rows with the same column value are in the same partition, which can be useful for `join` or `group-by` operations.

    To partition a DataFrame by a column using hash partitioning, you can use the `repartition()` method:

    ```
    df_hash = large_df.repartition(10, "id")
    ```

6. **Repartition a DataFrame by a column using range partitioning**: Another way to partition a DataFrame by a column is to use range partitioning. Range partitioning assigns rows to partitions based on the range of the column value. This ensures that rows with similar column values are in the same partition, which can be useful for sorting or ordering operations.

    To repartition a DataFrame by a column using range partitioning, you can use the `repartitionByRange()` method:

    ```
    df_range = large_df.repartitionByRange(10, "id")
    ```

7. **Coalesce a DataFrame to reduce the number of partitions**: Sometimes, you may want to reduce the number of partitions in a DataFrame to avoid having too many small partitions that can cause overhead and inefficiency. One way to do this is to use coalescing. Coalescing combines adjacent partitions into larger ones without shuffling data across nodes. This can improve performance and reduce resource consumption.

    To coalesce a DataFrame to reduce the number of partitions, you can use the `coalesce()` method:

    ```
    df_coalesce = df_range.coalesce(4)
    ```

8. **Write a partitioned DataFrame to disk using partitionBy**: Finally, you may want to write a partitioned DataFrame to disk for later use or analysis. One way to do this is to use the `partitionBy()` method. This method takes one or more column names as arguments and writes the DataFrame to disk in a hierarchical directory structure based on the column values. This can improve the performance and efficiency of reading data from a disk.

To write a partitioned DataFrame to a disk using `partitionBy`, you can use the `write` method with the `partitionBy` option:

```
(large_df.write
    .format("parquet")
    .partitionBy("id")
    .mode("overwrite")
    .save("/data/tmp/partitioned_output"))
```

This will create a directory named `/data/tmp/partitioned_output` that contains subdirectories named after the `id` values, such as `/data/tmp/partitioned_output/id=1` and `/data/tmp/partitioned_output/id=2`. Each subdirectory will contain one or more Parquet files that store the rows with the corresponding `id` value.

9. **Stop the Spark session**: Finally, we need to stop the Spark session to release the resources used by Spark:

```
spark.stop()
```

There's more...

Partitioning can improve the performance and scalability of Spark applications by enabling parallel processing, reducing data shuffling, and balancing the workload across the cluster. However, partitioning also involves some trade-offs and challenges that needs to be considered and addressed. In this section, we will discuss how to choose the appropriate partitioning method and number of partitions for different scenarios and how to deal with common issues and challenges related to partitioning:

- **How to choose the appropriate partitioning method**: Choosing the appropriate partitioning method depends on the type of operation and the data distribution. Here are some general guidelines:

  - If you are performing `join` or `group-by` operations that require data co-location, you should use hash partitioning by using the `join` or `group-by` key. This will reduce the amount of shuffling and improve the performance of the operation.

  - If you are performing sorting or ordering operations that require data locality, you should use range partitioning by using the `sort` or `order` key. This will reduce the amount of shuffling and improve the performance of the operation.

  - If you are not performing any operation that requires data co-location or locality, you can use the default partitioning method or any other method that suits your use case.

- **How to choose the appropriate number of partitions**: Choosing the appropriate number of partitions depends on the size of the dataset and the resources available on the cluster. Here are some general guidelines:

  - The number of partitions should be large enough to fit in memory and small enough to avoid too much overhead. A good rule of thumb is to have two to four partitions per core per node.

  - The number of partitions should be proportional to the size of the dataset. A larger dataset may require more partitions to achieve better parallelism and load balancing. A smaller dataset may require fewer partitions to avoid wasting resources and creating empty partitions.

  - The number of partitions should be adjustable according to the dynamic nature of the cluster. You may need to increase or decrease the number of partitions depending on the availability or demand of resources on the cluster.

- **How to deal with common issues and challenges related to partitioning**: Partitioning is not a silver bullet that can solve all performance and scalability issues in Spark applications. There are some common issues and challenges that you may encounter when using partitioning, such as the following:

  - **Data skewness**: Data skewness occurs when some partitions have significantly more or less data than others. This can cause performance degradation, resource wastage, or even job failure. To deal with data skewness, you can try to use a different partitioning key, increase or decrease the number of partitions, or apply some data preprocessing techniques, such as filtering, sampling, or hashing.

  - **Data spilling**: Data spilling occurs when some partitions are too large to fit in memory and have to be written to disk temporarily. This can cause performance degradation, disk I/O contention, or even job failure. To avoid data spilling, you can try to reduce the size of partitions, increase the memory available on nodes, or tune some Spark configuration parameters, such as `spark.sql.shuffle.partitions`, `spark.shuffle.memoryFraction`, or `spark.shuffle.spill.compress`.

  - **Data duplication**: Data duplication occurs when some rows are present in more than one partition due to shuffling or replication. This can cause performance degradation, memory consumption, or incorrect results. To prevent data duplication, you can try to use a unique partitioning key, avoid unnecessary shuffling or replication, or apply some data deduplication techniques such as `distinct`, `dropDuplicates`, or `groupBy`.

Optimizing join strategies

In this recipe, we will explore how to optimize join strategies in Apache Spark using various techniques and configurations. Joining data is one of the most common and expensive operations in Apache Spark. Depending on the size and distribution of the data, different join strategies can have a significant impact on the performance and resource utilization of your Spark applications.

How to do it...

1. **Import the required libraries**: Start by importing the necessary libraries for working with Delta Lake. In this case, we need the `delta` module and the `SparkSession` class from the `pyspark.sql` module:

   ```
   # Import modules
   from pyspark.sql import SparkSession
   from pyspark.sql.functions import broadcast, col, rand, skewness
   ```

2. **Create a SparkSession object**: To interact with Spark and Delta Lake, you need to create a `SparkSession` object:

   ```
   # Create a Spark session
   spark = SparkSession.builder.appName("Optimizing Join
   Strategies").getOrCreate()


   # Set the log level to WARN to reduce noise
   spark.sparkContext.setLogLevel("WARN")
   ```

3. **Create a sample DataFrame**: We will create a large, small, and skewed DataFrame to test the techniques to optimize the joins:

   ```
   # Create some sample data frames
   # A large data frame with 1 million rows and two columns: id and
   value
   large_df = spark.range(0, 1000000).withColumn("value",
   rand(seed=42))


   # A small data frame with 10000 rows and two columns: id and
   name
   small_df = spark.range(0, 10000).withColumn("name", col("id").
   cast("string"))


   # A skewed data frame with 1 million rows and two columns: id
   and value
   skewed_df = spark.range(0, 1000000).withColumn("value",
       rand(seed=42)).withColumn("id", col("id") ** 4)
   ```

4. **Choosing the right join type**: The first thing to consider when joining data is the type of join that we want to perform. Spark supports various types of joins, such as inner, outer, left, right, cross, semi-, and anti-joins. Each join type has different semantics and implications for performance and resource utilization.

In general, we should choose the join type that best matches our business logic and minimizes the amount of data that needs to be shuffled across the network. For example, if we only need to keep the matching rows from both tables, we should use an inner join instead of an outer join. If we only need to keep the rows from one table that have a matching row in another table, we should use a semi-join instead of an inner join.

To demonstrate how different join types affect performance, let's try to join our `large_df` and `small_df` using different join types and measure the execution time:

```
# Join large_df and small_df using an inner join on id column
measure_time(large_df.join(small_df, "id"))

# Join large_df and small_df using a left outer join on id
column
measure_time(large_df.join(small_df, "id", "left"))

# Join large_df and small_df using a right outer join on id
column
measure_time(large_df.join(small_df, "id", "right"))

# Join large_df and small_df using a full outer join on id
column
measure_time(large_df.join(small_df, "id", "full"))

# Join large_df and small_df using a left semi join on id column
measure_time(large_df.join(small_df, "id", "left_semi"))

# Join large_df and small_df using a left anti join on id column
measure_time(large_df.join(small_df, "id", "left_anti"))
```

The output should look like this:

```
Execution time: 1.0291528701782227 seconds
Execution time: 25.628353357315063 seconds
Execution time: 5.767467021942139 seconds
Execution time: 18.269603490829468 seconds
Execution time: 0.5185227394104004 seconds
Execution time: 6.6788036823272705 seconds
```

As we can see, the inner join and the left semi-join are the fastest, as they only keep the matching rows from both tables. The left outer join and the left anti-join are slightly slower, as they also keep the non-matching rows from the left table. The right outer join and the full outer join are the slowest, as they also keep the non-matching rows from both tables.

5. **Broadcasting small tables**: Another important factor that affects join performance is the size of the tables that we are joining. If one of the tables is much smaller than the other, we can use a technique called broadcasting to optimize the join.

Broadcasting is sending a copy of a small table to each executor node in the cluster so that it can be cached in memory and used for local join operations with the larger table. This can reduce the amount of data that needs to be shuffled across the network and improve performance.

Spark can automatically decide whether to broadcast a table or not based on its size and the `spark.sql.autoBroadcastJoinThreshold` configuration parameter, which is set to 10 MB by default. However, we can also explicitly hint at Spark to broadcast a table using the `broadcast` function.

To demonstrate how broadcasting can improve performance, let's try to join our `large_df` and `small_df` using an inner join with and without broadcasting:

```
spark.conf.set("spark.sql.adaptive.enabled", "false")
spark.conf.set("spark.sql.autoBroadcastJoinThreshold", -1)


# Join large_df and small_df using an inner join without
broadcasting
measure_time(large_df.join(small_df, "id"))


# Join large_df and small_df using an inner join with
broadcasting
measure_time(large_df.join(broadcast(small_df), "id"))
```

The output should look like this:

```
Execution time: 3.1970551013946533 seconds
Execution time: 0.20557928085327148 seconds
```

As we can see, broadcasting the small table reduces the execution time by more than half.

6. **Using join hints**: Sometimes, Spark may not choose the optimal join strategy for our query, or we may want to override its default behavior for some reason. In such cases, we can use join hints to provide Spark with hints about which join strategy to use.

Join hints are specified using special comments in SQL queries or using hint methods in the DataFrame API. For example, we can use the following join hints to instruct Spark to use different join strategies:

*   **/*+ BROADCAST */ or .hint("broadcast") for broadcast hash join**: A broadcast `hash` join broadcasts the smaller table to each executor node and builds a hash table in memory. Then, it scans the larger table and probes the hash table for matches. A broadcast `hash` join can improve performance by avoiding shuffling data across the network, but it requires sufficient memory to store the hash table.

*   **/*+ SHUFFLE_HASH */ or .hint("shuffle_hash") for shuffle hash join**: A shuffle `hash` join shuffles both tables based on the join keys and partitions them into buckets. Then, it builds a hash table for each bucket of one table and probes it with each bucket of another table. A shuffle `hash` join can handle large tables that cannot fit in memory, but it involves shuffling data across the network and requires sufficient memory to store each hash table.

- /*+ SHUFFLE_REPLICATE_NL */ or .hint("shuffle_replicate_nl") for shuffle replicate nested loop join: A shuffle replicate nested loop join shuffles and replicates the entire right table to each partition of the left table. Then, it performs a nested loop join for each pair of partitions. A shuffle replicate nested loop join can handle unequal-sized tables that cannot be broadcasted, but this involves shuffling and replicating a large amount of data across the network and may cause data imbalance and out-of-memory errors.

- /*+ MERGE */ or .hint("merge") for sort-merge join: A sort-merge join shuffles both tables based on the join keys and sorts them in ascending order. Then, it performs a merge operation to find matching rows from both tables. A sort-merge join can handle large and sorted tables efficiently, but it involves shuffling and sorting data across the network and may require additional disk space for spill files.

Using join hints can help us optimize performance for some specific scenarios, such as the following:

- When we know that one of the tables is much smaller than the other and can fit in memory, we can use the broadcast hint to avoid shuffling data across the network

- When we know that both tables are large and have a high cardinality of join keys, we can use the shuffle hash hint to avoid sorting data and reduce disk I/O

- When we know that both tables are large and have a low cardinality of join keys, we can use the sort-merge hint to avoid building hash tables and reduce memory consumption

- When we know that one of the tables is much larger than the other and cannot be broadcasted, we can use the shuffle replicate nested loop hint to avoid skewness and out-of-memory errors

To demonstrate how join hints can affect performance, let's try to join our `large_df` and `small_df` using an inner join with different join hints:

```
# Join large_df and small_df using an inner join with broadcast
hash join hint
inner_join_broadcast_hint = \
    large_df.hint("broadcast").join(small_df, "id")
measure_time(inner_join_broadcast_hint)


# Join large_df and small_df using an inner join with shuffle
hash join hint
inner_join_shuffle_hash_hint = \
    large_df.hint("shuffle_hash").join(small_df, "id")
measure_time(inner_join_shuffle_hash_hint)


# Join large_df and small_df using an inner join with shuffle
replicate nested loop join hint
inner_join_shuffle_replicate_nl_hint = \
    large_df.hint("shuffle_replicate_nl").join(small_df, "id")
measure_time(inner_join_shuffle_replicate_nl_hint)
```

```
# Join large_df and small_df using an inner join with sort-merge
join hint
inner_join_merge_hint = large_df.hint("merge").join(small_df,
"id")
measure_time(inner_join_merge_hint)
```

The output should look like this:

```
Execution time: 1.8980967998504639 seconds
Execution time: 2.253967046737671 seconds
Execution time: 761.0224421024323 seconds
Execution time: 2.357747793197632 seconds
```

As we can see, the broadcast `hash` join is the fastest, as it avoids shuffling data across the network. The shuffle `hash` join is slightly slower, as it shuffles data based on the join keys. The shuffle replicate nested loop join is the slowest, as it shuffles and replicates the entire right table to each partition of the left table. The `sort-merge` join is somewhere in between, as it shuffles and sorts data based on the join keys.

Using join hints may also have some drawbacks, such as the following:

- Join hints may not always guarantee that Spark will use the specified join strategy, as Spark may still override them based on some internal logic or constraints

- Join hints may not always result in the best performance, as Spark may have a better knowledge of the data and the environment than us and may choose a more optimal join strategy without hints

- Join hints may become obsolete or incompatible with future versions of Spark, as Spark may introduce new features or changes that affect join behavior

Therefore, we should use join hints with caution and test them thoroughly before applying them to production.

7. **Enabling adaptive query execution**: Spark 3.x introduced a new feature called **adaptive query execution** (**AQE**), which allows Spark to dynamically optimize the execution plan of a query based on the runtime statistics of the data. AQE can automatically adjust the following aspects of a query:

- **Coalesce shuffle partitions**: AQE can reduce the number of shuffle partitions based on the size of the data, which can improve performance and reduce resource consumption

- **Switch join strategies**: AQE can switch between different join strategies based on the size and skewness of the data, which can improve performance and avoid out-of-memory errors

- **Optimize skew joins**: AQE can handle skewed data by splitting skewed partitions into smaller sub-partitions and performing a local shuffle before joining them with other partitions, which can improve performance and avoid data imbalance

To enable AQE, we must set the `spark.sql.adaptive.enabled` configuration parameter to `true`. We can also tune some other parameters related to AQE, such as `spark.sql.adaptive.coalescePartitions.enabled`, `spark.sql.adaptive.skewJoin.enabled`, and `spark.sql.adaptive.skewJoin.skewedPartitionFactor`.

Enabling AQE can help us optimize performance for some dynamic scenarios, such as the following:

* When we have queries that involve multiple joins with different table sizes, AQE can coalesce shuffle partitions and switch join strategies to achieve better performance and resource utilization

* When queries involve skewed data that may cause data imbalance and out-of-memory errors, AQE can optimize skew joins to achieve better performance and resource utilization

To demonstrate how AQE can improve performance, let's try to join our `large_df` and `skewed_df` using an inner join with and without AQE:

```
# Join large_df and skewed_df using an inner join without AQE
spark.conf.set("spark.sql.adaptive.enabled", "false")
inner_join_no_aqe = large_df.join(skewed_df, "id")
measure_time(inner_join_no_aqe)

# Join large_df and skewed_df using an inner join with AQE
spark.conf.set("spark.sql.adaptive.enabled", "true")
inner_join_aqe = large_df.join(skewed_df, "id")
measure_time(inner_join_aqe)
```

The output should look like this:

```
Execution time: 8.188302278518677 seconds
Execution time: 2.7499380111694336 seconds
```

As we can see, AQE reduces the execution time by almost one-third.

8. **Stop the Spark session**: Finally, we need to stop the Spark session to release the resources used by Spark:

```
spark.stop()
```

See also

* Spark SQL programming guide – a comprehensive guide on how to use Spark SQL for structured and semi-structured data processing: `https://document360.com/blog/technical-documentation/`

* Spark SQL tuning guide – a detailed guide on how to tune the performance of Spark SQL applications and queries: `https://plan.io/blog/technical-documentation/`

- Spark SQL configuration – a reference of all the configuration parameters related to Spark SQL and AQE: `https://clevertap.com/blog/technical-documentation/`

- Spark SQL built-in functions – a reference of all the built-in functions supported by Spark SQL, including string functions, date functions, and math functions: `https://en.wikipedia.org/wiki/Technical_documentation`

- Spark SQL join types – a blog post that explains the different types of joins supported by Spark SQL and how they work: `https://whatfix.com/blog/types-of-technical-documentation/`

Performance Tuning in Delta Lake

Delta Lake is an open source data lake that supports ACID transactions and provides reliable data versioning and schema evolution capabilities. This chapter covers several techniques to optimize query performance in Delta Lake, including optimizing table partitioning, caching tables for fast query response, organizing data with Z-ordering, skipping data for faster query execution, reducing table size and I/O cost with compression, and boosting query performance.

We will cover the following recipes in this chapter:

- Optimizing Delta Lake table partitioning for query performance
- Organizing data with Z-ordering for efficient query execution
- Skipping data for faster query execution
- Reducing Delta Lake table size and I/O cost with compression

By the end of this chapter, you will have a solid understanding of how to tune Delta Lake tables for optimal performance and how to avoid or solve performance problems. You will also learn some useful tips and tricks to make your tables more efficient and elegant.

Technical requirements

Before starting, make sure that your `docker-compose` images are up and running, and open the JupyterLab server running on the localhost (`http://127.0.0.1:8888/lab`). Also, ensure that you have cloned the Git repo for this book and have access to the notebook and data used in this chapter.

Remember to stop all services defined in the `docker-compose` file for this book when you are done running the code examples. You can do this by executing this command:

```
$ docker-compose stop
```

You can find the notebooks and data for this chapter at `https://github.com/PacktPublishing/Data-Engineering-with-Databricks-Cookbook/tree/main/Chapter07`.

Optimizing Delta Lake table partitioning for query performance

Partitioning is a technique that improves the performance of Delta Lake queries by reducing the amount of data that needs to be scanned, filtered, or shuffled. Partitioning works by dividing a large table into smaller, more manageable parts, called partitions, based on a column or a set of columns. Each partition contains only the rows that match the partitioning criteria.

In this recipe, we will use PySpark to read the CSV file into a Spark DataFrame and write it to a Delta Lake table. We will then partition the table by specific columns and compare the query performance of the partitioned table with the non-partitioned table.

How to do it...

1. **Import the required libraries**: Start by importing the necessary libraries for working with Delta Lake. In this case, we need the `delta` module and the `SparkSession` class from the `pyspark.sql` module:

```
from delta import configure_spark_with_delta_pip
from pyspark.sql import SparkSession
from pyspark.sql.functions import when, rand
import timeit
```

2. **Create a SparkSession object**: To interact with Spark and Delta Lake, you need to create a `SparkSession` object:

```
builder = (SparkSession.builder
    .appName("optimize-table-partitions-delta")
    .master("spark://spark-master:7077")
    .config("spark.executor.memory", "512m")
    .config("spark.sql.extensions",
        "io.delta.sql.DeltaSparkSessionExtension")
    .config("spark.sql.catalog.spark_catalog",
```

```
                    "org.apache.spark.sql.delta.catalog.DeltaCatalog"))

    spark = configure_spark_with_delta_pip(builder).getOrCreate()
    spark.sparkContext.setLogLevel("ERROR")
```

3. **Generate a large DataFrame**: In this example, we will generate a randomized 1-million-record DataFrame with salary, gender, and country code:

```
# Create some sample data frames
# A large data frame with 1 million rows
large_df = (spark.range(0, 1000000)
    .withColumn("salary", 100*(rand() * 100).cast("int"))
    .withColumn("gender", when((rand() * 2).cast("int") == 0,
"M").otherwise("F"))
    .withColumn("country_code",
        .when((rand() * 4).cast("int") == 0, "US")
        .when((rand() * 4).cast("int") == 1, "CN")
        .when((rand() * 4).cast("int") == 2, "IN")
        .when((rand() * 4).cast("int") == 3, "BR")
        .otherwise('RU')))
large_df.show(5)
```

The output should look something like this:

```
+---+------+------+------------+
| id|salary|gender|country_code|
+---+------+------+------------+
|  0|  3100|     M|          US|
|  1|  4300|     M|          CN|
|  2|  3000|     M|          IN|
|  3|  4500|     F|          US|
|  4|  5900|     F|          RU|
+---+------+------+------------+
only showing top 5 rows
```

4. **Write the DataFrame to a Delta Lake**: Write a Delta table using the `df.write.format("delta").save` method. Specify the path where you want to save the table. The following code will create a Delta Lake table named `large_delta` in the specified path. By default, the table will not be partitioned:

```
(large_df.write
 .format("delta")
 .mode("overwrite")
 .save("../data/tmp/large_delta"))
```

5. **Write a partitioned Delta table**: To partition the table by the country column, you can use the partitionBy option in the df.write.format("delta").save method. The following code will create a partitioned Delta Lake table named large_delta_partitioned in the specified path. The table will be divided into partitions, one for each distinct value of the country column. Each partition will be stored as a subdirectory under the table path, with the name country_code=value:

```
(large_df.write
 .format("delta")
 .mode("overwrite")
 .partitionBy("country_code")
 .option("overwriteSchema", "true")
 .save("../data/tmp/large_delta_partitioned"))
```

6. **Compare query performance**: To compare the query performance of the partitioned and non-partitioned tables, you can use the spark.sql method to run SQL queries on the tables. The code and output with execution time for the query run on the non-partitioned table are as follows:

```
non_partitioned_query = "spark.sql(\"SELECT country_code,
gender, COUNT(*) AS employees FROM delta.`/opt/workspace/
data/tmp/large_delta` GROUP BY ALL ORDER BY employees DESC\").
show()"non_partitioned_time = timeit.timeit(non_partitioned_
query, number=1, globals=globals())

print(f"Non-partitioned query time: {non_partitioned_time}
seconds")
```

The output should look something like this:

```
+------------+------+---------+
|country_code|gender|employees|
+------------+------+---------+
|          RU|     F|   158374|
|          RU|     M|   158085|
|          US|     M|   124924|
|          US|     F|   124921|
|          CN|     F|    94448|
|          CN|     M|    93464|
|          IN|     F|    70416|
|          IN|     M|    70189|
|          BR|     F|    52726|
|          BR|     M|    52453|
+------------+------+---------+

Non-partitioned query time: 1.854158850002881 seconds
```

Here is the code and output with execution time for the query run on the partitioned table:

```
Partitioned_query = "spark.sql(\"SELECT country_code, gender,
COUNT(*) AS employees FROM delta.`/opt/workspace/data/tmp/large_
delta_partitioned` GROUP BY ALL ORDER BY employees DESC\").
show()"
partitioned_time = timeit.timeit(partitioned_query, number=1,
globals=globals())
print(f"Partitioned query time: {partitioned_time} seconds")
```

The output should look something like this:

```
+------------+------+---------+
|country_code|gender|employees|
+------------+------+---------+
|          RU|     F|   158374|
|          RU|     M|   158085|
|          US|     M|   124924|
|          US|     F|   124921|
|          CN|     F|    94448|
|          CN|     M|    93464|
|          IN|     F|    70416|
|          IN|     M|    70189|
|          BR|     F|    52726|
|          BR|     M|    52453|
+------------+------+---------+

Partitioned query time: 0.9437477390001732 seconds
```

This shows that the partitioned query is twice as fast as the non-partitioned query.

7. **Stop the Spark session**: Finally, we need to stop the Spark session to release the resources used by Spark:

```
Spark.stop()
```

There's more...

Partitioning has the following benefits:

- It reduces the I/O cost of reading data from a disk or network, as only the relevant partitions need to be accessed for a query

- It reduces the memory and CPU cost of processing data, as only the relevant partitions need to be scanned, filtered, or shuffled for a query

- It improves the parallelism and concurrency of query execution, as different partitions can be processed by different executors or threads simultaneously

Partitioning has the following drawbacks:

- It increases the metadata cost of managing data, as each partition adds an entry to the Delta Lake transaction log or metastore

- It may introduce data skew or imbalance if some partitions are much larger or smaller than others or if some partitions are accessed more frequently or less frequently than others

Therefore, partitioning should be done carefully and selectively, based on the characteristics and usage patterns of the data and queries. Some best practices for partitioning are presented here:

- Choose a partitioning column or columns that have high cardinality, low skew, and high selectivity. Avoid partitioning by a column or columns that have low cardinality, high skew, low selectivity. Low cardinality means that the column has few distinct values, which can create fewer partitions and increase the size of each partition; for example, categorical columns such as `gender`, `segment`, and `customer_type` are greater candidates. High skew means that the column has an imbalanced distribution of values, which can create too large or too small partitions; for example, timestamps and IDs, are bad candidates for partitioning. Low selectivity means that the column is rarely used as a filter or join condition, which can increase the amount of data that needs to be scanned or shuffled for a query.

- Use a partitioning scheme that matches the query patterns. For example, if most of the queries filter by a single column, then use a single-column partitioning scheme. If most of the queries filter by a combination of columns, then use a multi-column partitioning scheme. If most of the queries filter by a range of values, then use a range partitioning scheme. If most of the queries filter by a list of values, then use a list partitioning scheme.

- Monitor and optimize the partitioning scheme over time. As the data and queries change, the partitioning scheme may become suboptimal or outdated. You can use the Delta Lake utilities to analyze, repair, optimize, or compact the partitions. For example, you can use the `ANALYZE TABLE` command to collect statistics on partitions, the `VACUUM` command to remove obsolete files from partitions, the `OPTIMIZE` command to coalesce small files into larger ones within partitions, or the `ZORDER` command to reorder the data within partitions based on a column or columns.

See also

- Delta Lake optimizations: `https://docs.delta.io/latest/optimizations-oss.html`

- `Delta Lake best practices:` `https://docs.delta.io/latest/best-practices.html`

Organizing data with Z-ordering for efficient query execution

Data skipping in Delta Lake can speed up query performance by avoiding unnecessary data reads. Data skipping works best when the data is organized in a way that collocates related information in the same set of files. This can be achieved by using Z-ordering, a technique that reorders the data based on one or more columns. In this recipe, you will learn how to use Z-ordering with Delta Lake for efficient query execution.

In this recipe, we will use PySpark to read the CSV file into a Spark DataFrame and write it to a Delta Lake table. We will then Z-order the table by specific columns and compare the query performance of the optimized table with the non-optimized table.

How to do it...

1. **Import the required libraries**: Start by importing the necessary libraries for working with Delta Lake. In this case, we need the `delta` module and the `SparkSession` class from the `pyspark.sql` module:

```
from delta import configure_spark_with_delta_pip, DeltaTable
from pyspark.sql import SparkSession
from pyspark.sql.functions import when, rand
import timeit
```

2. **Create a SparkSession object**: To interact with Spark and Delta Lake, you need to create a `SparkSession` object:

```
builder = (SparkSession.builder
    .appName("z-order-delta-table")
    .master("spark://spark-master:7077")
    .config("spark.executor.memory", "512m")
    .config("spark.sql.extensions",
        "io.delta.sql.DeltaSparkSessionExtension")
    .config("spark.sql.catalog.spark_catalog",
        "org.apache.spark.sql.delta.catalog.DeltaCatalog"))

spark = configure_spark_with_delta_pip(builder).getOrCreate()
spark.sparkContext.setLogLevel("ERROR")
```

3. **Create a Delta table from the sample dataset**: You can use the `spark.read` and `spark.write` methods in Python to do this. For example, you can run the following commands in a notebook cell to create a Delta table named `online_retail` in the default database:

```
# Read the CSV file into a Spark DataFrame
df = (spark.read.format("csv")
    .option("header", "true")
    .option("inferSchema", "true")
    .load("../data/Online_Retail.csv"))
# Write the DataFrame into a Delta table

(df.write.format("delta")
.mode("overwrite")
.save("../data/delta_lake/online_retail"))
```

4. **Run a query on the non-Z-ordered table**: You can use the `spark.sql` method in Python or the SQL syntax to do this. For example, you can run the following commands in a notebook cell to query the total quantity sold for each stock code and customer ID, and measure the time taken for each query:

```
# Query the original table
query = "spark.sql(\"SELECT StockCode, CustomerID, SUM(Quantity)
AS TotalQuantity FROM delta.`/opt/workspace/data/delta_lake/
online_retail` GROUP BY StockCode, CustomerID\").show()"query_
time = timeit.timeit(query, number=1, globals=globals())
print(f"Time taken for original table: {query_time} seconds")
```

This will result in the following output:

```
+---------+----------+-------------+
|StockCode|CustomerID|TotalQuantity|
+---------+----------+-------------+
|    22637|     15311|           15|
|    22141|     17920|            4|
|    22242|     17920|            5|
|    21257|     14849|            8|
|    21670|     17841|           85|
|   85123A|     15235|           12|
|    21042|     13715|            3|
|    22752|     12471|           11|
|    22726|     13418|           72|
|   85099B|     14388|           70|
|    22094|     18041|           12|
|    21211|     16916|            1|
|    22335|     14449|           15|
|    84347|     15061|         1200|
```

```
|     20752|     15574|             1|
|     22593|     16546|          -144|
|     71477|     13295|            -1|
|    85014B|     18239|             3|
|     21916|     15093|            24|
|     79321|     12841|            33|
+---------+----------+-------------+
only showing top 20 rows

Time taken for original table: 5.887500449000072 seconds
```

5. **Identify columns for Z-ordering**: Identify columns that are commonly used in query predicates and have high cardinality (that is, a large number of distinct values). For example, in the `online_retail` table, the `StockCode` and `CustomerID` columns are good candidates for Z-ordering as they are frequently used to filter or join the data, and they have many unique values.

6. **Apply Z-ordering on the Delta table**: Run the `OPTIMIZE` command with the `ZORDER BY` clause to reorder the data based on the selected columns. You can use the SQL syntax or the `delta.tables` module in Python to do this. For example, you can run the following command in a notebook cell to optimize the `online_retail` table with Z-ordering on `StockCode` and `CustomerID`:

```
# Get the DeltaTable object for the online_retail table
deltaTable = DeltaTable.forPath(spark, "/opt/workspace/data/
delta_lake/online_retail")

# Optimize the table with Z-Ordering on StockCode and CustomerID
deltaTable.optimize().executeZOrderBy("StockCode", "CustomerID")
```

7. **Run the same queries now on the optimized table**: You can use the `spark.sql` method in Python or the SQL syntax to do this. For example, you can run the following commands in a notebook cell to query the total quantity sold for each stock code and customer ID, and measure the time taken for each query:

```
# Query the Z-Ordered table
query = "spark.sql(\\\"SELECT StockCode, CustomerID,
SUM(Quantity) AS TotalQuantity FROM delta.`/opt/workspace/data/
delta_lake/online_retail` GROUP BY StockCode, CustomerID\\\").
show()"
query_time = timeit.timeit(query, number=1, globals=globals())
print(f"Time taken for original table: {query_time} seconds")
```

This will result in the following output:

```
+---------+----------+-------------+
|StockCode|CustomerID|TotalQuantity|
+---------+----------+-------------+
|    22637|     15311|           15|
|    22141|     17920|            4|
|    22242|     17920|            5|
|    21257|     14849|            8|
|    21670|     17841|           85|
|   85123A|     15235|           12|
|    21042|     13715|            3|
|    22752|     12471|           11|
|    22726|     13418|           72|
|   85099B|     14388|           70|
|    22094|     18041|           12|
|    21211|     16916|            1|
|    22335|     14449|           15|
|    84347|     15061|         1200|
|    20752|     15574|            1|
|    22593|     16546|         -144|
|    71477|     13295|           -1|
|   85014B|     18239|            3|
|    21916|     15093|           24|
|    79321|     12841|           33|
+---------+----------+-------------+
only showing top 20 rows

Time taken for z-ordered table: 1.648877622999862 seconds
```

> **Note**
>
> The same query after Z-ordering now completes three times faster than the query on the original table. Your results may vary between the example shown here and the execution on your own machine.

8. **Stop the Spark session**: Finally, we need to stop the Spark session to release the resources used by Spark:

```
spark.stop()
```

How it works...

Z-ordering is a technique that reorders data based on one or more columns, such that rows with similar values for those columns are stored close to each other in the same set of files. This reduces the number of files that needs to be scanned when querying those columns, as Delta Lake can skip files that do not contain the relevant values. This improves query performance and reduces I/O cost.

The `OPTIMIZE` command is used to compact and reorder data in a Delta table. It merges small files into larger files and optionally applies Z-ordering on the specified columns. The `ZORDER BY` clause is used to indicate the columns to use for Z-ordering. The order of the columns in the clause matters, as the first column has the highest priority and the last column has the lowest priority. The `OPTIMIZE` command is not idempotent, meaning that running it multiple times on the same table may produce different results. However, if no new data was added to a partition that was just Z-ordered, another Z-ordering of that partition will not have any effect.

The query performance improvement depends on the query workload and the data distribution. In general, Z-ordering is more effective when the query predicates are selective and the data is skewed. For example, in the `online_retail` table, some stock codes and customer IDs are more popular than others, and queries often filter or join on those columns. By applying Z-ordering on those columns, the data is reordered such that rows with the same stock code and customer ID are grouped together in fewer files, and files that do not contain the queried values are skipped. This reduces the amount of data that needs to be read and processed and hence improves query performance.

See also

- Delta Lake documentation: `https://docs.delta.io/latest/optimizations-oss.html`

- *Data skipping with Z-order indexes* – Databricks documentation: `https://docs.databricks.com/en/delta/data-skipping.html`

- *Optimizations* – Delta Lake documentation: `https://docs.delta.io/latest/optimizations-oss.html`

Skipping data for faster query execution

In this recipe, we will show you how to perform tasks to optimize and vacuum Delta tables. Delta Lake enables you to create, manage, and query data as tables, which are stored as Parquet files in a directory. However, without proper optimization, Delta Lake tables can suffer from performance issues such as slow query execution, high I/O costs, and inefficient data layout.

To optimize Delta Lake tables for efficient read query execution, you need to perform the following tasks:

- Use the OPTIMIZE command to compact small files into larger ones and sort the data within each file by one or more columns

- Use the ZORDER clause to cluster data by one or more columns that are frequently used in filter predicates

- Use the VACUUM command to remove stale files that are no longer referenced by the table

How to do it...

1. **Import the required libraries**: Start by importing the necessary libraries for working with Delta Lake. In this case, we need the delta module and the SparkSession class from the pyspark.sql module:

```
from delta import configure_spark_with_delta_pip, DeltaTable
from pyspark.sql import SparkSession
from pyspark.sql.functions import when, rand
import timeit
```

2. **Create a SparkSession object**: To interact with Spark and Delta Lake, you need to create a SparkSession object:

```
builder = (SparkSession.builder
    .appName("data-skipping-delta-table")
    .master("spark://spark-master:7077")
    .config("spark.executor.memory", "512m")
    .config("spark.sql.extensions", "io.delta.sql.
DeltaSparkSessionExtension")
    .config("spark.sql.catalog.spark_catalog", "org.apache.
spark.sql.delta.catalog.DeltaCatalog"))

spark = configure_spark_with_delta_pip(builder).getOrCreate()
spark.sparkContext.setLogLevel("ERROR")
```

3. **Generate a large DataFrame**: In this example, we will generate a randomized 1-million-record DataFrame with salary, gender, and country code:

```
# Create some sample data frames
# A large data frame with 1 million rows
df = (spark.range(0, 1000000)
    .withColumn("salary", 100*(rand() * 100).cast("int"))
    .withColumn("gender", when((rand() * 2).cast("int") == 0,
"M").otherwise("F"))
    .withColumn("country_code",
        .when((rand() * 4).cast("int") == 0, "US")
```

```
            .when((rand() * 4).cast("int") == 1, "CN")
            .when((rand() * 4).cast("int") == 2, "IN")
            .when((rand() * 4).cast("int") == 3, "BR")
            .otherwise('RU')))
    df.show(5)
```

The output should look something like this:

```
+---+------+------+------------+
| id|salary|gender|country_code|
+---+------+------+------------+
|  0|  3100|     M|          US|
|  1|  4300|     M|          CN|
|  2|  3000|     M|          IN|
|  3|  4500|     F|          US|
|  4|  5900|     F|          RU|
+---+------+------+------------+
only showing top 5 rows
```

4. **Write the DataFrame to a Delta Lake**: Write to the Delta table using the `df.write.format("delta").save` method. Specify the path where you want to save the table. The following code will create a Delta Lake table named `employee_salary` in the specified path. By default, the table will not be partitioned:

```
(df.write
 .format("delta")
 .mode("overwrite")
 .save("../data/tmp/employee_salary"))
```

5. **Append the new data to the table**: Using the `write` method of the `df` DataFrame, specify the `format` option as `delta` and the `mode` option as `append`. This will create some small files in the `../data/tmp/events` directory, which can affect the read performance:

```
df = (spark.range(0, 1000)
      .withColumn("salary", 100*(rand() * 100).cast("int"))
      .withColumn("gender", when((rand() * 2).cast("int") == 0,
"M").otherwise("F"))
      .withColumn("country_code",
          .when((rand() * 4).cast("int") == 0, "US")
          .when((rand() * 4).cast("int") == 1, "CN")
          .when((rand() * 4).cast("int") == 2, "IN")
          .when((rand() * 4).cast("int") == 3, "BR")
          .otherwise('RU')))
(df.write
```

```
.format("delta")
.mode("append")
.save("../data/tmp/employee_salary"))
```

6. **Check the number and size of files in the table**: Check the history using the `DESCRIBE` `HISTORY` command and the `operationMetrics` column. You can use the `numFiles` value in each operation:

```
%%sparksql
DESCRIBE HISTORY delta.`/opt/workspace/data/tmp/employee_salary`
```

The output should look something like this:

ok	clusterId	readVersion	isolationLevel	isBlindAppend	operationMetrics	userMetadata	engineInfo
ull	null	0	Serializable	True	{'numOutputRows': '1000', 'numOutputBytes': '9597', 'numFiles': '2'}	null	Apache-Spark/3.4.1 Delta-Lake/2.4.0
ull	null	null	Serializable	False	{'numOutputRows': '1000000', 'numOutputBytes': '5402017', 'numFiles': '2'}	null	Apache-Spark/3.4.1 Delta-Lake/2.4.0

Figure 7.1 – Delta table history

7. **Compact the Delta files**: Use the `OPTIMIZE` command to compact and sort the data in the Delta table. You can use the `WHERE` clause to specify a predicate to filter the files to optimize. You can use the `ZORDER BY` clause to cluster the data by one or more columns that are frequently used in filter predicates. The following code snippet shows how to use the `OPTIMIZE` command:

```
%%sparksql
OPTIMIZE delta.`/opt/workspace/data/tmp/employee_salary`
```

This will result in the following output:

path	metrics
file:/opt/workspace/data/tmp/employee_salary	Row(numFilesAdded=1, numFilesRemoved=4, filesAdded=Row(min=5406862, max=5406862, avg=5406862.0, totalFiles=1, totalSize=5406862), filesRemoved=Row(min=4431, max=2701371, avg=1352903.5, totalFiles=4, totalSize=5411614), partitionsOptimized=1, zOrderStats=None, numBatches=1, totalConsideredFiles=4, totalFilesSkipped=0, preserveInsertionOrder=False, numFilesSkippedToReduceWriteAmplification=0, numBytesSkippedToReduceWriteAmplification=0, startTimeMs=1696086587316, endTimeMs=0, totalClusterParallelism=2, totalScheduledTasks=0, autoCompactParallelismStats=None, deletionVectorStats=None, numTableColumns=4, numTableColumnsWithStats=4)

Figure 7.2 – Optimized operation metrics

8. **Delete unused files**: Use the VACUUM command to remove stale files that are no longer referenced by the Delta table. You can use the RETAIN clause to specify the number of hours to retain the history of the table. The default value is 168 hours (7 days). You can set it to a lower value to reduce the storage space used by the table. The following code snippet shows how to use the VACUUM command:

```
%%sparksql
VACUUM delta.`/opt/workspace/data/tmp/employee_salary` RETAIN 0
HOURS
```

This results in the following output:

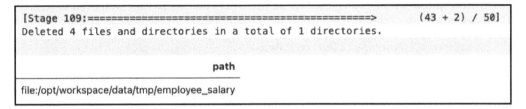

Figure 7.3 – VACUUM command details

9. **Stop the Spark session**: Finally, we need to stop the Spark session to release the resources used by Spark:

```
spark.stop()
```

See also

- Delta Lake documentation: https://docs.delta.io/latest/index.html

- Delta Lake quickstart guide: https://docs.delta.io/latest/quick-start.html

- Delta Lake best practices: https://docs.delta.io/latest/best-practices.html

Reducing Delta Lake table size and I/O cost with compression

Delta Lake tables are stored as Parquet files in a directory, along with a transaction log that tracks changes to the table. One of the benefits of using Delta Lake is that it supports various compression codecs for Parquet files, such as gzip, snappy, lzo, zstd, and brotli. Compression can help reduce the size of the table on disk and the amount of data transferred over the network, which can improve performance and save costs.

In this recipe, we will learn how to use compression with Delta Lake tables and how to measure the impact of compression on table size and I/O cost.

How to do it...

1. **Import the required libraries**: Start by importing the necessary libraries for working with Delta Lake. In this case, we need the `delta` module and the `SparkSession` class from the `pyspark.sql` module:

```
from delta import configure_spark_with_delta_pip, DeltaTable
from pyspark.sql import SparkSession
from pyspark.sql.functions import when, rand
import timeit
```

2. **Create a SparkSession object**: To interact with Spark and Delta Lake, you need to create a `SparkSession` object:

```
builder = (SparkSession.builder
    .appName("data-skipping-delta-table")
    .master("spark://spark-master:7077")
    .config("spark.executor.memory", "512m")
    .config("spark.sql.extensions", "io.delta.sql.
DeltaSparkSessionExtension")
    .config("spark.sql.catalog.spark_catalog", "org.apache.
spark.sql.delta.catalog.DeltaCatalog"))


spark = configure_spark_with_delta_pip(builder).getOrCreate()
spark.sparkContext.setLogLevel("ERROR")
```

3. **Generate a large DataFrame**: In this example, we will generate a randomized 1-million-record DataFrame with salary, gender, and country code:

```
# Create some sample data frames
# A large data frame with 1 million rows
df = (spark.range(0, 1000000)
    .withColumn("salary", 100*(rand() * 100).cast("int"))
    .withColumn("gender", when((rand() * 2).cast("int") == 0,
"M").otherwise("F"))
    .withColumn("country_code",
        .when((rand() * 4).cast("int") == 0, "US")
        .when((rand() * 4).cast("int") == 1, "CN")
        .when((rand() * 4).cast("int") == 2, "IN")
        .when((rand() * 4).cast("int") == 3, "BR")
        .otherwise('RU')))
df.show(5)
```

The output should look something like this:

```
+---+------+------+------------+
| id|salary|gender|country_code|
+---+------+------+------------+
|  0|  3100|     M|          US|
|  1|  4300|     M|          CN|
|  2|  3000|     M|          IN|
|  3|  4500|     F|          US|
|  4|  5900|     F|          RU|
+---+------+------+------------+
only showing top 5 rows
```

4. **Write a Delta table with default compression**: Write the DataFrame to a Delta Lake table with the default compression codec (`snappy`):

```
# Write the DataFrame to a Delta Lake table with the default
compression codec (snappy)
(df.write.format("delta")
    .mode("overwrite")
    .save("../data/tmp/employee_salary_snappy"))
```

5. **Get the query time on the Delta table**: We can get the execution time for reading the Delta table with `snappy` compression by using `spark.read` to read the Delta table and `noop` to write to it:

```
# Check the size of the table on disk
query = "(spark.read.format(\\"delta\\").load(\\"../data/
tmp/employee_salary_snappy\\").write.mode(\\"overwrite\\").
format(\\"noop\\").save())"
snappy_time = timeit.timeit(query, number=1, globals=globals())
print(f"Snappy Compression query time: {snappy_time} seconds")
```

This will result in the following output:

```
Snappy Compression query time: 2.3627631140006997 seconds
```

6. **Write to the Delta table with zstd compression**: To reduce the Delta Lake table size and I/O cost with compression, we need to do the following:

   - **Choose a compression codec that suits our data and workload**: Different compression codecs have different trade-offs between compression ratio, speed, and CPU usage. For example, `gzip` has a high compression ratio but is slow and CPU-intensive, while `snappy` has a low compression ratio but is fast and CPU-efficient. `zstd` and `brotli` are newer codecs that offer better compression ratios and speeds than `gzip` and `snappy`, respectively, but may not be supported by all platforms and tools.

- **Rewrite the Delta Lake table with the chosen compression codec**: We can use the `compression` option when writing the table to specify the codec. We can also use the `overwriteSchema` option to update the table schema with the new codec:

```
# Write the DataFrame to a Delta Lake table with the default
compression codec (zstd)
(df.write.format("delta")
 .mode("overwrite")
 .option("compression", "zstd")
 .save("../data/tmp/employee_salary_zstd"))
```

7. **Compare the size and I/O cost of the table before and after compression**: We can use the same methods as in the previous section to measure the size and I/O cost of the table. The following code shows how to do these steps with `zstd` compression:

```
# Check the size of the table on disk
query = "(spark.read.format(\\"delta\\").load(\\"../data/
tmp/employee_salary_zstd\\").write.mode(\\"overwrite\\").
format(\\"noop\\").save())"
zstd_time = timeit.timeit(query, number=1, globals=globals())
print(f"zstd Compression query time: {zstd_time} seconds")
```

This will result in the following output:

```
zstd Compression query time: 1.8800181449987576 seconds
```

> **Note**
>
> The `zstd`-compressed Delta table was read faster. If you look at the Parquet files associated with the Delta table, you will also notice that the storage size is much lower compared to the `snappy`-compressed Delta table.

8. **Stop the Spark session**: Finally, we need to stop the Spark session to release the resources used by Spark:

```
spark.stop()
```

How it works...

Delta Lake supports various compression codecs for Parquet files, such as `gzip`, `snappy`, `lzo`, `zstd`, and `brotli`. Each codec has its own advantage and disadvantage, and the choice of codec depends on the data and workload characteristics. Delta Lake allows users to specify the compression codec when writing the table, using the `compression` option. The codec can be specified at the table level or column level, depending on the data and workload requirements. For example, if some columns have more compressible data than others, users can choose different codecs for different columns to optimize the compression performance and cost.

Delta Lake also allows users to update the table schema with the new codec, using the `overwriteSchema` option. This option can be useful when users want to change the codec of an existing table without affecting existing queries or applications that read the table. This option can also be useful when users want to migrate a table from one platform or tool to another that supports a different codec.

See also

- Delta Lake storage configuration: `https://docs.delta.io/2.1.1/delta-storage.html`

- Delta Lake optimizations: `https://docs.delta.io/latest/optimizations-oss.html`

Part 2 – Data Engineering Capabilities within Databricks

In this part, we will comprehensively explore advanced Databricks features, guiding you through orchestrating and scheduling data pipelines, building dynamic pipelines with Delta Live Tables, enforcing data governance with Unity Catalog, and integrating DataOps and DevOps practices. It offers practical, hands-on guidance for enhancing data engineering processes, ensuring data quality and compliance, and promoting a culture of continuous improvement within the Databricks ecosystem.

This part contains the following chapters:

- *Chapter 8, Orchestration and Scheduling Data Pipeline with Databricks Workflows*
- *Chapter 9, Building Data Pipelines with Delta Live Tables*
- *Chapter 10, Data Governance with Unity Catalog*
- *Chapter 11, Implementing DataOps and DevOps on Databricks*

8

Orchestration and Scheduling Data Pipeline with Databricks Workflows

Databricks Workflows is a way to automate and orchestrate data processing tasks on the Databricks platform. A **workflow** is a sequence of tasks that can be defined using the Databricks Workflow API or the Databricks UI. Workflows can also include conditional logic, loops, and branching to handle complex scenarios.

Databricks Workflows can help you achieve various goals, such as the following:

- Running data pipelines or ETL processes on a regular basis or in response to events
- Training and deploying machine learning models in a scalable and reproducible way
- Performing batch or streaming analytics on large datasets
- Testing and validating data quality and integrity
- Generating reports and dashboards for business insights

In this chapter, you will learn how to orchestrate and schedule Databricks Workflows. We will cover the following recipes:

- Building Databricks Workflows
- Running and managing Databricks Workflows
- Passing task and job parameters within a Databricks Workflow

- Conditional branching in Databricks Workflows

- Triggering jobs based on file arrival

- Setting up workflow alerts and notifications

- Troubleshooting and repairing failures in Databricks Workflows

By the end of this chapter, you will have a solid understanding of how to use Databricks Workflows to operationalize tasks in the Databricks Lakehouse Platform.

Technical requirements

To follow along with the examples in this chapter, you will need a Databricks workspace with the premium plan.

You can find the notebooks and data for this chapter at `https://github.com/PacktPublishing/Data-Engineering-with-Databricks-Cookbook/tree/main/Chapter08`.

Building Databricks workflows

Databricks workflows are a way to automate and orchestrate your data processing tasks on the Databricks platform. You can use workflows to create pipelines, ETL processes, machine learning models, and more.

A task is a unit of work that you can run on a schedule or on-demand. A Databricks workflow is a sequence of tasks that are linked by dependencies. A workflow can be defined using the Databricks UI, the Databricks CLI, or the Databricks REST API. A workflow can also have parameters that can be passed to it at runtime. A workflow can have one or more runs, which are the instances of the workflow execution.

In this recipe, you will learn how to use the Databricks Workflows UI to create a multi-task workflow in Databricks.

How to do it...

1. **Create a new workflow**: Choose one of the two ways to access the job creation page:

   - Click on **Workflows** in the sidebar, then click on the **Create job** button as shown:

Figure 8.1 – The Create job option in Databricks Workflows

- Click on **New** in the sidebar, then select **Job** from the menu:

Figure 8.2 – The new job panel

2. **Enter job name and task details**: On the job creation page, you will see the **Tasks** tab with a dialog box to create a **task**. A task is a unit of work that runs on a cluster and performs a specific action, such as running a notebook or a JAR file. In the dialog box, enter the following information for your task:

    - **Job name**: Enter the job name at the top of the page. Give your job a descriptive name that will help you identify it later. In our case, it will be `On-Shelf-Availability Workflow`.

    - **Task name**: Give your task a name that reflects what it does. In our case, it will be `Setup`.

    - **Type**: Choose the type of task that you want to run, such as **Notebook**, **JAR**, **Python**, or **Spark Submit**. Depending on the type of task, you may need to provide additional information, such as the path to the notebook or the JAR file, the main class name, the arguments, and the Spark configuration. We will choose **Notebook**.

    - **Source**: Choose the source of the notebook. **Workspace** for a workspace folder or Databricks Repo; **Git provider** for a remote repository. We will choose **Workspace**.

    - **Path**: This is the path to a notebook. For a local notebook, it will be the full path. For a Git repository, the path is relative to the repository root. We will choose the repo's location: `https://github.com/PacktPublishing/Data-Engineering-with-Databricks-Cookbook/blob/main/Chapter08/Setup.sql`.

    - **Cluster**: Specify the cluster where you want to run your task. You can either create a new job cluster or use an existing all-purpose cluster. A job cluster is a cluster that is created and terminated automatically for each job run, while an all-purpose cluster is a cluster that you can use for multiple purposes and manage manually. To create a new job cluster, click on **Add new job cluster** in the **Cluster** drop-down menu and configure the cluster settings, such as the cluster mode, the node type, the number of workers, the autoscaling options, and the termination conditions. Using a job cluster to run the workflow is recommended. Here is a screenshot of the job cluster settings we will be using:

Figure 8.3 – Job cluster configuration UI

- **Dependent libraries**: If your task depends on any external libraries, such as Python packages, Maven packages, or JAR files, you can add them to your task by clicking on + **Add** next to **Dependent libraries**. You can either upload the libraries from your local machine, or specify the library source, such as PyPI, Maven, or DBFS.

- **Parameters**: If your task accepts any parameters, such as notebook parameters or `spark-submit` parameters, you can pass them to your task by entering them in the **Parameters** field. In our case, we have two parameters: `catalog`, which in our case will have the value `main`, and `schema`, which in our case will have the value `on_shelf_availability`.

- **Notifications**: If you want to get email alerts when your task begins, completes, or fails, click on + **Add** next to **Emails**. You will receive failure alerts for the first and any following attempts to re-run the task. To avoid getting too many emails, you can check the boxes for **Mute notifications for skipped runs**, **Mute notifications for canceled runs**, or **Mute notifications until the last retry**. We will skip setting this up for now.

- **Retries**: If you want to set a policy that controls the number and frequency of retries for failed task runs, click on + **Add** next to **Retries**. The system measures the retry interval in milliseconds from the start of the failed run to the next retry run. We will skip setting this up for now.

- **Duration threshold**: With duration thresholds, Databricks enables users to specify how long a workflow can run, enhancing the ability to track job performance. The system sends an alert when a job runs longer than the set time limit, allowing for faster resolution of possible issues. We will skip this for now.

The following screenshot shows all these details filled in:

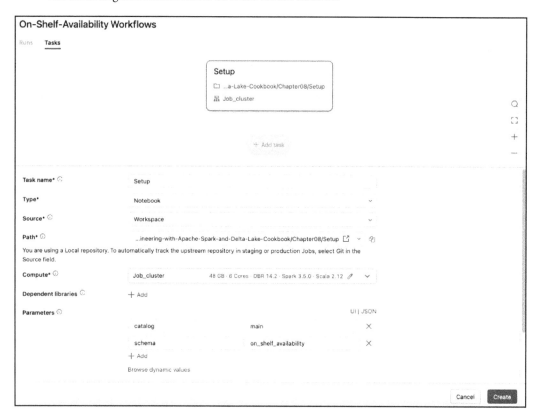

Figure 8.4 – Job task details view

3.   **Create task**: Click on **Create task** to create your task and add it to the job.

4. **Add more tasks**: If you want to add more tasks to your job, you can click on the **Add Task** button in the **DAG** view. A **directed acyclic graph** (**DAG**) is a graphical representation of the dependencies and execution order of your tasks. You can also use the **DAG** view to edit, delete, or reorder your tasks. When you add a new task, you can either use the same cluster as the previous task or configure a different cluster for each task. To use the same cluster, select **Shared Cluster** from the cluster drop-down menu. To configure a different task, follow the same steps as in *step 2*. The details for the additional task we will create are as follows:

- **Task name**: `Download_Inventory_Data`.

- **Type**: **Notebook**.

- **Source**: **Workspace**.

- **Path**: This is the path to the notebook. We will choose the repo's location: `https://github.com/PacktPublishing/Data-Engineering-with-Databricks-Cookbook/blob/main/Chapter08/Download%20Inventory%20Data.sql`.

- **Cluster**: We will use the same job cluster as the previous task.

- **Parameters**: We have two parameters: `catalog`, which in our case will have the value `main`, and `schema`, which in our case will have the value `on_shelf_availability`.

- **Depends on**: To specify the execution order of tasks in a job, use the **Depends on** drop-down menu. You can select one or more tasks in the job that must run before the current task. This creates a DAG of task execution, which is a standard way of showing the execution order in job schedulers. We will set this task to run depending on the `Setup` task.

- **Run if dependencies**: A condition that determines whether the task should be run once its dependencies have been completed. We will set this to **All succeeded**.

- **Notifications**: We will skip setting this up for now.

- **Retries**: We will skip setting this up for now.

- **Duration threshold**: We will skip this for now.

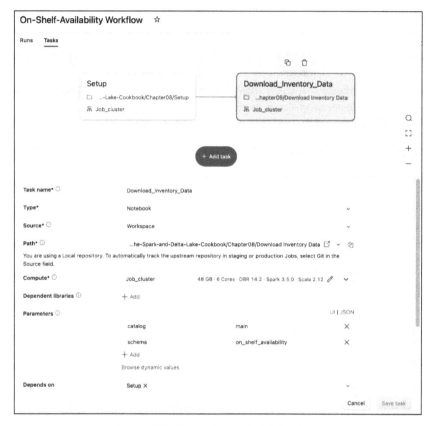

Figure 8.5 – Dependent task details view

5. **Create task**: Click on **Create task** to create your task and add it to the job.

6. **Set job schedules and triggers**: If you want to schedule your job to run automatically at a certain frequency or time, you can create a trigger by clicking on **Add trigger**. You can either create a periodic trigger, which runs your job at a fixed interval, triggers the job on file arrival, which will watch a location, and executes the workflows when a new file is received, or execute the job continuously.

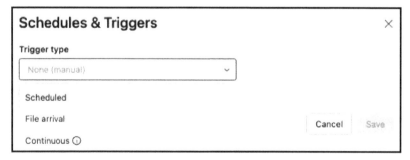

Figure 8.6 – The Schedules & Triggers popup

We will choose to run our workflow daily at 00:00:00 UTC:

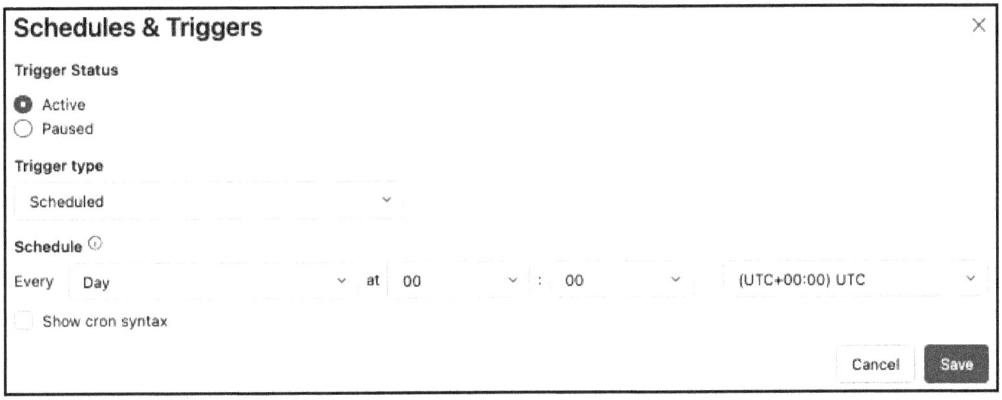

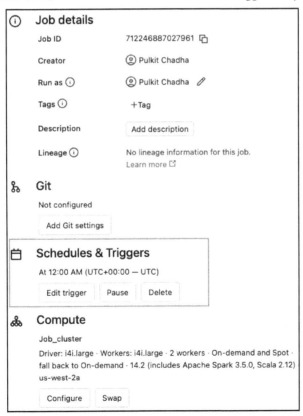

Figure 8.7 – Daily schedule cron setting

Once you click on **Save**, the workflow will be scheduled to trigger every day at 00:00:00 UTC.

Figure 8.8 – Workflow scheduled

See also

- Introduction to Databricks Workflows: `https://docs.databricks.com/en/workflows/index.html`

- Introduction to Azure Databricks Workflows: `https://learn.microsoft.com/en-us/azure/databricks/workflows/`

- Implement data processing and analysis workflows with Jobs: `https://docs.databricks.com/en/workflows/jobs/how-to/index.html`

- Create and run Databricks Jobs: `https://docs.databricks.com/en/workflows/jobs/create-run-jobs.html`

Running and managing Databricks Workflows

With Databricks Workflows, you can see the current and past runs of any jobs you can access, even if they were started by other tools. Databricks keeps your job runs for up to 60 days. If you want to save them longer, Databricks suggests exporting them before expiration.

In this recipe, you will learn how to use the Databricks UI to see the jobs you can access, the past runs of a job, and the run details.

How to do it...

1. **Go to the workflow:** Click on **Workflows** in the sidebar, then click on a job name in the **Name** column:

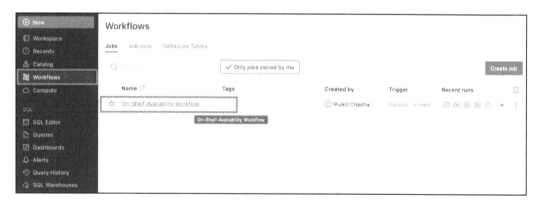

Figure 8.9 – Databricks Workflows list view

You will see the **Runs** tab with two views: matrix and list, as shown:

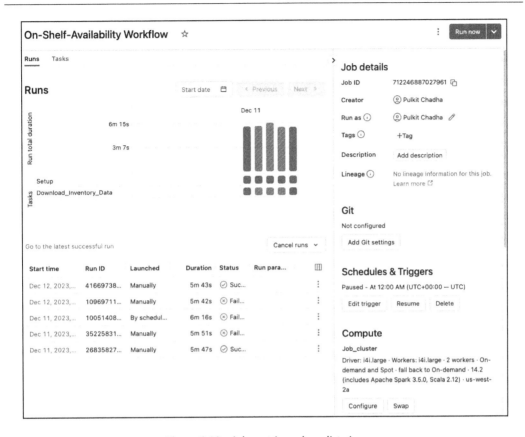

Figure 8.10 – Job matrix and run list view

2. **Matrix view**: The matrix view shows you the previous runs of the job and its tasks.

    ▪ The **Run total duration** row shows you how long each run took and what state it was in:

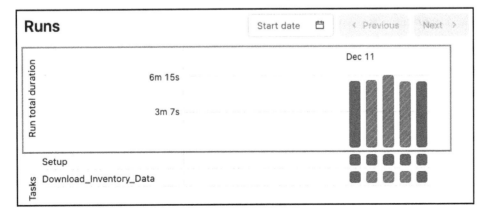

Figure 8.11 – Matrix view run duration

- To see more details about a run, such as when it started, how long it lasted, and what status it had, hover over its bar in the **Run total duration** row, which will show information in a popup as shown:

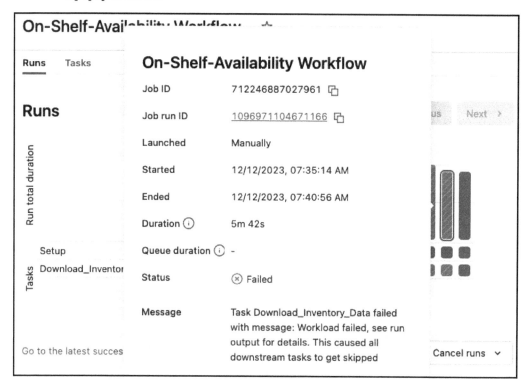

Figure 8.12 – Matrix view job details

- The **Tasks** row has cells for each task and its status. To see more details about a task, such as when it started, how long it lasted, what cluster it used, and what status it had, hover over its cell in the **Tasks** row, which will show the details like this:

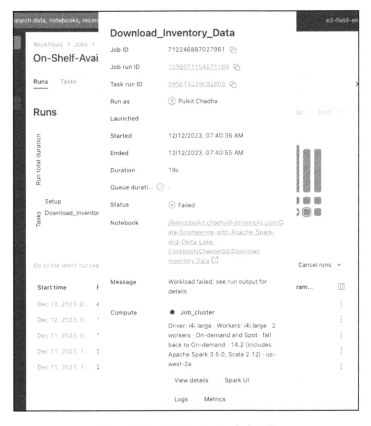

Figure 8.13 – Matrix view task details

- The bars for the job runs and task runs have different colors to show their status: green means successful, red means unsuccessful, and pink means skipped. The height of the bars shows how long the runs took.

- If you set an expected completion time, the matrix view will warn you if a run takes longer than that.

- **Run list view**: The runs list view shows you by default the following fields:

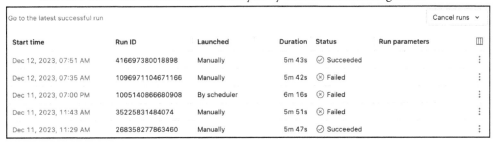

Figure 8.14 – Job run list view

- **Start time**: When the run started.

- **Run ID**: The unique run ID.

- **Launched**: How the run was triggered: by a job schedule, an API request, or manually.

- **Duration**: How much time has passed for a running job or how much time it took for a completed job. You will see a warning if the time is longer than the expected completion time.

- **Status**: The run status can be **Queued**, **Pending**, **Running**, **Skipped**, **Succeeded**, **Failed**, **Terminating**, **Terminated**, **Internal Error**, **Timed Out**, **Canceled**, **Canceling**, or **Waiting for Retry**.

3. **View job run details**: To see more details about a run, click on its link in the **Start time** column in the runs list view. The job run details page shows you the job output links to logs and tells you whether each task in the job run succeeded or failed.

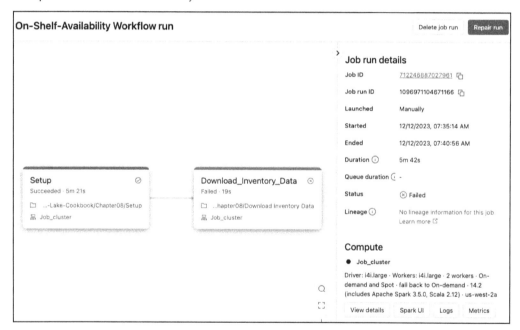

Figure 8.15 – Job run task details view

If the job has more than one task, you can click a task to see more details about it, such as the following details:

- The cluster that ran the task

- The Spark UI for the task

- Logs for the task

- Metrics for the task

The following screenshot shows the view you get for the task details, showing the error and an executed notebook:

Figure 8.16 – Job run task details notebook error view

See also

- Introduction to Databricks Workflows: `https://docs.databricks.com/en/workflows/index.html`

- Implement data processing and analysis workflows with Jobs: `https://docs.databricks.com/en/workflows/jobs/how-to/index.html`

- View and manage job runs: `https://docs.databricks.com/en/workflows/jobs/create-run-jobs.html`

Passing task and job parameters within a Databricks Workflow

Parameters are a convenient way to run the same code with different values in Databricks. When you create a task in a Databricks Workflow, you can set parameters for that task only. Job-level parameters are linked to the job, so they have the same runtime value for all tasks in the job that use them. If you change the job parameters at runtime, all tasks will use the new value. However, the default values of the parameters are specific to each task, so they can be different even if the parameter name is the same. This does not matter if you use task parameters, as they are not shared with other tasks.

In this recipe, we will learn how to pass task and job parameters within Workflows and share information from one job task to another.

How to do it...

We will be looking at two ways to share parameters in Databricks Workflows:

- Passing task parameters to tasks within a workflow
- Share information from one task to another in a workflow

To pass task parameters within Workflows

1. **Define your parameters in your notebook, script, or JAR file**: You can use the `dbutils.widgets` library to create input widgets that accept different types of parameters, such as text, dropdown, combo box, or multi-select. For example, you can create a text widget named `catalog` and `schema` with a default value of `main` and `on_shelf_availability` using the following code:

    ```
    CREATE WIDGET TEXT catalog DEFAULT "main";
    CREATE WIDGET TEXT schema DEFAULT "on_shelf_availability";
    ```

    In Python, this would be as follows:

    ```
    dbutils.widgets.text("catalog", "main")
    dbutils.widgets.text("schema", "on_shelf_availability")
    ```

2. **Retrieve parameters in the notebook**: Use the `dbutils.widgets.get` function to access the value of the parameter in your code, or just use `${<parameter_name>}` in SQL:

    ```
    USE CATALOG ${catalog};
    CREATE SCHEMA IF NOT EXISTS ${schema};
    ```

    In Python, use the following code:

    ```
    catalog = dbutils.widgets.get("catalog")
    schema = dbutils.widgets.get("schema")
    ```

3. **Configure your job or workflow to pass the parameter values when you run it**: You can do this in different ways depending on how you create and manage your job or workflow:

    - If you use the Databricks UI to create a job, you can go to the **Tasks** tab and click on the **Edit** button of the task that you want to pass the parameter to. Then, you can enter the parameter name and value in the **Parameters** section. For example, you can enter `catalog` as the parameter name and `main` as the parameter value to override the default value of the catalog widget:

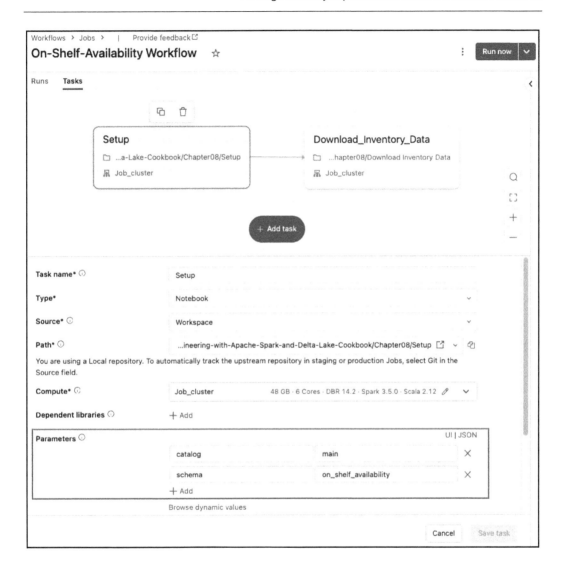

Figure 8.17 – Job task parameter UI

- If you use the REST API to create a job, you can specify the parameter name and value in the `notebook_params` field of the `notebook_task` object. For example, you can use the following JSON payload to create a job with a notebook task that passes the date parameter:

```
{
  "name": "On-Shelf-Availability Workflow",
  ...,
  "tasks": [
    {
```

```
              "task_key": "Setup",
              ...,
              "notebook_task": {
                ...,
                "base_parameters": {
                  "catalog": "main",
                  "schema": "on_shelf_availability"
                }
              }
            },
            {
              "task_key": "Download_Inventory_Data",
              ...,
              "notebook_task": {
                ...,
                "base_parameters": {
                  "catalog": "main",
                  "schema": "on_shelf_availability"
                }
              }
            }
          ],
          ...,
        }
```

4. **Set parameters for all tasks in a job**: You can set key-value parameters on a job that applies
    to any tasks in the job that can take such parameters, including Python wheels that can take
    keyword arguments. These job-level parameters are combined with any task-level parameters
    that are already set. You can see the job parameters that are passed to each task in the task
    configuration, along with the task parameters. You can also use job parameters for tasks that
    do not have key-value parameters, such as JAR or Spark Submit tasks. To do this, use the
    format {{job.parameters.[name]}} for the arguments, where [name] is the key of
    the parameter.

    To set job parameters, click on **Edit** parameters in the **Job details** side panel and enter the key
    and default value for each parameter.

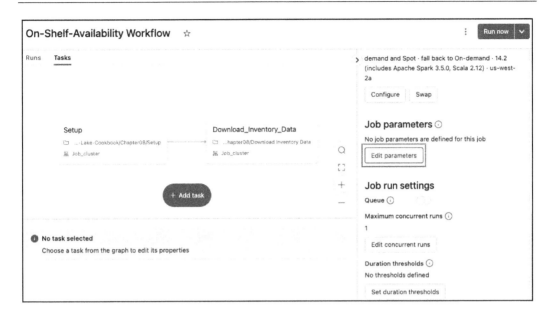

Figure 8.18 – Job parameters detail view

Note

Job parameters have higher priority than task parameters. If a job parameter and a task parameter have the same key, the value of the job parameter is used instead of the task parameter. We get a warning that job parameters with the same name as task parameters will be overwritten:

Job parameters   ✕

Add job parameters that are automatically pushed down to tasks that take key/value-based parameters. The job parameters can also be referenced explicitly in individual tasks.

Browse dynamic values

Key	Value	
catalog	main	🗑

⚠ This parameter overrides the task parameter "catalog" defined in the task(s) "Setup", "Download_Inventory_Data"

| schema | on_shelf_availability | 🗑 |

⚠ This parameter overrides the task parameter "schema" defined in the task(s) "Setup", "Download_Inventory_Data"

                                                                        🗑

                                                     Cancel   Save

Figure 8.19 – Job parameters UI

To share information from one job task to another

1. **Share values between tasks**: You can use the `taskValues` sub-utility to set and get values between different notebook tasks. For example, if you have two notebook tasks, `Setup` and `Download_Inventory_Data`, and you want to pass `volumeName` from the first task to the second task, you can do this:

```
dbutils.jobs.taskValues.set(key = 'volumeName', value = 'data')
```

- `key` is a unique name for the value you want to set
- `value` is the value you want to set for the key

2. **Retrieve task value**: To get the values in the `Download_Inventory_Data` task, you can use this example:

```
dbutils.jobs.taskValues.get(taskKey = "Setup", key =
"volumeName", default = "data", debugValue = "data")
```

- `taskKey` is the name of the task that sets the value. It raises `ValueError` if it is not found.
- `key` is the name of the value you want to get. It raises `ValueError` if it is not found (unless you provide a default value).
- `default` is an optional value that is used if the key is missing. It cannot be `None`.
- `debugValue` is an optional value that is used if you run the notebook outside of a job. This can help you debug your notebook. It cannot be `None`.

See also

- Introduction to Databricks Workflows: `https://docs.databricks.com/en/workflows/index.html`
- Implement data processing and analysis workflows with Jobs: `https://docs.databricks.com/en/workflows/jobs/how-to/index.html`
- View and manage job runs: `https://docs.databricks.com/en/workflows/jobs/create-run-jobs.html`
- Configure settings for Databricks jobs: `https://docs.databricks.com/en/workflows/jobs/settings.html`
- Pass context about job runs into job tasks: `https://docs.databricks.com/en/workflows/jobs/parameter-value-references.html`
- Share information between tasks in a Databricks job: `https://docs.databricks.com/en/workflows/jobs/share-task-context.html`

Conditional branching in Databricks Workflows

Tasks in a Databricks workflow can run based on the run status of their dependencies or the outcome of a Boolean expression. You can use these methods to run tasks conditionally in a Databricks workflow:

- **Run if dependencies**: This method allows you to run a task only when certain conditions are met by the task's dependencies. For instance, this method can help your job resume running and recover from failures by running a task even when some or all of its dependencies have failed.

- **If/else condition task**: This method lets you run a part of a job DAG based on a Boolean expression. The if/else task allows you to incorporate branching logic into your job. For instance, you can use this method to run data processing tasks only if the upstream ingestion task has added new data.

To run a part of a job DAG based on a Boolean expression, use the `if/else` task. The expression consists of a Boolean operator and two operands, which refer to a job or task state using job and task parameter variables or task values.

In this recipe, we will look at how to set up tasks with `run/if` dependencies and `if/else` conditions.

> **Note**
>
> Part of this recipe needs a Delta Live Table created in *Chapter 9, Building Data Pipelines with Delta Live Tables*.

How to do it...

We will look at two ways to set up the conditional execution of Databricks Workflows:

- **Run/if conditional task**: To execute task depending on the status of other tasks
- **If/else condition task**: To execute a branch of tasks based on a specific condition

To configure run/if conditional tasks

1. **Create a run/if task**: Click on the **Add Task** button in the **DAG** view. When you add a new task, you can either use the same cluster as the previous task or configure a different cluster for each task. To use the same cluster, select **Shared Cluster** from the cluster drop-down menu. The details for the additional task we will create are as follows:

   - **Task name**: `Cleanup`.
   - **Type**: **Notebook**.
   - **Source**: **Workspace**.

- **Path**: The path to the notebook. We will choose the repo's location: `https://github.com/PacktPublishing/Data-Engineering-with-Databricks-Cookbook/blob/main/Chapter08/Clean%20Up.sql`.

- **Cluster**: We will use the same job cluster as the previous task.

- **Parameters**: We have two parameters: `catalog`, which in our case will have the value `main`, and `schema`, which will have the value `on_shelf_availability`.

- **Depends on**: To specify the execution order of tasks in a job, use the **Depends on** drop-down menu. You can select one or more tasks in the job that must run before the current task. This creates a DAG of task execution, which is a standard way of showing the execution order in job schedulers. We will set this task to run depending on the `Setup` task and `Download_Inventory_Data`.

- **Run if dependencies**: A condition that determines whether the task should be run once its dependencies have been completed. We will set this to **At least one failed**.

- **Run if condition options**: You can use the following `run/if` conditions to run a task based on its dependencies:

  - **All succeeded**: Run the task when all its dependencies have run and succeeded

  - **At least one succeeded**: Run the task when at least one dependency has run and succeeded

  - **None failed**: Run the task when no dependent task has failed and at least one dependency has run

  - **All done**: Run the task when all its dependencies have run, irrespective of their run status

  - **At least one failed**: Run the task when at least one dependency has failed

- **Notifications**: We will skip setting this up for now.

- **Retries**: We will skip setting this up for now.

- **Duration threshold**: We will skip this for now.

Figure 8.20 – Run if, Depends on, and dependencies setting

2. **Create the task:** Click on **Create task** to create your task and add it to the job.

To create an if/else condition task

1. **Create an if/else task:** click the **Add Task** button in the **DAG** view. Choose the **If/else condition** from the **Type** drop-down menu:

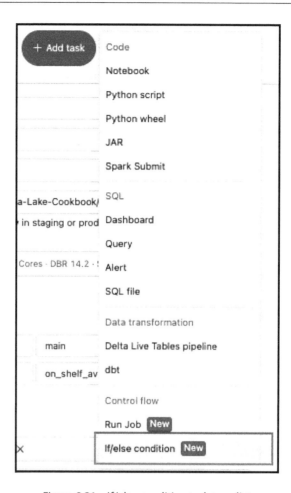

Figure 8.21 – If/else condition task type list

3. **Enter task details**: In the screen that follows, input your task details as shown:

    *   **Task name**: A unique identifier to refer to your task. Whitespace is not allowed and is automatically replaced by underscores. In our case, it is Check_File_Size.

    *   **Type**: Select the operand to be tested in the first **Condition** box. The operand can refer to a job or task parameter variable or a task value.

    *   **Condition**: You have to do the following configurations:

        *   Choose a Boolean operator from the drop-down menu. We will choose the greater than operator >.

        *   Type the value for testing the condition in the second **Condition** box. We want to check if the size of the file is more than 0.

- **Depends on**: Set up dependencies on an `if/else` condition task. In our case, the check will depend on the **Download_Inventory_Data** task.

Figure 8.22 – Check file size task details

4. **Create the task**: Click on **Create task** to create your task and add it to the job.

5. **Create post-check tasks**: Click on the **Add Task** button in the **DAG** view. The details for the additional task we will create are as follows:

- **Task name**: `OSA_Data_Preparation_DLT`.

- **Type**: **Delta Live Tables pipeline**.

- **Pipeline**: **OSA Data Preparation**.

- **Depends on**: To specify the execution order of tasks in a job, use the **Depends on** drop-down menu. We want this task to execute when the check of file size is true (that is, the file size is more than 0). We will set this task to run depending on `Check_File_Size (true)`.

- **Run if dependencies**: A condition that determines whether the task should be run once its dependencies have been completed. We will set this to **All succeeded**.

- **Notifications**: We will skip setting this up for now.

- **Retries**: We will skip setting this up for now.

- **Duration threshold**: We will skip this for now.

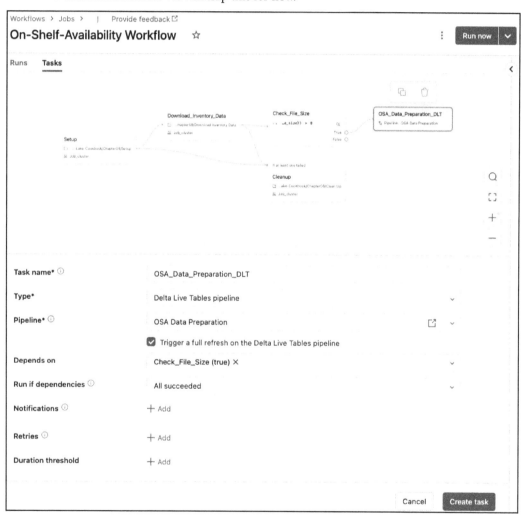

Figure 8.23 – Check_File_Size (true) branch

6. **Create the task**: Click on **Create task** to create your task and add it to the job.

7. **Create post-check tasks**: Click on **Cleanup** task and update the **Depends on** to add Check_ File_Size (false):

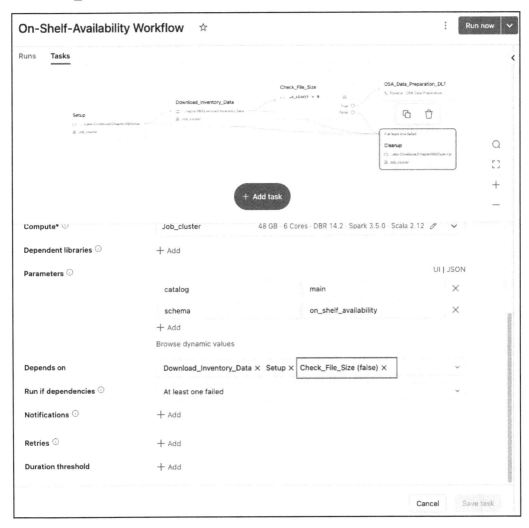

Figure 8.24 – The Check_File_Size (false) branch task

See also

- Introduction to Databricks Workflows: `https://docs.databricks.com/en/workflows/index.html`

- Implement data processing and analysis workflows with jobs: `https://docs.databricks.com/en/workflows/jobs/how-to/index.html`

- Run tasks conditionally in a Databricks job: `https://docs.databricks.com/en/workflows/jobs/conditional-tasks.html`

Triggering jobs based on file arrival

File arrival triggers can be used to start your Databricks Workflow when new files show up in an external location such as Amazon S3 or Azure storage. This feature is useful when new data does not come regularly and a scheduled job would not work well. Every minute, file arrival triggers look for new files. They do not have any extra cost except for the cloud provider costs for listing files in the storage location.

In this recipe, you will learn how to set up file arrival triggers to run Databricks Workflows.

Getting ready

Before you can use file arrival triggers, you need to meet the following conditions:

- Your workspace should have Unity Catalog activated. Unity Catalog is a unified governance solution for data and AI assets on the lakehouse.

- An external location that belongs to the Unity Catalog metastore is recommended. This is an object that has both a cloud storage path and a storage credential that allows access to that path. To create an external location, follow the instructions in **Create an external location**.

- You should have the ability to read the external location and manage the job.

How to do it...

1. **Go to the workflow**: Click on **Workflows** in the sidebar, then click on a job name in the **Name** column that you want to add a trigger to on the **Jobs** tab.

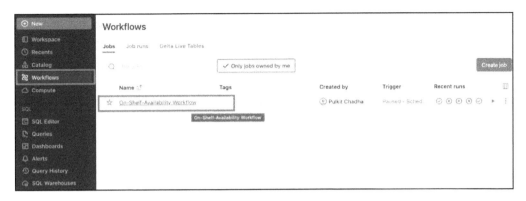

Figure 8.25 – Databricks Workflows jobs list view

2. **Add a trigger to the job**: On the right side of the screen, you will see the **Job details** panel. Click on the **Add trigger** button.

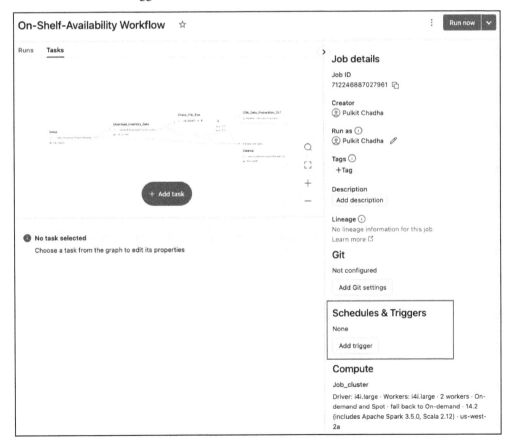

Figure 8.26 – Jobs' triggers panel

3.   **Choose file arrival trigger**: In the **Trigger type** drop-down menu, choose **File arrival**.

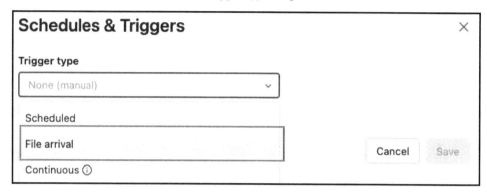

Figure 8.27– File arrival trigger drop-down view

4.   **Enter the file trigger details**:

- **Storage location**: In the **Storage location** field, type the URL of the external location where you want to monitor for new files. You can also specify a subdirectory within that location. In our case, this will be `s3://databricks-file-arrival-trigger/`.

- **Advanced**: You can adjust some advanced options for your trigger:

  - **Minimum time between triggers in seconds**: This is the shortest amount of time that the trigger will wait before starting a new run after the previous run finishes. If new files arrive during this time, they will not trigger a run until the time is up. You can use this option to limit how often your job runs. We will set this to 10 mins, that is, `600`.

  - **Wait after the last change in seconds**: This is the amount of time the trigger will wait after a new file arrives before starting a run. If another new file arrives within this time, the timer will reset. You can use this option to make sure that your job runs only after all the files in a batch have arrived. We will set this to 10 mins, that is, `600`.

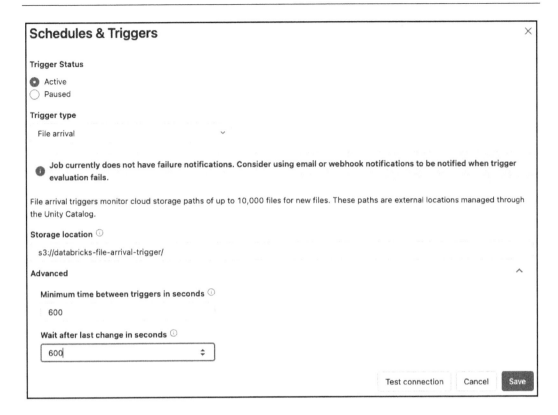

Figure 8.28 – File arrival trigger details

5. **Check the configuration**: To check if your trigger configuration is correct, click on the **Test connection** button. You will see a message indicating whether the connection was successful or not.

    To save your trigger, click on the **Save** button.

See also

- Introduction to Databricks Workflows: `https://docs.databricks.com/en/workflows/index.html`

- Implement data processing and analysis workflows with jobs: `https://docs.databricks.com/en/workflows/jobs/how-to/index.html`

- Trigger jobs when new files arrive: `https://docs.databricks.com/en/workflows/jobs/file-arrival-triggers.html`

Setting up workflow alerts and notifications

Alerts and notifications are essential features for monitoring and troubleshooting your workflows. They enable you to get notified when a workflow run starts, completes successfully, or fails. You can also set up custom conditions to trigger alerts based on the results of your workflow queries. You can send alerts and notifications to multiple email addresses or system destinations, for example, webhook destinations or Slack.

In this recipe, you will learn how to set up alerts and notifications for your workflows using the Databricks UI.

How to do it...

1. **Go to the workflow**: Click on **Workflows** in the sidebar, then click on a job name in the **Name** column.

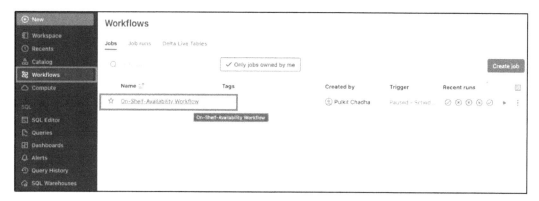

Figure 8.29 – Databricks Workflows job list

2. **Add a trigger to the job**: On the right side of the screen, you will see the **Job details** panel. Click on the **Edit notifications** button.

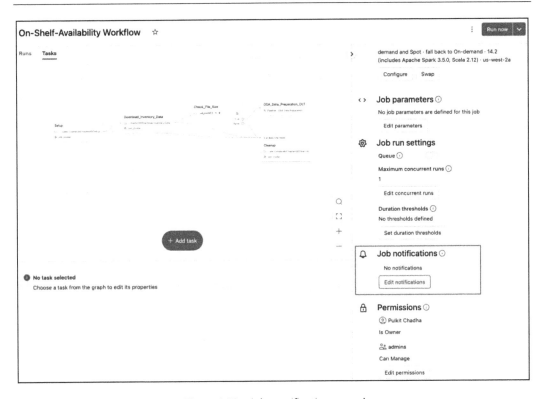

Figure 8.30 – Jobs notifications panel

3. **Add notifications**: Click on the + **Add notification** button in the pop-up window.

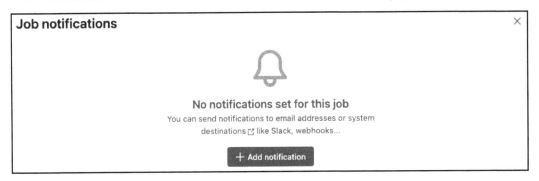

Figure 8.31 – Add notification popup

4. **Add job notification details**: In the next screen, input the following details:

    - **Choose the type of notification that you want to create**: The options are **Email** or **System destination**. Depending on your choice, you will need to enter the email address or the system destination URL that you want to receive notifications from. We will select the email address and enter the email address to be notified.

    - **Choose the events that you want to trigger notifications for**: The options for events are **On Start**, **On Success**, or **On Failure.** You can also choose to send notifications only when the workflow run status changes from the previous run. We will choose the **Failure** event.

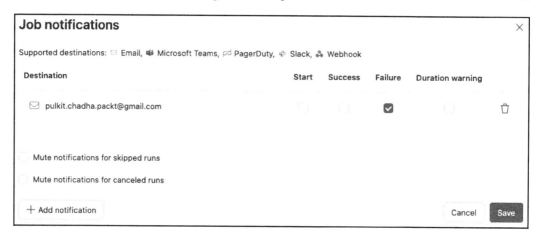

Figure 8.32 – Notification details

5. **Save job notifications**: Click on the **Save** button to save your notification settings.

You can repeat these steps to add more notifications to your workflow. You can also edit or delete existing notifications by clicking on the **Edit** or **Delete** buttons next to each notification.

There's more…

Setting up alerts and notifications for your workflows is just one of the many features that Databricks Workflows offers to help you monitor and troubleshoot your workflows. Here are some more features that you can explore:

- **Job Runs**: This is a new dashboard that gives you an aggregated overview of all your workflow runs in real time. You can see the start time, duration, status, and other relevant information of each workflow run. You can also see the trends and the top error types of your workflow runs. You can access the **Job Runs** dashboard by clicking on the **Job Runs** tab on the **Workflows** page in the Databricks UI.

- **The matrix view**: This is a new view that shows you the health and the performance of each task in your workflow across multiple runs. You can see the duration, status, and output of each task in a matrix format. You can also drill down into each task to see the details and the logs. You can access the matrix view by clicking on the **Matrix View** button on the top right corner of the workflow page in the Databricks UI.

- **The Workflow API**: This is a RESTful API that allows you to programmatically create, update, delete, run, and monitor your workflows. You can use the Workflow API to integrate your workflows with your existing tools and workflows. You can also access the Workflow API documentation by clicking on the **API Docs** button on the top-right corner of the **Workflows** page in the Databricks UI.

See also

- Introduction to Databricks Workflows: `https://docs.databricks.com/en/workflows/index.html`

- Implement data processing and analysis workflows with jobs: `https://docs.databricks.com/en/workflows/jobs/how-to/index.html`

- Better Alerts: Announcing New Workflows Notifications: `https://www.databricks.com/blog/2023/01/03/better-alerts-announcing-new-workflows-notifications.html`

- Add email and system notifications for job events: `https://docs.databricks.com/en/workflows/jobs/job-notifications.html`

- Manage notification destinations: `https://docs.databricks.com/en/administration-guide/workspace-settings/notification-destinations.html`

Troubleshooting and repairing failures in Databricks Workflows

With Databricks Workflows, you can create, manage, and track multi-task workflows for ETL, analytics, and machine learning pipelines. Your data teams can leverage various task types, rich observability features, and high dependability to improve the automation and coordination of any pipeline and boost their efficiency. However, sometimes things can go wrong and your workflows may fail due to various reasons, such as data quality issues, code errors, resource constraints, or external dependencies.

In this recipe, you will learn how to repair common failures in Databricks Workflows, using the tools and best practices provided by the Databricks platform.

How to do it...

1. **Go to the workflow**: Click on **Workflows** in the sidebar, then click on a job name in the **Name** column.

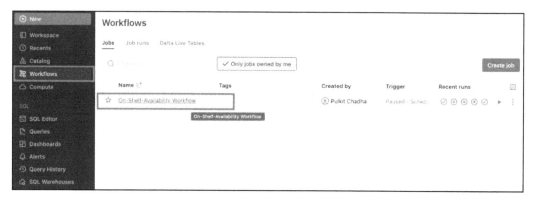

Figure 8.33 – The Databricks Workflows jobs list view

2. **Job run details**: The **Runs** tab displays the current and past runs, along with their outcomes. The matrix view in the **Runs** tab provides a history of runs for the job, showing which tasks were successful and which were not. A task may fail or be skipped if it depends on another task that failed. The matrix view helps you spot the problematic tasks for your job run.

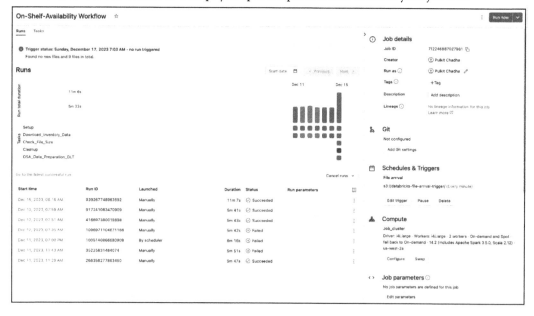

Figure 8.34 – The Databricks On-Shelf-Availability Workflow matrix and run list view

3. **Matrix view of job runs**: Move your cursor over a failed task to view its metadata. This metadata contains the start and end dates, status, duration, cluster details, and sometimes an error message.

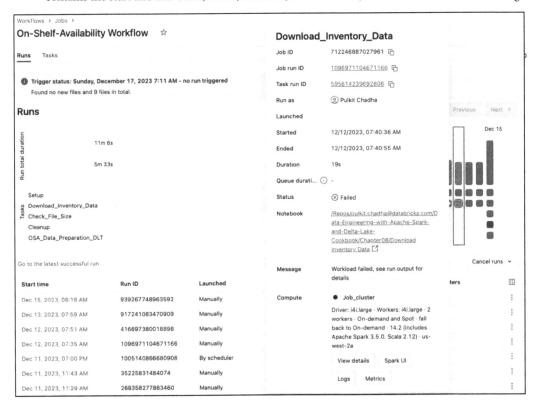

Figure 8.35 – Matrix view job run details popup

4. **Investigate the reasons for the failure**: To see more details about a run, click on its link in the **Start time** column in the runs list view. The job run details page shows you the job output links to logs and tells you if each task in the job run succeeded or failed.

Start time	Run ID	Launched	Duration	Status	Run parameters	
Dec 15, 2023, 08:18 AM	939267748963592	Manually	11m 7s	⊘ Succeeded		⋮
Dec 13, 2023, 07:59 AM	917241083470909	Manually	5m 41s	⊘ Succeeded		⋮
Dec 12, 2023, 07:51 AM	416697380018898	Manually	5m 43s	⊘ Succeeded		⋮
Dec 12, 2023, 07:35 AM	1096971104671166	Manually	5m 42s	⊗ Failed		⋮
Dec 11, 2023, 07: Dec 12, 2023, 07:35 AM 5140866680908		By scheduler	6m 16s	⊗ Failed		⋮
Dec 11, 2023, 11:43 AM	35225831484074	Manually	5m 51s	⊗ Failed		⋮
Dec 11, 2023, 11:29 AM	268358277863460	Manually	5m 47s	⊘ Succeeded		⋮

Figure 8.36 – Job run list view

5. **Resume interrupted and skipped tasks**: Once you find the source of the problem, you can fix failed or stopped multi-task jobs by executing only the portion of tasks that did not succeed and any tasks that depend on them by clicking on **Repair run**. This saves time and resources by avoiding re-running tasks that were successful and any tasks that rely on them.

Figure 8.37 – Job run details task view

6. **Click on Repair job run**: The **Repair job run** dialog will pop up, showing all the tasks that failed and any tasks that depend on them that will be re-run. You also have the option to pass parameters to the repair job run. To change or add parameters for the tasks to fix, type the parameters in the **Repair job run** dialog. The parameters you type in the **Repair job run** dialog will replace the old values.

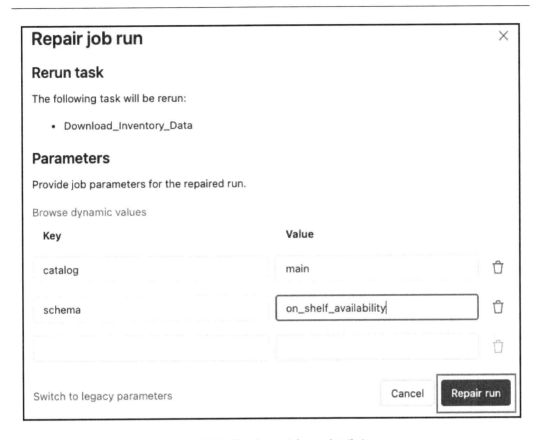

Figure 8.38 – The Repair job run detail view

Click on **Repair run** in the **Repair job run** dialog.

When the repair run is done, the matrix view will have a new column for the repaired run. Any red tasks should be green now, meaning your whole job ran successfully.

See also

- Introduction to Databricks Workflows: `https://docs.databricks.com/en/workflows/index.html`

- Implement data processing and analysis workflows with jobs: `https://docs.databricks.com/en/workflows/jobs/how-to/index.html`

- Troubleshoot and repair job failures: `https://docs.databricks.com/en/workflows/jobs/repair-job-failures.html`

9

Building Data Pipelines with Delta Live Tables

Data pipelines are essential for transforming raw data into useful insights. However, building and managing data pipelines can be challenging, especially when dealing with complex, large-scale, and streaming data sources. Data engineers often have to write and maintain multiple Spark jobs, handle cluster provisioning and scaling, monitor data quality and performance, and troubleshoot errors and failures.

With Delta Live Tables, you can easily build data processing pipelines on the Databricks Lakehouse platform that are reliable, maintainable, and testable. You only need to specify the data transformations you want to apply and Delta Live Tables will take care of the rest, such as task orchestration, cluster management, monitoring, data quality, and error handling. Delta Live Tables leverages the power of Delta Lake to provide ACID transactions, schema enforcement, time travel, and unified batch and streaming processing.

In this chapter, you will learn how to build data pipelines with Delta Live Tables. We will cover the following recipes:

- Creating a multi-hop medallion architecture data pipeline with Delta Live Tables in Databricks
- Building a data pipeline with Delta Live Tables on Databricks
- Implementing data quality and validation rules with Delta Live Tables in Databricks
- Quarantining bad data with Delta Live Tables in Databricks
- Monitoring Delta Live Tables pipelines
- Deploying Delta Live Table pipelines with Databricks Access Bundles
- Applying changes (CDC) to Delta tables with Delta Live Tables

By the end of this chapter, you will have a solid understanding of how to use Delta Live Tables pipelines to implement a medallion architecture.

Technical requirements

To follow along with the examples in this chapter, you will need to have the following:

- A Databricks workspace with the Premium plan. Delta Live Tables requires the Premium plan. Contact your Databricks account team for more information.

- Clone the repo associated with this book to your Databricks workspace. For details, take a look at the *Using Databricks repos* recipe in *Chapter 11, Implementing DataOps and DevOps on Databricks*, which tells you how to store code in Git.

You can find the notebooks and data for this chapter at `https://github.com/PacktPublishing/Data-Engineering-with-Databricks-Cookbook/tree/main/Chapter09`.

Creating a multi-hop medallion architecture data pipeline with Delta Live Tables in Databricks

The multi-hop medallion architecture is a data design pattern that organizes data in a lakehouse into multiple layers: bronze, silver, and gold. The bronze layer contains raw data from various sources, the silver layer contains validated and conformed data, and the gold layer contains curated and enriched data for analytics and AI.

In this recipe, you will learn how to use Delta Live Tables to create a multi-hop medallion architecture data pipeline in Databricks. You will use SQL to define your datasets and pipelines.

How to do it...

1. **Incremental ingestion with an autoloader**: Create a bronze dataset from the `iot-stream` device JSON data and `iot-stream user` CSV data.

    I. **Create a streaming table for device data**: We are defining a streaming table to denote that this is an incremental append-only load from the JSON file that lands in a folder to the `device_data` Delta Lake table:

    ```
    CREATE
    OR REFRESH STREAMING TABLE device_data AS
    SELECT
      *
    FROM
      cloud_files(
        "/databricks-datasets/iot-stream/data-device",
        "json",
        map("cloudFiles.inferColumnTypes", "true")
      )
    ```

II. **Create a streaming table for user data**: We are defining a streaming table to denote that this is an incremental append-only load from the CSV file that lands in a folder to the user_data Delta table:

```
CREATE
OR REFRESH STREAMING TABLE user_data AS
SELECT
  *
FROM
  cloud_files(
    "/databricks-datasets/iot-stream/data-user",
    "csv",
    map("cloudFiles.inferColumnTypes", "true")
  )
```

2. **Create the silver datasets**: We will apply some transformation validation and enrichments to the bronze datasets. You can use the CREATE OR REPLACE STREAMING LIVE TABLE command to define a streaming table that processes each record exactly once. We will read the bronze live table with STREAM() to denote that we only want to process the incremental data and not the full table.

Create the device_data_prepared Delta table with the following code:

```
CREATE
OR REFRESH STREAMING TABLE device_data_prepared (
  CONSTRAINT valid_timestamp EXPECT (timestamp IS NOT NULL)
) -- COMMENT ""
AS
SELECT
  device.id,
  device.device_id,
  device.user_id,
  device.calories_burnt,
  device.miles_walked,
  device.num_steps,
  CAST(device.timestamp as TIMESTAMP) AS timestamp
FROM
  STREAM(live.device_data) device
```

The code for the user_data_prepared Delta table is as follows:

```
CREATE
OR REFRESH STREAMING TABLE user_data_prepared (
  CONSTRAINT valid_user EXPECT (user_id IS NOT NULL) ON
VIOLATION DROP ROW
) -- COMMENT ""
```

```
AS
SELECT
  users.userid as user_id,
  CASE
    WHEN users.gender = 'F' THEN 'Female'
    WHEN users.gender = 'M' THEN 'Male'
  END AS gender,
  users.age,
  users.height,
  users.weight,
  CAST(users.smoker as BOOLEAN) AS isSmoker,
  CAST(users.familyhistory as BOOLEAN) AS hasFamilyHistory,
  users.cholestlevs AS cholestrolLevels,
  users.bp AS bloodPressure,
  users.risk
FROM
  STREAM(live.user_data) users;
```

3. **Create a gold dataset**: We will now apply aggregations to the silver dataset to create a gold dataset. You can use the CREATE OR REFRESH LIVE TABLE command to define a materialized view that processes records as required to return accurate results for the current data state:

```
CREATE
OR REFRESH LIVE TABLE user_metrics AS
SELECT
  users.user_id,
  users.gender,
  users.age,
  users.height,
  users.weight,
  users.isSmoker,
  users.hasFamilyHistory,
  users.cholestrolLevels,
  users.bloodPressure,
  users.risk,
  SUM(devices.calories_burnt) AS totalCaloriesBurnt,
  SUM(devices.miles_walked) AS totalMilesWalked,
  SUM(devices.num_steps) AS totalNumberOfSteps
FROM
  live.user_data_prepared users
  LEFT OUTER JOIN LIVE.device_data_prepared devices on devices.
user_id = users.user_id
GROUP BY
  ALL;
```

How it works...

Delta Live Tables allows users to write SQL queries that can perform data transformations in a declarative manner. This is achieved by extending the standard SQL syntax with a unified declarative way of specifying **data definition language** (**DDL**) and **data manipulation language** (**DML**) operations.

Delta Live Tables supports two kinds of persistent tables that are stored using the Delta Lake protocol, which ensures ACID transactions, versioning, and other benefits. The following are two types of abstractions we can use while working with Delta Live Tables:

- **Materialized views**, which are similar to materialized views in relational databases. They store the results of a query and can be refreshed periodically or on demand to reflect the latest state of the data.

- **Streaming tables**, which are designed for processing streaming data incrementally and in near-real time. They continuously update the results of a query based on new data arriving in the source tables.

To create or update a live table or a streaming live table, Delta Live Tables uses a modified version of the **create table as select** (**CTAS**) statement. Users only need to specify the query that defines the transformation logic and Delta Live Tables will handle the rest of the tasks, such as schema inference, dependency tracking, and execution planning.

The general syntax for a Delta Live Tables query in SQL is `CREATE OR REFRESH [STREAMING] LIVE TABLE table_name AS select_statement`.

See also

- *What is Delta Live Tables?*: `https://docs.databricks.com/en/delta-live-tables/index.html`

- *Delta Live Tables*: `https://www.databricks.com/product/delta-live-tables`

- *What is a medallion architecture?*: `https://www.databricks.com/glossary/medallion-architecture`

Building a data pipeline with Delta Live Tables on Databricks

In this recipe, you will learn how to use Delta Live Tables to build a simple data pipeline that ingests streaming data from a Kafka topic, performs some transformations and aggregations, and writes the results to a Delta table. You will also learn how to define data quality expectations, deploy and trigger your pipeline, and monitor its execution.

How to do it...

1. **Go to the Workflows UI**: Click on **Workflows** from the left navigation panel:

Figure 9.1 – The Workflows navigation panel

2. **Go to Delta Live Tables**: Click on the **Delta Live Tables** tab:

Figure 9.2 – The Delta Live Tables tab

3. **Create a Delta Live Tables pipeline**: Click on the **Create pipeline** button at the top right of the page:

Figure 9.3 – The Create pipeline button

4. **General Delta Live Tables pipeline details**: Provide some general information about the Delta Live Tables pipeline:

   • **Name**: This is the name of the Delta Live Tables pipeline. We will name it `IOT Stream Pipeline`.

- **Product edition**: Delta Live Tables comes in three flavors:

  - **Core**: This has capabilities for basic integration.

  - **Pro**: This has capabilities for **change data capture** (**CDC**) and upserts or merges.

  - **Advanced**: This has capabilities for data profiling, including data quality rules. We will choose the **Advanced** edition.

- **Pipeline mode**: This specifies how the pipeline will be run. Choose the mode based on latency and cost requirements:

  - **Triggered** pipelines run once and then shut down until the next manual or scheduled update

  - **Continuous** pipelines run continuously, ingesting new data as it arrives

We will choose **Triggered**:

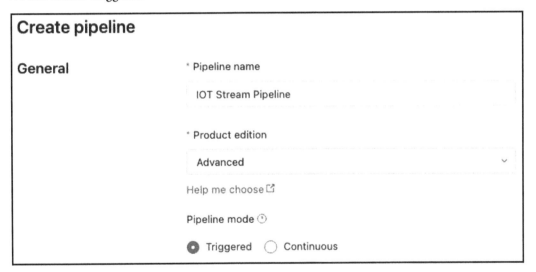

Figure 9.4 – The General section of Create pipeline

1. **Choose the source code paths**: You can have one or more source code files that build the Delta Live Tables pipeline. We will choose the source code from the *Creating a multi-hop medallion architecture data pipeline with Delta Live Tables in Databricks* recipe:

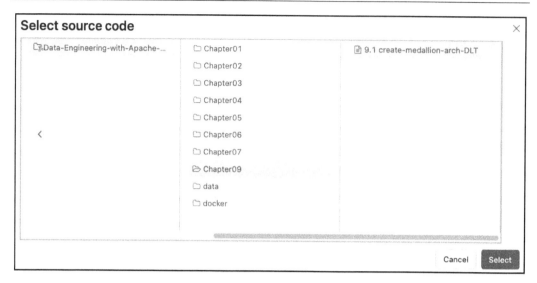

Figure 9.5 – Selecting the source code

2. **Choose destination information**: You will need to choose the appropriate storage options:

    *   **Hive Metastore**: If chosen, the tables in the pipeline will be registered in the Hive Metastore. You will need to provide the following details:

        *   **Storage location**: This optional field allows the user to specify a location to store logs, tables, and other information related to pipeline execution. If not specified, Delta Live Tables will automatically generate a directory.

        *   **Target**: Specify the target schema/database if you want to publish your table to the Metastore.

    *   **Unity Catalog**: If chosen, the tables in the pipeline are registered and managed by Unity Catalog. You will need to provide the following details:

        *   **Catalog**: A catalog where tables and other metadata will be stored

        *   **Target**: Specify the target schema/database if you want to publish your table to the metastore

    We will choose to register our tables in the Hive Metastore and let Delta Live Tables use the default storage location. For **Target schema**, we will use the `iot_stream` database, as shown here:

Destination

Storage options

⦿ Hive Metastore   ◯ Unity Catalog   Preview

Storage location ⓘ

Target schema ⓘ

iot_stream

Figure 9.6 – The Destination section of Create pipeline

3. **Choose your compute**: This section controls the worker configuration for the underlying cluster processing the pipeline. Here, we set the number of minimum workers to 1 and the number of maximum workers to 2. We also choose **Enhanced autoscaling** for **Cluster mode**:

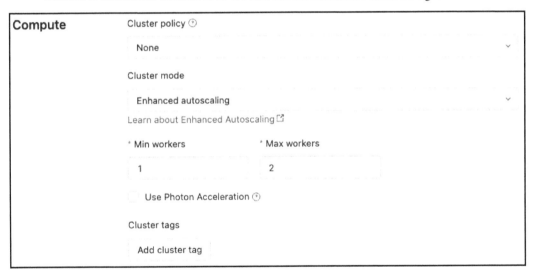

Compute

Cluster policy ⓘ

None

Cluster mode

Enhanced autoscaling

Learn about Enhanced Autoscaling ⬀

* Min workers   * Max workers

1   2

☐ Use Photon Acceleration ⓘ

Cluster tags

Add cluster tag

Figure 9.7 – The Compute section of Create pipeline

4. **Create the pipeline**: Click on the **Create** button in the bottom-right corner:

Figure 9.8 – The Create button

5. **Run the pipeline**: To execute the pipeline, follow these steps:

    I.   Choose **Development** to run the pipeline in development mode. This mode allows you to develop faster by reusing the cluster (instead of creating a new one for each run) and disabling retries so that you can quickly find and fix errors.

    II. Click on **Start**:

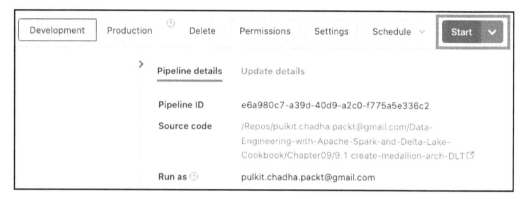

Figure 9.9 – The Pipeline details section

6. **Explore the Delta Live Tables graph**: After waiting a few minutes for the cluster to spin and the pipeline to start, you will see that the DAG generated by Delta Live Tables is based on the declarations we provided in the source code:

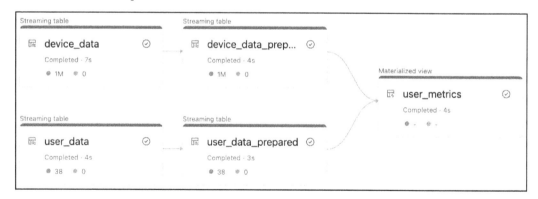

Figure 9.10 – Delta Live Tables DAG

See also

- *What is Delta Live Tables?*: https://learn.microsoft.com/en-us/azure/databricks/delta-live-tables/

- *Run your first Delta Live Tables pipeline*: `https://docs.databricks.com/en/delta-live-tables/tutorial-pipelines.html`

- *Continuous versus triggered pipeline execution*: `https://docs.databricks.com/en/delta-live-tables/updates.html#continuous-vs-triggered-pipeline-execution`

Implementing data quality and validation rules with Delta Live Tables in Databricks

One of the key features of Delta Live Tables is the ability to define and enforce data quality rules on your datasets using expectations. You can add optional clauses called expectations to your dataset declarations that check the quality of each record in a query. You can specify what action to take when a record fails an expectation, such as warning, dropping, failing, or quarantining the record.

In this recipe, you will learn how to use Delta Live Tables expectations to implement data quality and validation rules in your data pipelines. You will also learn how to monitor and troubleshoot your pipelines using the Delta Live Tables event log and UI.

How to do it...

1. **Create a live table for customer data**: We are defining a live table to create a materialized view from the `samples.tpch.customer` Delta table. We can now define some data expectation rules. The `CONSTRAINT valid_customer_key EXPECT (c_custkey IS NOT NULL) ON VIOLATION DROP ROW` rule expects that the `c_custkey` column is not null for every row in the live table. If this expectation is violated, the row will be dropped from the live table.

   ```
   CREATE
   OR REFRESH LIVE TABLE customers (
     CONSTRAINT valid_customer_key EXPECT (c_custkey IS NOT NULL)
   ON VIOLATION DROP ROW
   ) AS
   SELECT
     *
   FROM
     samples.tpch.customer
   ```

2. **Deduplicate records data quality check for the customers table**: We are defining a temporary live table from the `live.customer` Delta table; this will not create a table in the schema. The table contains two columns: `c_custkey`, which is the customer key, and `cnt`, which is the count of how many times the customer key appears in `live.customers`. We can now define the data expectation rules, `CONSTRAINT unique_customer_key EXPECT (cnt = 1) ON VIOLATION DROP ROW`, which expect that the count of each customer

key is equal to 1. This means that there are no duplicate customer keys in the source table. If this expectation is violated, the code specifies that the row with the duplicate customer key should be dropped from the live table.

```
CREATE TEMPORARY LIVE TABLE duplicate_customers_test (
  CONSTRAINT unique_customer_key EXPECT (cnt = 1) ON VIOLATION
DROP ROW
) AS
SELECT
  c_custkey, count(*) as cnt
FROM
  live.customers
GROUP BY ALL;
```

3. **Create a live table for orders data**: We are defining a live table to create a materialized view from the samples.tpch.orders Delta table. We can now define some data expectation rules with the following constraints:

   - valid_order_key: This constraint expects that the o_orderkey column is not null for every row in the live table. If this expectation is violated, the row is dropped from the live table.

   - valid_customer_key: This constraint expects that the o_custkey column is not null for every row in the live table. If this expectation is violated, the row is dropped from the live table.

   - valid_reference_customer: This constraint expects that the c_custkey column, which is joined from another live table named customers, is not null for every row in the live table. If this expectation is violated, the row is dropped from the live table.

```
CREATE
OR REFRESH LIVE TABLE orders (
  CONSTRAINT valid_order_key EXPECT (o_orderkey IS NOT NULL) ON
VIOLATION DROP ROW,
  CONSTRAINT valid_customer_key EXPECT (o_custkey IS NOT NULL)
ON VIOLATION DROP ROW,
  CONSTRAINT valid_reference_customer EXPECT (cust.c_custkey IS
NOT NULL) ON VIOLATION DROP ROW
) AS
SELECT
  ord.*,
  cust.c_custkey
FROM
  samples.tpch.orders ord
  LEFT OUTER JOIN live.customers cust on cust.c_custkey = ord.o_
custkey
```

4. **Deduplicate records data quality check for the orders table**: We are defining a temporary live table from the `live.orders` Delta table; this will not create a table in the schema. The table contains two columns: `o_orderkey`, which is the order key, and `cnt`, which is the count of how many times the order key appears in `live.orders`. We can now define the data expectation rules, `CONSTRAINT unique_order_key EXPECT (cnt = 1) ON VIOLATION DROP ROW`, which expect that the count of each order key is equal to 1. This means that there are no duplicate order keys in the source table. If this expectation is violated, the code specifies that the row with the duplicate order key should be dropped from the live table.

```
CREATE TEMPORARY LIVE TABLE duplicate_orders_test (
  CONSTRAINT unique_order_key EXPECT (cnt = 1) ON VIOLATION DROP
ROW
) AS
SELECT
  o_orderkey, count(*) as cnt
FROM
  live.orders
GROUP BY ALL;
```

5. **Create a Delta Live Tables pipeline**: Provide the following information:

   - **Name**: This is the name of the Delta Live Tables pipeline. We will name it `IOT Stream Pipeline (Quality Check)`.

   - **Product edition**: Delta Live Tables comes in three flavors. We will choose the **Advanced** edition.

   - **Pipeline mode**: We will choose **Triggered**.

   - **Choose the source code paths**: We will choose the source code path for this recipe, which is `Chapter09/9.3 data-quality-and-validation`.

   - **Choose destination information**: We will choose to register our tables in Unity Catalog:

     - **Catalog**: **main** or the catalog you want to use.

     - **Target schema**: We will use the `tpch_dlt` database. You might need to create the schema if it has not been created.

   - Choose your compute:

     - **Cluster mode**: We will choose **Enhanced autoscaling**

     - **Min workers**: 1

     - **Max workers**: 2

6. **Create the pipeline**: Click on the **Create** button in the bottom-right corner:

Figure 9.11 – The Create button

7. **Run the pipeline**: To execute the pipeline, click on **Start**. This will cause the Delta Live Tables pipeline to start executing. Once this happens, the following Delta Live Tables dependency graph will be generated from the source code:

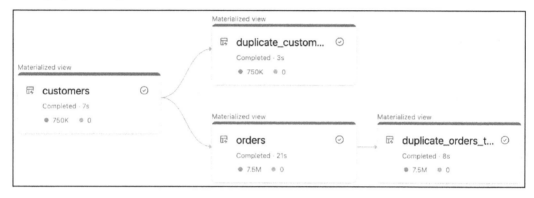

Figure 9.12 – Delta Live Tables DAG

How it works...

Delta Live Tables datasets allow you to add optional expectations that check the quality of each record in a query. An expectation has three parts:

*   A unique name
*   A Boolean expression that evaluates to true or false based on a condition
*   An action to be performed when a record does not meet the expectation (that is, when the Boolean is false)

The following table shows the three options you can choose from for handling invalid records, along with the Python and SQL syntax to implement them:

Action	Result	Python Syntax	SQL Syntax

Warn (default)	Invalid records are written to the target; the dataset reports the failure as a metric.	`@dlt. expect("description", "constraint")`
		
		`@dlt.expect_ all(expectations)`
			`CONSTRAINT expectation_ name_1 EXPECT (expectation_expr1)`

Drop	Invalid records are not written to the target; the dataset reports the failure as a metric.	`@dlt.expect_or_drop(« description », « constraint »)`	`CONSTRAINT expectation_ name_1 EXPECT (expectation_expr1) ON VIOLATION DROP ROW`

Fail	Invalid records stop the update from completing. You need to fix the issue before re-running the query.	`@dlt.expect_or_ fail("description", "constraint")`
		
		`@dlt.expect_all_or_ fail(expectations)`
			`CONSTRAINT expectation_ name_1 EXPECT (expectation_expr1) ON VIOLATION FAIL UPDATE`

Table 9.1 – Types of exception handling actions and syntax

See also

- *Delta Live Tables Python language reference*: https://docs.databricks.com/en/ delta-live-tables/python-ref.html

- *Delta Live Tables SQL language reference*: https://docs.databricks.com/en/ delta-live-tables/sql-ref.html

- *Manage data quality with Delta Live Tables*: https://docs.databricks.com/en/ delta-live-tables/expectations.html

Quarantining bad data with Delta Live Tables in Databricks

Quarantining bad data means sending it to a separate location for further inspection and correction. You can also replay the quarantined data once you have fixed the issues. With Delta Live Tables, you can quarantine bad data that does not meet your expectations or requirements. You can define expectations for your data using SQL or Python expressions, and specify how to handle records that fail those expectations. You can choose to fail, drop, alert, or quarantine the bad data.

In this recipe, you will learn how to use Delta Live Tables to quarantine bad data in your data pipelines, and how to backfill the quarantined data after resolving the problems.

How to do it...

1. **Incremental ingestion with an autoloader**: Create a dataset from the `farmers_markets_geographic_data` CSV data by creating a streaming table. We are defining a streaming table to denote that this is an incremental append-only load from the CSV file that lands in a folder to the `raw_farmers_market` Delta Lake table:

    ```
    CREATE
    OR REFRESH STREAMING LIVE TABLE raw_farmers_market AS
    SELECT
      *
    FROM
      cloud_files(
        "/databricks-datasets/data.gov/farmers_markets_geographic_
    data/data-001/",
        "csv",
        map(
          "cloudFiles.inferColumnTypes",
          "true"
        )
      )
    ```

2. **Create a streaming live table for the cleansed data**: We are defining a streaming live table from the `live.raw_farmers_market` Delta table. Use `STREAM()` to load only incremental data. We can now define some data expectation rules with the following constraints:

    *   `valid_website`: This constraint expects that the `Website` column is not null for every row in the live table. If this expectation is violated, the row is dropped from the live table.

    *   `valid_location`: This constraint expects that the `Location` column is not null for every row in the live table. If this expectation is violated, the row is dropped from the live table.

    ```
    CREATE
    OR REFRESH STREAMING LIVE TABLE farmers_market_clean (
      CONSTRAINT valid_website EXPECT (Website IS NOT NULL) ON
    ```

```
VIOLATION DROP ROW,
  CONSTRAINT valid_location EXPECT (Location IS NOT NULL) ON
VIOLATION DROP ROW
) AS
SELECT
  *
FROM
  STREAM(live.raw_farmers_market)
```

3. **Define the quarantine table:** We are defining a streaming live table from the `live.farmers_market_quarantine` Delta table. Use `STREAM()` to load only incremental data. We can now define some data expectation rules with the following constraints:

   - `valid_website`: This constraint expects that the `Website` column is `null` for every row in the live table. If this expectation is violated, the row is dropped from the live table.

   - `valid_location`: This constraint expects that the `Location` column is `null` for every row in the live table. If this expectation is violated, the row is dropped from the live table.

```
CREATE
OR REFRESH STREAMING LIVE TABLE farmers_market_quarantine (
  CONSTRAINT valid_website EXPECT (NOT(Website IS NOT NULL)) ON
VIOLATION DROP ROW,
  CONSTRAINT valid_location EXPECT (NOT(Location IS NOT NULL))
ON VIOLATION DROP ROW
) AS
SELECT
  *
FROM
  STREAM(live.raw_farmers_market)
```

> **Note**
>
> The `CONSTRAINTS` keyword, when added to the quarantine table, is reversed to make sure we capture the dropped rows from the cleansed dataset.

4. **Create a Delta Live Tables pipeline:** Provide the following information:

   - **Name:** This is the name of the Delta Live Tables pipeline. We will name it `Farmers Market DLT Pipeline`.

   - **Product edition:** Delta Live Tables comes in three flavors. We will choose the **Advanced** edition.

- **Pipeline mode**: We will choose **Triggered**:

  - **Choose the source code paths**: We will choose the source code for this recipe, which is `Chapter09/9.4 quarantine-bad-data-dlt`

  - **Choose destination information**: We will choose to register our tables in Unity Catalog

- **Catalog: main**:

  - **Target schema**: We will use the `farmers_market` database

- **Choose the compute**:

  - **Cluster mode**: We will choose **Enhanced autoscaling**

  - **Min workers**: 1

  - **Max workers**: 2

5. **Create the pipeline**: Click on the **Create** button in the bottom-right corner:

Figure 9.13 – The Create button

6. **Run the pipeline**: To execute the pipeline, click on **Start**. This will trigger the Delta Live Tables pipeline to start executing. Once the run is triggered, the following Delta Live Tables dependency graph will be generated from the source code:

Figure 9.14 – Delta Live Tables DAG

See also

- *Delta Live Tables SQL language reference*: `https://docs.databricks.com/en/` `delta-live-tables/sql-ref.html`

- *Manage data quality with Delta Live Tables*: `https://docs.databricks.com/en/` `delta-live-tables/expectations.html`

Monitoring Delta Live Tables pipelines

Monitoring and observability are essential aspects of data engineering as they allow you to track, understand, and troubleshoot the state of your data pipelines. Delta Live Tables provides built-in features for monitoring and observability in the form of an event log that contains all the information about a pipeline, such as audit logs, data quality checks, pipeline progress, and data lineage.

In this recipe, you will learn how to use the event logs captured by Delta Live Tables to monitor Delta Live Tables pipelines in Databricks.

How to do it...

1. **Create a Delta Live Tables pipeline**: Provide the following information:

   - **Name**: This is the name of the Delta Live Tables pipeline. We will name it `Farmers Market DLT Pipeline`.

   - **Product edition**: Delta Live Tables comes in three flavors. We will choose the **Advanced** edition.

   - **Pipeline mode**: We will choose **Triggered**:

     - **Choose the source code paths**: We will choose the source code for this recipe, which is `9.4 quarantine-bad-data-dlt`

     - **Choose destination information**: We will choose to register our tables in Unity Catalog

   - **Catalog**: **main**:

     - **Target schema**: We will use the `farmers_market` database

   - Choose your compute:

     - **Cluster mode**: We will choose **Enhanced autoscaling**

     - **Min workers**: 1

     - **Max workers**: 2

2. **Create the pipeline**: Click on the **Create** button in the bottom-right corner:

Figure 9.15 – The Create button

3. **Run the pipeline**: To execute the pipeline, click on **Start**. Delta Live Tables will execute the pipeline and generate the graph shown here:

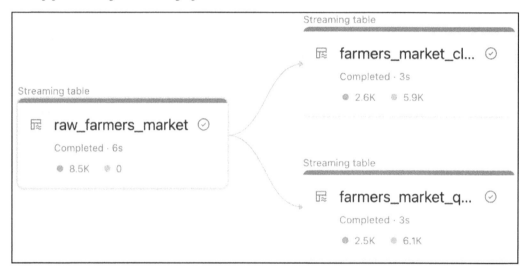

Figure 9.16 – Delta Live Tables DAG

4. **Get the Delta Live Tables pipeline ID**: From the **Delta Live Tables** tab, get the **Pipeline ID** value from the **Pipeline details** panel:

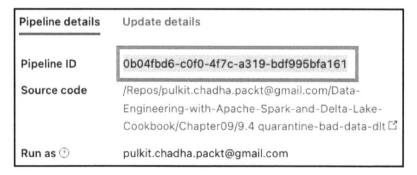

Figure 9.17 – Delta Live Tables Pipeline ID

5. **Create a temporary view of the event logs**: Since our pipeline is publishing tables in Unity Catalog, we can use the `event_log` **table-valued function** (**TVF**) to fetch the event log for the pipeline. To retrieve the event log for a pipeline, pass the pipeline ID to `event_log()`. We will create a view so that we can run the subsequent queries against it.

```
CREATE VIEW main.farmers_market.event_log_raw AS SELECT * FROM
event_log("<pipeline-ID>");
```

6. **Report on data quality metrics**: The result of this query is a table that shows the number of records that passed or failed each data quality expectation for each dataset and `update_id`. This can be useful for monitoring the data quality of different datasets over time.

```
SELECT
  update_id,
  row_expectations.dataset as dataset,
  row_expectations.name as expectation,
  SUM(row_expectations.passed_records) as passing_records,
  SUM(row_expectations.failed_records) as failing_records
FROM
  (
    SELECT
      origin.update_id,explode(
        from_json(
          details :flow_progress :data_quality :expectations,
          "array<struct<name: string, dataset: string, passed_
records: int, failed_records: int>>"
        )
      ) row_expectations
    FROM
      main.pkc_farmers_market.event_log_raw
    WHERE
      event_type = 'flow_progress'
  )
GROUP BY
  update_id,
  row_expectations.dataset,
  row_expectations.name
```

7. **Report on cluster utilization**: In the events log, the `cluster_resources` events show how many task slots are available and used in the cluster, and how many tasks are waiting for a slot.

When **Enhanced autoscaling** is enabled, the events related to `cluster_resources` include some metrics that the autoscaling algorithm uses, such as the number of executors that were recently requested and the optimal number of executors. The events also indicate the status of the algorithm, such as whether the cluster has reached the desired size, whether it is scaling up and waiting for more executors, or whether it is prevented from scaling down by the configuration. The result of this query is a table that shows the metrics and state of the cluster resources. This can be useful for analyzing the performance and behavior of the cluster, especially when enhanced autoscaling is enabled.

```
SELECT
  origin.update_id,
  timestamp,
  Double(details :cluster_resources.avg_num_queued_tasks) as
queue_size,
  Double(
    details :cluster_resources.avg_task_slot_utilization
  ) as utilization,
  Double(details :cluster_resources.num_executors) as current_
executors,
  Double(
    details :cluster_resources.latest_requested_num_executors
  ) as latest_requested_num_executors,
  Double(details :cluster_resources.optimal_num_executors) as
optimal_num_executors,
  details :cluster_resources.state as autoscaling_state
FROM
  main.pkc_farmers_market.event_log_raw
WHERE
  event_type = 'cluster_resources'
ORDER BY
  origin.update_id,
  timestamp DESC
```

8. **Analyze user actions**: The event log lets you track events, such as what users do. Events that contain information about user actions are of the user_action type.

    The user_action object in the details field contains information about the action. Use this query to make an audit log of user events.

    The result of this query is a table that shows the timestamp, action, and user name of each user action event. This can be useful for auditing the user activities on the database.

```
SELECT
  timestamp,
  details :user_action :action,
  details :user_action :user_name
FROM
  main.pkc_farmers_market.event_log_raw
WHERE
  event_type = 'user_action'
```

See also

- *Monitor Delta Live Tables pipelines*: `https://docs.databricks.com/en/delta-live-tables/observability.html`

Deploying Delta Live Tables pipelines with Databricks Asset Bundles

Databricks Asset Bundles is a new feature that allows you to deploy your data, analytics, and machine learning projects as a collection of source files. You can use a YAML file to specify the resources and settings for your project, such as jobs, pipelines, endpoints, experiments, and models. This way, you can easily manage your code and infrastructure in a consistent and automated way.

The following are some of the benefits of using Databricks Asset Bundles:

- You can use best practice tools and processes to work with source code, such as source control, code review, testing, and CI/CD

- You can streamline your local development with IDEs and run your resources before deploying them to production

- You can configure your deployments across multiple workspaces, regions, and clouds

In this recipe, you will learn how to package your pipeline as a Databricks Access Bundle and deploy it to a production workspace.

Getting ready

Before you start, you will need the following:

- A Databricks account with access to a workspace and a cluster.

- The Databricks CLI installed on your local machine. Please make sure you follow the instructions here: `https://docs.databricks.com/en/dev-tools/cli/install.html`.

- Authentication with a token for the Databricks CLI: `https://docs.databricks.com/en/dev-tools/cli/authentication.html`.

How to do it...

1. **Create a Databricks Asset Bundle manually**: A Databricks Asset Bundle is a YAML file with the name `databricks.yml`. The file has three main sections:

   - The `bundle` section specifies the name of the bundle, which is `dlt_dabs_cicd` in this case.

- The `include` section lists the source files that are part of the bundle. In this case, there is only one file: `dlt_dabs_cicd_pipeline.yml`. This is a YAML file that defines a Delta Live Tables pipeline.

- The `targets` section defines the workspaces where the bundle can be deployed. There are two targets: dev and prod.

  - The dev target is used for development purposes. It has the following properties:

    - It uses the mode development setting, which adds a prefix such as [dev my_user_name] to everything that is deployed to this target. This prevents clashes with other developers who are working on the same project.

    - It also turns off any schedules and automatic triggers for jobs and turns on development mode for Delta Live Tables pipelines. This lets developers execute and test their code without impacting production data or resources.

    - It sets the default true setting, which means that this is the default target for deploying the bundle. If no target is specified, the bundle will be deployed to the dev target.

    - It specifies the workspace property, which contains the host URL of the workspace where the bundle will be deployed. In this case, it is `https://adb-7637940272361795.15.azuredatabricks.net`.

  - The prod target is used for production deployment. It has the following properties:

    - It uses the mode production setting, which means that everything deployed to this target does not get any prefix and uses the original names of the source files.

    - It overrides the `workspace.root_path` property, which is the default path where the bundle will be deployed in the workspace. By default, it is `/Users/${workspace.current_user.userName}/.bundle/${bundle.target}/${bundle.name}`, which is specific to the current user. In this case, it has been changed to `/Shared/.bundle/prod/${bundle.name}`, which is a shared path that is not specific to any user.

    - It specifies the same workspace property as the dev target, which means that both targets use the same workspace host URL.

    - It specifies the `run_as` property, which defines who will run the resources in the bundle. In this case, it uses the `user_name: pulkit.chadha.packt@gmail.com` setting, which means that everything will run as this user. Alternatively, a service principal name could be used here using the `service_principal_name` setting.

2. **Create resources within the Databricks Asset Bundle**: We will only define the bundle with a single Delta Live Tables pipeline. The YAML file has the following sections:

    *   `pipelines`: This section defines one or more pipelines that are part of the resource. Each pipeline has a unique name, a target schema, and other settings. In this case, there is only one pipeline: `dlt_dabs_cicd_pipeline`.

    *   `target`: This setting specifies the schema where the output tables of the pipeline will be written. The target schema can be different for different environments, such as development, testing, or production. In this case, the target schema is `dlt_dabs_cicd_${bundle.environment}`, where `${bundle.environment}` is a variable that can be replaced with the actual environment name.

    *   `continuous`: This setting controls whether the pipeline runs in continuous mode or not. Continuous mode means that the pipeline will automatically update whenever there is new data available in the input sources. In this case, the pipeline runs in non-continuous mode, which means that it will only update when triggered manually or by a schedule.

    *   `channel`: This setting specifies the channel of Delta Live Tables that the pipeline uses. A channel is a version of Delta Live Tables that has different features and capabilities. In this case, the pipeline uses the **CURRENT** channel, which is the latest and most stable version of Delta Live Tables.

    *   `photon`: This setting controls whether the pipeline uses Photon or not. Photon is a vectorized query engine that can speed up data processing and reduce resource consumption. In this case, the pipeline does not use Photon.

    *   `libraries`: This section lists the source code libraries that are required for the pipeline. In this case, there is only one library, which is a notebook named `9.6 create-medallion-arch-DLT.sql`.

    *   `clusters`: This section defines the cluster that is used to run the pipeline. In this case, there is only one cluster with a label of `default`.

    *   `autoscale`: This setting enables or disables autoscaling for the cluster. In this case, autoscaling is enabled with a minimum of 0 workers and a maximum of 1 worker.

    *   `mode`: This setting specifies the mode of autoscaling. There are two modes: `STANDARD` and `ENHANCED`. `ENHANCED` mode can scale up and down faster and more efficiently than `STANDARD` mode. In this case, the mode is set to `ENHANCED`.

    ```
    # The main pipeline for dlt_dabs_cicd
    resources:
      pipelines:
        dlt_dabs_cicd_pipeline:
          name: dlt_dabs_cicd_pipeline
          target: dlt_dabs_cicd_${bundle.environment}
    ```

```
continuous: false
channel: CURRENT
photon: false
libraries:
  - notebook:
      path: 9.6 create-medallion-arch-DLT.sql
clusters:
  - label: default
    autoscale:
      min_workers: 1
      max_workers: 1
      mode: ENHANCED
```

3. **Validate the Databricks Asset Bundle**: Run the `bundle validate` command from the root directory of the bundle using the Databricks CLI. The validation is successful if you see a JSON representation of the bundle configuration. Otherwise, correct the errors and try again.

```
databricks bundle validate
```

4. **Deploy the bundle in the dev environment**: Run the `bundle deploy` command with the Databricks CLI to deploy the bundle in a specific environment. Once run, the local notebooks and SQL code will be moved to your Databricks workspace on the cloud and the Delta Live Tables pipeline will be set up.

```
databricks bundle deploy -t dev
```

The output should look like this:

```
% databricks bundle deploy -t dev

Starting upload of bundle files
Uploaded bundle files at /Users/pulkit.chadha.packt@gmail.com/.
bundle/dlt_dabs_cicd/dev/files!

Starting resource deployment
Resource deployment completed!
```

5. **See if the local notebook was moved**: Go to **Workspace** in your Databricks workspace's sidebar and navigate to **Users** | `<your-username>` | `<.bundle` > `<project-name>` | **dev** | **files** | **src**. The notebook should be there:

Figure 9.18 – Databricks Bundle Working Directory

6. **See if the pipeline was set up**: Go to **Delta Live Tables** in your Databricks workspace's sidebar. Click on [dev <your-username>] <project-name>_pipeline on the **Delta Live Tables** tab.

Figure 9.19 – Databricks Asset Bundle dev pipeline

7. **Run the deployed bundle**: You will run the Delta Live Tables pipeline in your workspace in this step. Go to the root directory and run the bundle run command with the Databricks CLI. Use your project name instead of <project-name>.

    ```
    databricks bundle run -t dev
    ```

    The output from the run command will list the pipeline events:

    ```
    sh-3.2$ databricks bundle run -t dev
    Update URL: <https://adb-7637940272361795.15.azuredatabricks.
    net/#joblist/pipelines/1a3434c7-df77-41b9-b423-ff8bdcd002a2/
    updates/839e9f65-6cb3-4e46-bf1e-ef0339b50421>
    2023-10-23T21:24:07.574Z update_progress INFO "Update 839e9f is
    WAITING_FOR_RESOURCES."
    2023-10-23T21:27:29.924Z update_progress INFO "Update 839e9f is
    INITIALIZING."
    2023-10-23T21:27:44.805Z update_progress INFO "Update 839e9f is
    SETTING_UP_TABLES."
    2023-10-23T21:28:01.566Z update_progress INFO "Update 839e9f is
    RUNNING."
    2023-10-23T21:28:01.571Z flow_progress   INFO "Flow 'device_
    data' is QUEUED."
    . . . .
    ```

```
2023-10-23T21:28:23.066Z flow_progress    INFO "Flow 'user_
metrics' is RUNNING."
2023-10-23T21:28:28.102Z flow_progress    INFO "Flow 'user_
metrics' has COMPLETED."
2023-10-23T21:28:29.987Z update_progress INFO "Update 839e9f is
COMPLETED."
sh-3.2$
```

When the pipeline finishes successfully, you can view the details in your Databricks workspace:

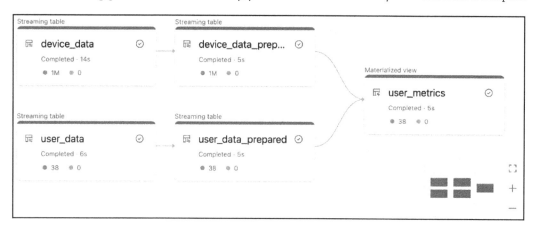

Figure 9.20 – Delta Live Tables DAG

There's more...

Within Databricks Asset Bundles, resources are a collection of entities that represent a data engineering or data science task. Each resource has a unique identifier and a set of properties that can be configured using the Databricks REST API. The YAML file has the following sections:

- `resources`: This section specifies the top-level mapping of resources and their default settings.

- `experiments`: This section defines one or more experiments that are part of the resource. An experiment is a collection of runs that are associated with a machine learning model. Each experiment has a unique identifier and a set of properties that can be configured using the Experiments API.

- `jobs`: This section defines one or more jobs that are part of the resource. A job is a scheduled or on-demand execution of a notebook, JAR, or Python script. Each job has a unique identifier and a set of properties that can be configured using the Jobs API.

- `models`: This section defines one or more models that are part of the resource. A model is a machine learning artifact that can be registered, versioned, packaged, and deployed. Each model has a unique identifier and a set of properties that can be configured using the Models API.

- `pipelines`: This section defines one or more pipelines that are part of the resource. A pipeline is a Delta Live Tables entity that consists of one or more datasets that are updated by applying transformations to the input data. Each pipeline has a unique identifier and a set of properties that can be configured using the Delta Live Tables API.

See also

- *Databricks Asset Bundles documentation*: `https://docs.databricks.com/en/dev-tools/bundles/index.html`

- *Databricks Asset Bundles product tour*: `https://www.databricks.com/resources/demos/tours/data-engineering/databricks-asset-bundles`

- *Databricks Asset Bundles YAML Settings*: `https://docs.databricks.com/en/dev-tools/bundles/settings.html`

- *Databricks Asset Bundles Product page*: `https://www.databricks.com/resources/demos/tours/data-engineering/databricks-asset-bundles`

- *Databricks Asset Bundles session at DAIS 2023*: `https://www.databricks.com/dataaisummit/session/databricks-asset-bundles-standard-unified-approach-deploying-data-products/`

Applying changes (CDC) to Delta tables with Delta Live Tables

Delta Live Tables also supports CDC, a technique that's used to identify and capture changes made to data in a source database and then deliver those changes in real time to a target system. CDC enables you to keep your data lake or data warehouse in sync with your operational databases and also supports real-time analytics and data science.

One of the challenges of CDC is how to handle **slowly changing dimensions** (**SCDs**), which are dimensions that store and manage both current and historical data over time in a data warehouse. For example, a customer dimension may have attributes such as name, address, and phone number that can change over time. Depending on your business requirements, you may want to track the history of these changes in different ways. Several types of SCDs define how to handle these changes, such as Type 1 (overwrite), Type 2 (add new row), Type 3 (add new column), and so on.

In this recipe, we will show you how to implement CDC with Delta Live Tables for both SCD Type 1 and Type 2 using Python and SQL. We will use a sample dataset of customers and orders that simulates changes in the customer dimension over time.

How to do it...

To apply changes to a Delta table, we will have to create a volume that will receive the incoming data files. We will upload our data to this volume, and then create the Delta table to ingest the data file there. Let's go through the steps.

Creating a volume and schema and uploading the data

Follow these steps:

1. **Create a managed schema**:

    I.   Select **Catalog** from the navigation panel:

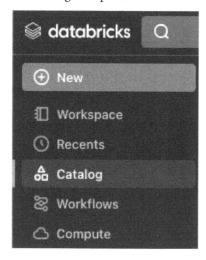

Figure 9.21 – Catalog

    II.   Search for the **main** catalog:

Figure 9.22 – The main catalog

    III.   Click on **Create Schema**:

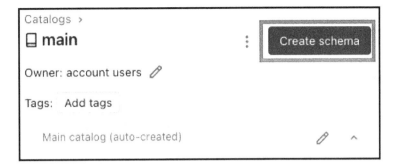

Figure 9.23 – The Create schema button

2. **Create a new schema**: Name the schema netflix. Optionally, you can give it a **Storage** location. We will skip this for now.

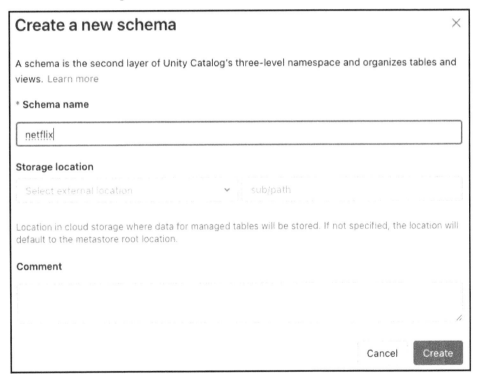

Figure 9.24 – Adding the schema's name

3. **Create a managed volume**:

   I.   Within the netflix schema, we can now click on **Create volume**:

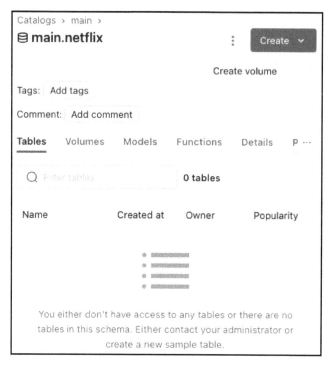

Figure 9.25 – Create volume

II.   Name the volume data, select **Managed volume**, and click on **Create**:

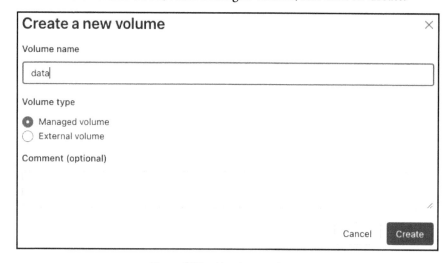

Figure 9.26 – Naming a volume

4.   **Upload data**: Click on **Upload to this volume**:

Figure 9.27 – Upload to this volume

Browse to the location where you cloned the repo, select the `data/netflix_titles.csv` file, and click on **Upload**:

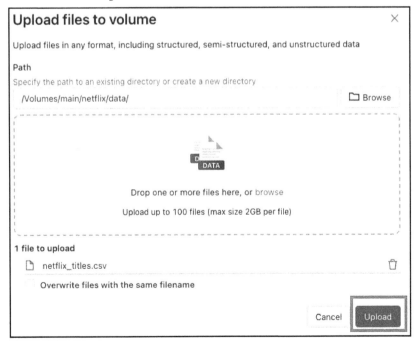

Figure 9.28 – Uploading the data file

Creating Delta Live Tables

Follow these steps:

1. **Create a live view for title data**: The following code creates a temporary streaming live view named `movie_and_show_titles` that reads data from the volume location in CSV format and also adds a `timestamp` column to each record. We will use this `timestamp` column later. A streaming live view is a type of dataset in Delta Live Tables that allows you to process a growing dataset and handle each row only once. Streaming live views are optimal for pipelines that require data freshness and low latency.

   ```
   CREATE TEMPORARY STREAMING LIVE VIEW movie_and_show_titles AS
   SELECT
     *,
     now() as ts
   FROM
     cloud_files(
       "/Volumes/main/netflix/data",
       "csv",
       map("header", "true")
     );
   ```

2. **Create a slowly changing dimension Type 1 table**: The following code applies changes from a streaming live view we created before to a live table named `live.movie_and_show_titles_scd_1`. A streaming table is a type of dataset in Delta Live Tables that allows you to process a growing dataset, handling each row multiple times. While using `APPLY CHANGES INTO`, we need to specify the following parameters for the streaming table:

   - `KEYS`: The columns that uniquely identify each row in the table. In this case, the keys are `(type, title, director)`, which means that each row is distinguished by a combination of these three columns.

   - `SEQUENCE BY`: The column that determines the order of the changes. In this case, the sequence column is `ts`, which is the timestamp column that was added by the streaming live view. This means that the changes are applied in the order of their arrival time.

   - `COLUMNS`: The columns that will be updated in the table. In this case, all the columns except the `ts` column will be updated.

   - `STORED AS`: This is the type of SCD that will be used to store the changes in the table. In this case, the SCD type is `1`, which means that the streaming table only keeps the latest version of each row and overwrites any previous versions.

     ```
     CREATE
     OR REFRESH STREAMING TABLE movie_and_show_titles_scd_1;
     ```

```
APPLY CHANGES INTO
  live.movie_and_show_titles_scd_1
FROM
  STREAM(LIVE.movie_and_show_titles)
KEYS
  (type, title, director)
SEQUENCE BY
  ts
COLUMNS * EXCEPT (ts)
STORED AS
  SCD TYPE 1;
```

3. **Create a slowly changing dimension Type 2 table**: The following code applies changes from a streaming live view we created before into a live table named `live.movie_and_show_titles_scd_2`. A streaming table is a type of dataset in Delta Live Tables that allows you to process a growing dataset, handling each row multiple times.

   This code specifies the same parameters as the previous code, except for the `STORED AS` parameter, which is different. The `STORED AS` parameter determines the type of SCD that is used to store the changes in the streaming table. In this case, the SCD type is 2, which means that the streaming table keeps both the current and historical versions of each row and adds validity period columns (`__START_AT` and `__END_AT`) to indicate the time range for which each version is valid.

```
CREATE
OR REFRESH STREAMING TABLE movie_and_show_titles_scd_2;

APPLY CHANGES INTO
  live.movie_and_show_titles_scd_2
FROM
  STREAM(LIVE.movie_and_show_titles)
KEYS
  (type, title, director)
SEQUENCE BY
  ts
COLUMNS * EXCEPT (ts)
STORED AS
  SCD TYPE 2;
```

4. **Create a Delta Live Tables pipeline**: Go to **Workflows** from the navigation panel. Then, go to the **Delta Live Tables** tab and click on **Create Pipeline**. Provide the following information:

   • **Name**: This is the name of the Delta Live Tables pipeline. We will name it `Netflix Titles Pipeline`.

   • **Product edition**: Delta Live Tables comes in three flavors. We will choose the **Advanced** edition.

- **Pipeline mode**: We will choose **Triggered**:

  - **Choose the source code paths**: We will choose the source code for this recipe, which is `9.7 apply-changes_into-dlt`

  - **Choose destination information**: We will choose to register our tables in Unity Catalog

- **Catalog: main**:

  - **Target schema**: We will use the `netflix` database

- Choose your compute:

  - **Cluster mode**: We will choose **Enhanced autoscaling**

  - **Min workers**: 1

  - **Max workers**: 2

5. **Create the pipeline**: Click on the **Create** button in the bottom-right corner:

Figure 9.29 – The Create button

6. **Run the pipeline**: To execute the pipeline, click on **Start**:

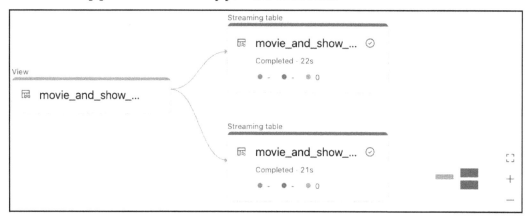

Figure 9.30 – Delta Live Tables DAG

Simulate an incremental load

Follow these steps:

1. **View the Catalog UI**: Go to **Catalog** from the navigation panel, search for the **main** catalog, and under the **netflix** schema, open the **data** volume.

2. **Upload the data**: Click on **Upload to this volume**:

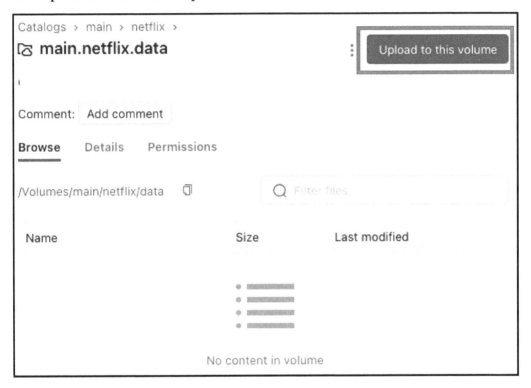

Figure 9.31 – Upload to this volume

Browse to the location where you cloned the repo, select the `data/netflix_titles_batch_2.csv` file, and click on **Upload**:

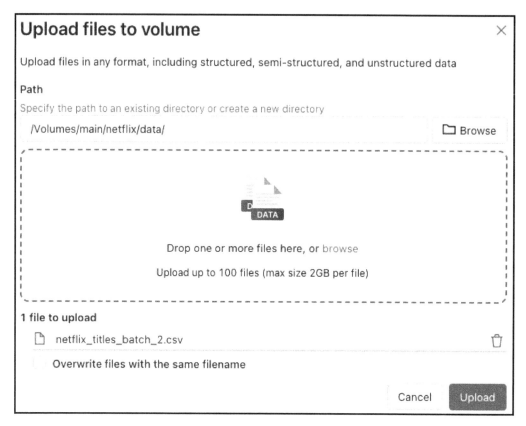

Figure 9.32 – Uploading the data file

3. **Execute the Delta Live Tables pipeline**: Click on **Start** in the **Netflix Titles Pipeline** section:

Figure 9.33 – Starting the Delta table

4. **Validate the data**: Once the pipeline has been executed, go to **Catalog** from the navigation panel. Then, go to the **main** catalog and the **Netflix** schema, select movie_and_show_titles_scd_2, click on the **Sample Data** tab, and scroll to the right to see the validity columns added by Delta Live Tables:

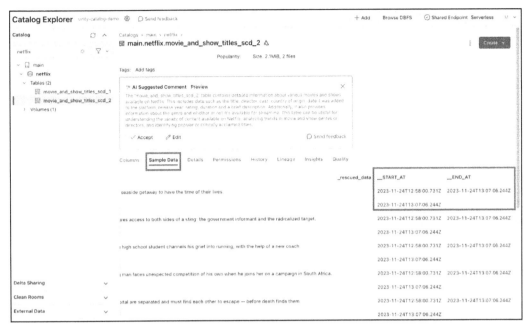

Figure 9.34 – Data Explorer UI

See also

- *Delta Live Tables documentation*: `https://docs.databricks.com/en/delta-live-tables/index.html`

- *Auto Loader SQL syntax*: `https://docs.databricks.com/en/delta-live-tables/sql-ref.html#auto-loader-sql-syntax`

- *Change data capture with SQL in Delta Live Tables*: `https://docs.databricks.com/en/delta-live-tables/sql-ref.html#change-data-capture-with-sql-in-delta-live-tables`

- *Create volumes*: `https://docs.databricks.com/en/data-governance/unity-catalog/create-volumes.html#create-a-managed-volume`

10

Data Governance with Unity Catalog

Databricks Unity Catalog is a unified governance solution for data and AI on a lakehouse. It allows you to securely discover, access, and collaborate on trusted data and AI assets from various platforms and clouds, leveraging AI to boost productivity and compliance.

Unity Catalog is a tool that helps you control, audit, trace, and discover data across Databricks workspaces. You can set up data access rules in one place and enforce them in all workspaces, following the ANSI SQL security standards. You can also see who created and used data assets, and how they did it, in any language. In addition, Unity Catalog allows you to label and describe data assets and offers a way to search for data that you need.

In this chapter, you will learn how to use Databricks Unity Catalog for data governance. We will cover the following recipes:

- Connecting to cloud object storage using Unity Catalog
- Creating and managing catalogs, schemas, volumes, and tables using Unity Catalog
- Defining and applying fine-grained access control policies using Unity Catalog
- Tagging, commenting, and capturing metadata about data and AI assets using Databricks Unity Catalog
- Filtering sensitive data with Unity Catalog
- Using Unity Catalogs lineage data for debugging, root cause analysis, and impact assessment
- Accessing and querying system tables using Unity Catalog

By the end of this chapter, you will have a solid understanding of how Databricks Unity Catalog can help you manage your data and AI assets in a lakehouse architecture.

Technical requirements

To follow along with the examples in this chapter, you will need a Databricks workspace with the premium plan.

You can find the notebooks and data for this chapter at `https://github.com/PacktPublishing/Data-Engineering-with-Databricks-Cookbook/tree/main/Chapter10`.

Connecting to cloud object storage using Unity Catalog

Databricks Unity Catalog allows you to manage and access data in cloud object storage using a unified namespace and a consistent set of APIs. With Unity Catalog, you can do the following:

*   Create and manage storage credentials, external locations, storage locations, and volumes using SQL commands or the Unity Catalog UI

*   Access data from various cloud platforms (AWS S3, Azure Blob Storage, or Google Cloud Storage) and storage formats (Parquet, Delta Lake, CSV, or JSON) using the same SQL syntax or Spark APIs

*   Apply fine-grained access control and data governance policies to your data using Databricks SQL Analytics or Databricks Runtime

In this recipe, you will learn what Unity Catalog is and how it integrates with AWS S3.

Getting ready

Before you start setting up and configuring Unity Catalog, you need to have the following prerequisites:

*   A Databricks workspace with administrator privileges

*   A Databricks workspace with the Unity Catalog feature enabled

*   A cloud storage account (such as AWS S3, Azure Blob Storage, or Google Cloud Storage) with the necessary permissions to read and write data

How to do it...

In this section, we will first create a storage credential, the IAM role, with access to an s3 bucket. Then, we will create an external location in Databricks Unity Catalog that will use the storage credential to access the s3 bucket.

Creating a storage credential

You must create a storage credential to access data from an external location or a volume. In this example, you will create a storage credential that uses an IAM role to access the S3 Bucket. The steps are as follows:

1. **Go to Catalog Explorer**: Click on **Catalog** in the left panel and go to **Catalog Explorer**.

2. **Create storage credentials**: Click on +**Add** and select **Add a storage credential**.

Figure 10.1 – Add a storage credential

3. **Enter storage credential details**: Give the credential a name, the IAM role ARN that allows Unity Catalog to access the storage location on your cloud tenant, and a comment if you want, and click on **Create**.

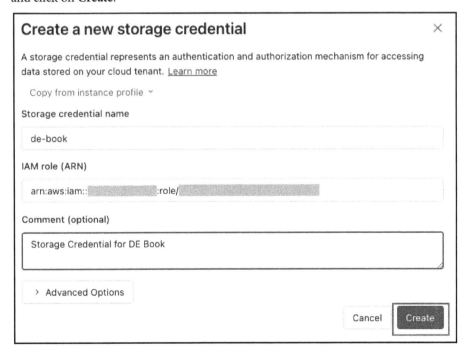

Figure 10.2 – Create a new storage credential

> **Important note**
>
> To learn more about IAM roles in AWS, you can reference the user guide here: `https://docs.aws.amazon.com/IAM/latest/UserGuide/introduction.html`.

4. **Get External ID**: In the **Storage credential created** dialog, copy the **External ID** value and click on **Done**.

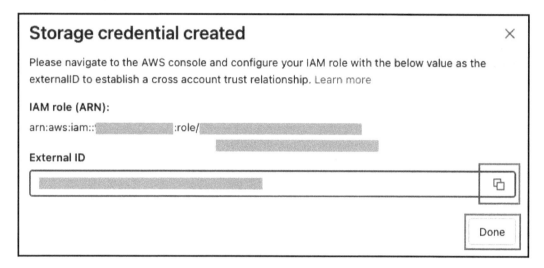

Figure 10.3 – External ID for the storage credential

5. **Update the trust policy with an External ID**: Update the trust policy associated with the IAM role and add the External ID value for `sts:ExternalId`:

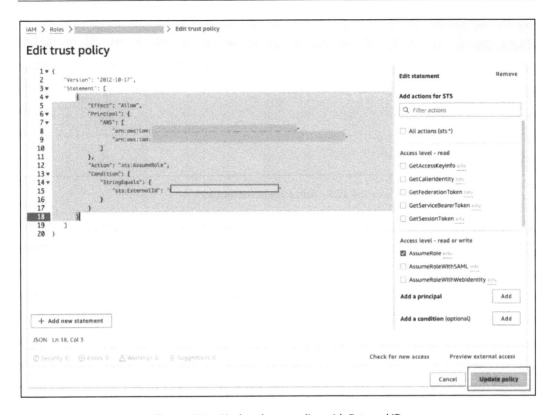

Figure 10.4 – Updated trust policy with External ID

Creating an external location

An external location contains a reference to a storage credential and a cloud storage path. You need to create an external location to access data from a custom storage location that Unity Catalog uses to reference external tables. In this example, you will create an external location that points to the de-book-ext-loc folder in an S3 bucket. To create an external location, you can follow these steps:

1. **Go to Catalog Explorer**: Click on **Catalog** in the left panel to go to **Catalog Explorer**.

2. **Create external location**: Click on +**Add** and select **Add an external location**:

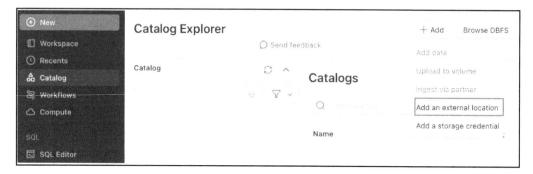

Figure 10.5 – Add an external location

3. **Pick an external location creation method**: Select **Manual** and then click on **Next**:

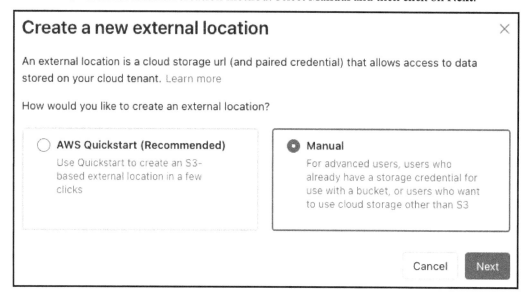

Figure 10.6 – Create a new external location

4. **Enter external location details**: Enter the external location name, select the storage credential, and enter the S3 URL; then, click on the **Create** button:

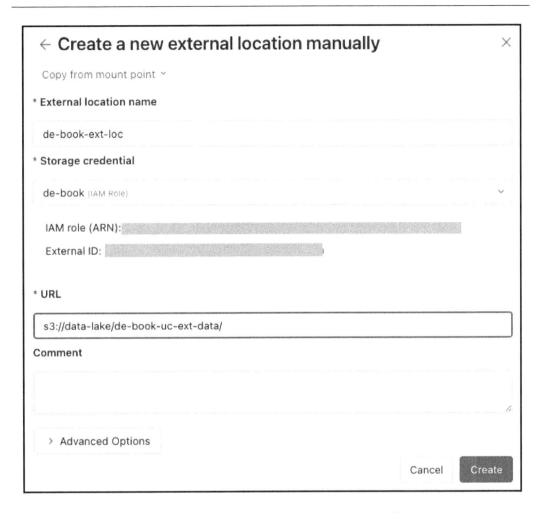

Figure 10.7 – Create a new external location manually

5. **Test connection**: Test the connection to make sure you have set up the credentials accurately and that Unity Catalog is able to access cloud storage:

Figure 10.8 – Test connection for external location

If everything is set up right, you should see a screen like the following. Click on **Done**:

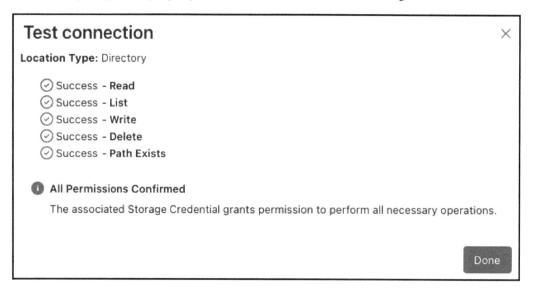

Figure 10.9 – Test connection results

See also

- Databricks Unity Catalog: `https://www.databricks.com/product/unity-catalog`

- What is Unity Catalog: `https://docs.databricks.com/en/data-governance/unity-catalog/index.html`

- Databricks Unity Catalog documentation: `https://docs.databricks.com/en/compute/access-mode-limitations.html`

- Databricks SQL documentation: `https://docs.databricks.com/en/data-governance/unity-catalog/create-tables.html`

- Databricks Unity Catalog: A Comprehensive Guide to Features, Capabilities, and Architecture: `https://atlan.com/databricks-unity-catalog/`

- Step By Step Guide on Databricks Unity Catalog Setup and its key Features: `https://medium.com/@sauravkum780/step-by-step-guide-on-databricks-unity-catalog-setup-and-its-features-1d0366c282b7`

Creating and managing catalogs, schemas, volumes, and tables using Unity Catalog

Unity Catalog introduces a hierarchy of data objects that organize your data assets:

- **Metastore**: A metadata storage that has a three-level structure (`catalog.schema.table` or `catalog.schema.volume`) to arrange your data.

- **Catalog**: An object that groups your data assets in the first level of the structure. A catalog can include schemas, tables, and volumes. A catalog can also specify a storage location that is used by default for its schemas, tables, and volumes.

- **Schema**: The second layer of the object hierarchy, used to group related tables and volumes. A schema can also have a managed storage location that serves as the default location for its tables and volumes.

- **Table**: The next layer of the object hierarchy is used to access tabular data stored in cloud object storage. A table can be either managed or external. A managed table is backed by a managed storage location and is automatically deleted when the table is dropped. An external table is backed by an external location and is not deleted when the table is dropped.

- **Volume**: The next layer of the object hierarchy is used to access non-tabular data stored in cloud object storage. A volume can be either managed or external. A managed volume is backed by a managed storage location and is automatically deleted when the volume is dropped. An external volume is backed by an external location and is not deleted when the volume is dropped.

The following diagram illustrates the hierarchy of objects in Unity Catalog:

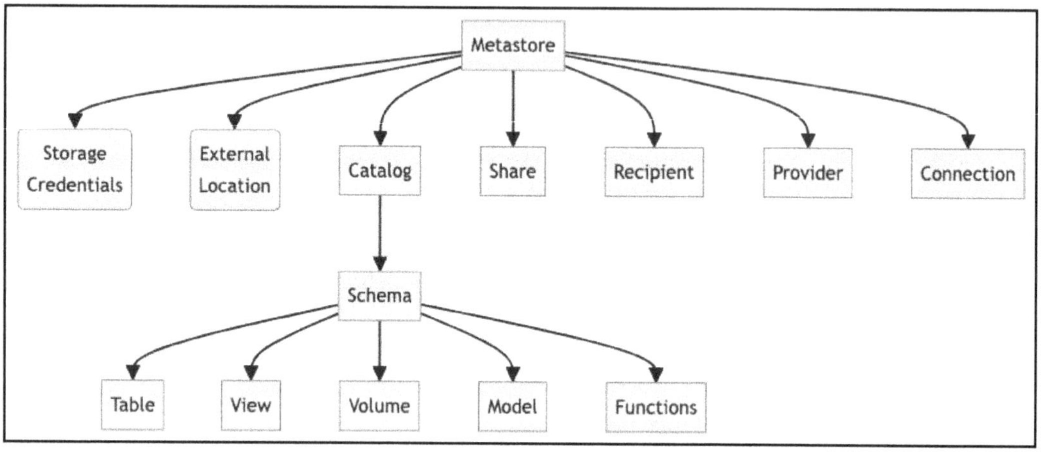

Figure 10.10 – Unity Catalog hierarchy of objects

In this recipe, you will learn the following:

- The hierarchy of data objects in Unity Catalog, such as metastore, catalog, schema, volume, and table
- How to create and manage these objects using SQL commands or the Unity Catalog UI

Getting ready

Before you start, you need to have the following prerequisites:

- A Databricks account with admin privileges
- A storage credential that provides access to your cloud storage
- An external location that contains a reference to your storage credential and a cloud storage path

We created the storage credential and external location in the previous recipe and will be using them in this recipe.

How to do it...

You can use SQL commands or the Unity Catalog UI to create and manage catalogs, schemas, tables, and volumes. The following examples show how to perform some common tasks using both methods.

Creating a catalog

To create a catalog named `sales`, you can use the following SQL command:

```
CREATE CATALOG de_book MANAGED LOCATION 's3://data-lake/de-book-ext-
data';
```

Alternatively, you can use the Unity Catalog UI to create a catalog. To do so, follow these steps:

1. **Go to Catalog Explorer**: Click on **Catalog** from the left navigation panel.

2. **Create catalog**: Click on the **Create catalog** button:

Figure 10.11 – The Create catalog button

3. **Enter catalog details**: Provide catalog details as follows:

   - **Catalog name**: de_book

   - Choose the type of catalog you want to create; select **Standard**

     - **Standard catalog**: This is a securable object that groups data assets under Unity Catalog. Use this for all scenarios except Lakehouse Federation.

     - **Foreign catalog**: This is a securable object in Unity Catalog that reflects a database from an external data system with Lakehouse Federation.

   - **Storage location**: Select the managed storage location for the catalog. We will use **de-book-ext-data**.

Click on **Create**:

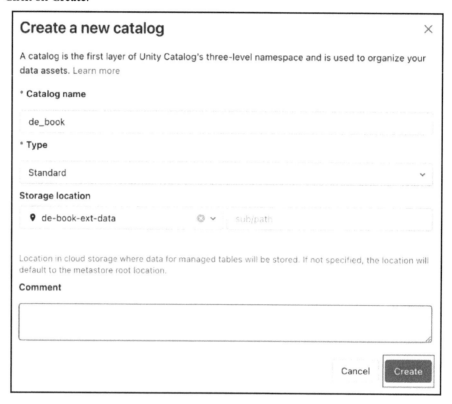

Figure 10.12 – Create a new catalog

4. **Set up catalog workspace binding**: Provide the workspace that the catalog is linked to. By default, the catalog is available to all workspaces that are attached to the current metastore. If the catalog has data that should be limited to certain workspaces, go to the **Workspaces** tab and add those workspaces.

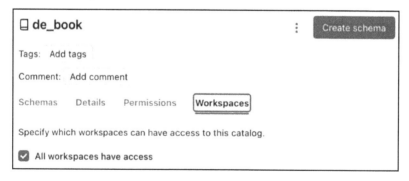

Figure 10.13 – Catalog workspace binding

Creating a schema

To create a schema named `credit_card` under the catalog `de_book`, you can use the following SQL command:

```
USE CATALOG de_book;
CREATE SCHEMA credit_card;
```

Alternatively, you can use the Unity Catalog UI to create a schema. To do so, follow these steps:

1. **Go to Catalog Explorer**: Click on **Catalog**.

2. **Search for a catalog**: In the left **Catalog** pane, select the catalog where you want to create the schema:

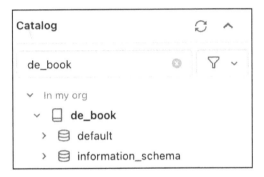

Figure 10.14 – Select a catalog

3. **Create schema**: Click on **Create schema**:

Figure 10.15 – The Create schema button

4.   **Enter schema details**: Enter a name for the schema and a comment that explains its purpose to users. Click on **Create**:

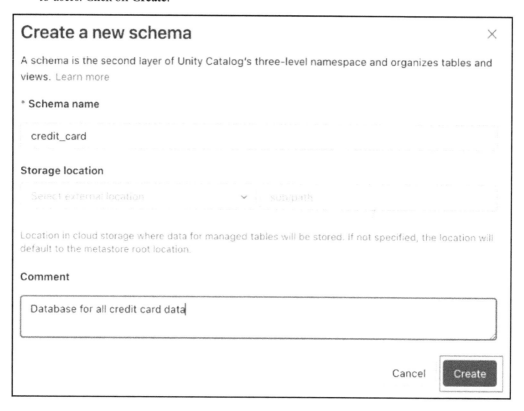

Create a new schema   ×

A schema is the second layer of Unity Catalog's three-level namespace and organizes tables and views. Learn more

* Schema name

credit_card

Storage location

Select external location   ∨   sub-path

Location in cloud storage where data for managed tables will be stored. If not specified, the location will default to the metastore root location.

Comment

Database for all credit card data

Cancel   Create

Figure 10.16 – Schema details

Creating a volume

To create a volume named `files` under the schema `de_book.credit_card`, you can use the following SQL command:

```
CREATE EXTERNAL VOLUME de_book.credit_card.files
LOCATION 's3://data-lake/de-book-ext-data/files';
```

Alternatively, you can use the Unity Catalog UI to create a volume. To do so, follow these steps:

1.   **Go to Catalog Explorer**: In your Databricks workspace, click on **Catalog**.

2.   **Search for schema**: Find and select the schema where you want to create the volume:

Figure 10.17 – Select schema

3. **Create volume**: Click on the **Create volume** button (you need enough privileges for this):

Figure 10.18 – The Create volume button

4. **Enter volume details**: Enter details for the volume:

    - Give a name to the volume

    - Pick an external location to create the volume

    - Change the path to match the sub-directory where you want the volume

    - Add a comment (optional)

Click on **Create**:

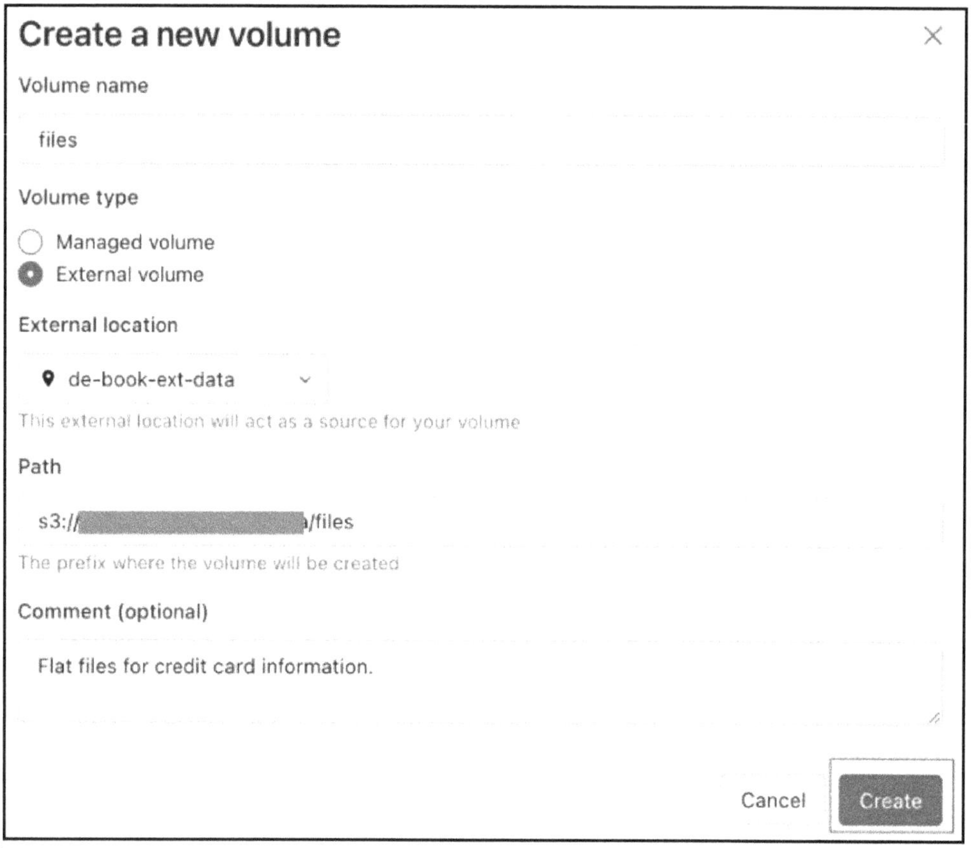

Create a new volume   ✕

Volume name

files

Volume type

◯ Managed volume
● External volume

External location

📍 de-book-ext-data   ⌄

This external location will act as a source for your volume

Path

s3://[_____]/files

The prefix where the volume will be created

Comment (optional)

Flat files for credit card information.

                                        Cancel   Create

Figure 10.19 – Volume details view

Creating a table

A table is the third layer of the object hierarchy in Unity Catalog, which is used to store data. A view is a named query on one or more tables. In this example, you will create a table named `transactions_table` and a view named `transactions_view` in the `credit_card` schema of the `data_eng_book` catalog. You can run the following code in an SQL to create the table:

```
USE CATALOG de_book;
USE SCHEMA credit_card;
CREATE TABLE IF NOT EXISTS transactions_table (
  Transaction_ID STRING,
  Transaction_Date STRING,
  Credit_Card_ID STRING,
  Transaction_Value FLOAT,
```

```
    Transaction_Segment STRING
);
INSERT INTO
    transactions_table
VALUES
    ('CTID28830551','24-Apr-16', '1629-9566-3285-2123', 23649, 'SEG25'),
    ('CTID45504917', '11-Feb-16', '3697-6001-4909-5350', 26726,
'SEG16'),
    ('CTID47312290', '1-Nov-16', '5864-4475-3659-1440', 22012, 'SEG14'),
    ('CTID25637718', '28-Jan-16', '5991-4421-8476-3804', 37637,
'SEG17');
```

You can run the following code in SQL to create the view:

```
CREATE OR REPLACE VIEW transactions_view (Credit_Card_ID, total_
Transaction_Value)
COMMENT 'A view that shows the total transaction value by credit card'
AS SELECT Credit_Card_ID, SUM(Transaction_Value) AS total_Transaction_
Value FROM data_eng_book.credit_card.transactions_table GROUP BY
Credit_Card_ID;
```

You can verify that the table and the view are created by going to **Data**, clicking on the **Unity Catalog** tab, and expanding the **Metastores**, **Catalogs**, and **Schemas** sections. You should see transactions_ table and transactions_view listed under the credit_card schema of the de_book catalog:

Figure 10.20 – Catalog explorer

See also

- What is Unity Catalog?: `https://learn.microsoft.com/en-us/azure/databricks/data-governance/unity-catalog/`

- Working with Unity Catalog in Azure Databricks: `https://techcommunity.microsoft.com/t5/fasttrack-for-azure/working-with-unity-catalog-in-azure-databricks/ba-p/3693781`

- Unity Catalog best practices: `https://learn.microsoft.com/en-us/azure/databricks/data-governance/unity-catalog/best-practices`

Defining and applying fine-grained access control policies using Unity Catalog

With Unity Catalog, you can use ANSI SQL to grant permissions to their existing data lake. A fine-grained access control method controls who can access certain data based on multiple conditions or entitlements. It allows you to specify the policies for each data item and attribute and apply them consistently across all workspaces and platforms. In this way, you can ensure data security and compliance while enabling data discovery and collaboration.

In this recipe, you will learn:

- How the security model of Unity Catalog allows fine-grained control over data and AI assets

- How to define and apply access policies using SQL commands or the Unity Catalog UI

Getting ready

Before you start defining and applying access policies using Unity Catalog, you need to have the following prerequisites:

- Access to a Databricks workspace enabled for Unity Catalog

- Data assets that you want to govern using Unity Catalog. It can be a table, a view, or a volume in your data lake or an external data source that supports Lakehouse Federation.

How to do it...

1. **Use GRANT and REVOKE statements**: To manage privileges in SQL, you need to use the GRANT and REVOKE statements. These statements allow you to grant or revoke privileges on securable objects to principals. A securable object is an object in the Unity Catalog that can have privileges, such as a catalog, a schema, a table, or a view. A principal is a user, a service principal, or a group that can have privileges.

The general syntax for the GRANT and REVOKE statements can be seen in the following:

```
GRANT <privilege_type> ON <securable_object> TO <principal>;
REVOKE <privilege_type> ON <securable_object> FROM <principal>;
```

where the following definitions apply:

- privilege_type is a Unity Catalog privilege type, such as ALL PRIVILEGES, APPLY TAG, CREATE SCHEMA, and USE CATALOG
- securable_object is a securable object in the Unity Catalog, such as a catalog, a schema, a table, a view, a volume, and so on
- principal is a user, a service principal, or a group that can have privileges

For a table of all the securable and relevant privileges, refer to this link: https://docs. databricks.com/en/data-governance/unity-catalog/manage-privileges/ privileges.html.

2. **Understand the inheritance model**: In Unity Catalog, securable objects have a hierarchy, and lower-level objects inherit privileges from higher-level objects. This means that when you give a privilege to an object, such as a catalog or a schema, all the objects within that catalog or schema, such as tables and views, get the same privilege automatically. For example, if you give the SELECT privilege to the main catalog of the finance group, the finance group can query any table or be viewed in any schema in the main catalog.

The inheritance model makes it easy to set up default access rules for your data. For example, if you want the machine learning team to create tables in a schema and read each other's tables, you can use these commands:

```
CREATE CATALOG ml;
CREATE SCHEMA ml.team_sandbox;
GRANT USE_CATALOG ON CATALOG ml TO ml_team;
GRANT USE_SCHEMA ON SCHEMA ml.team_sandbox TO ml_team;
GRANT CREATE TABLE ON SCHEMA ml.team_sandbox TO ml_team;
GRANT SELECT ON SCHEMA ml.team_sandbox TO ml_team;
```

3. **Unity catalog object ownership**: In Unity Catalog, every object has an owner, which can be a user, a service principal, or a group. Object owners get all privileges on their own objects by default. Moreover, object owners can give privileges to their own objects and any objects inside them. This means that schema owners do not automatically have all privileges over the tables in the schema, but they can give themselves privileges over the tables in the schema. For example, if you own the schema main.default, you can grant yourself the DROP TABLE privilege for the table de_book.credit_card.transaction_table by using the following command:

```
GRANT DROP TABLE ON TABLE de_book.credit_card.transaction_table
TO <your_user_name>;
```

4. **Grant privileges via SQL**: To grant privileges, you can use the GRANT statement with the appropriate privilege type, securable object, and principal. For example, the following command grants the CREATE TABLE privilege to the data-eng-team group for the de_book. credit_card schema:

    ```
    GRANT CREATE TABLE ON SCHEMA de_book.credit_card TO data-eng-
    team;
    ```

    You can also grant multiple privileges at once by separating them with commas. For example, the following command grants the SELECT and INSERT privileges to the user john.doe@ company.com for the de_book.credit_card.transaction_table table:

    ```
    GRANT SELECT, INSERT ON TABLE de_book.credit_card.transaction_
    table TO john.doe@company.com;
    ```

5. **Grant privileges via the Catalog Explore UI**: To give permissions over objects in Unity Catalog objects, do the following:

    I.   Click on **Catalog**.

    II.   Choose the object you want to give permissions to, such as a catalog, a schema, a table, or a view.

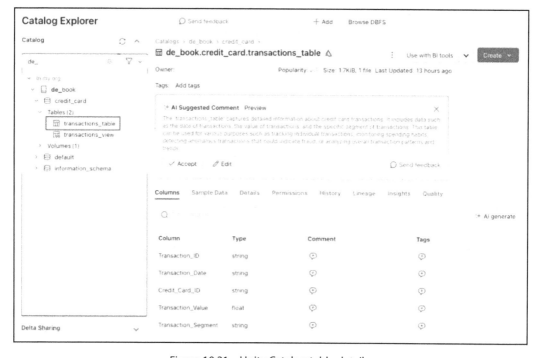

Figure 10.21 – Unity Catalog table details

    III.   Go to the **Permissions** tab and click on **Grant**:

Figure 10.22 – The Permissions tab

IV.   Type the email address of a user or the group name and pick the permissions you want to give them. Click on **Grant**:

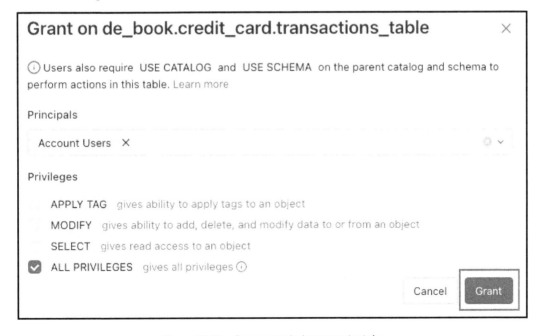

Figure 10.23 – Grant permissions to principles

6. **Revoke privileges via SQL**: To revoke privileges, you can use the REVOKE statement with the same syntax as the GRANT statement. For example, the following command revokes the DROP VIEW privilege from the service principal on de_book.credit_card.transaction_view:

```
REVOKE DROP VIEW ON VIEW de_book.credit_card.transaction_view
FROM app-1234-5678-9012;
```

You can also revoke all privileges from a principal on a securable object by using the ALL keyword. For example, the following command revokes all privileges from the ml_team group in the ml catalog:

```
REVOKE ALL ON CATALOG ml FROM ml_team;
```

7. **Revoke privileges via the catalog explore UI**: To take away permissions from objects in a Unity Catalog metastore, follow these steps:

    I.   Click on **Catalog**.

    II. Pick the object you want to take permissions away from, such as a catalog, a schema, a table, or a view.

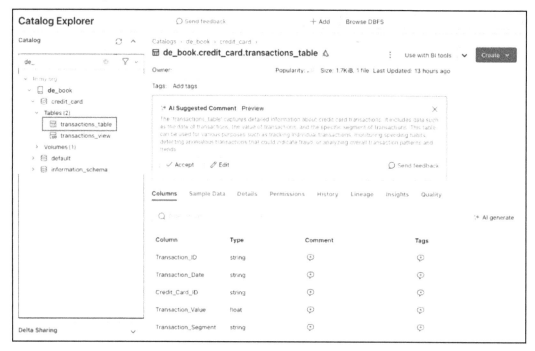

Figure 10.24 – Unity Catalog Table Details

    III. Go to the **Permissions** tab and choose a privilege that a user, a service principal, or a group has. Click on **Revoke**:

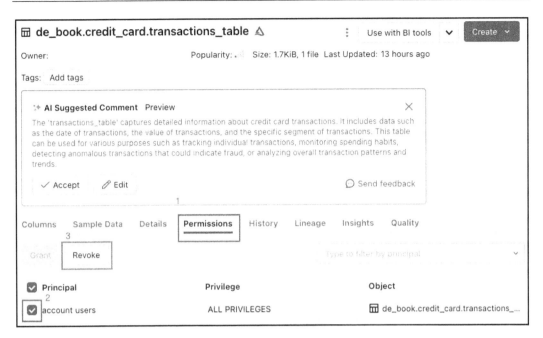

Figure 10.25 – Permissions and revoking privilege

- Confirm removing the privileges by clicking on **Revoke** again:

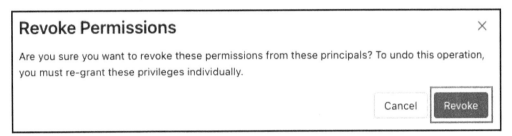

Figure 10.26 – Confirm revoke access

See also

- What is Unity Catalog?: https://learn.microsoft.com/en-us/azure/databricks/data-governance/unity-catalog/

- Databricks Unity Catalog: https://www.databricks.com/product/unity-catalog

- Working with Unity Catalog in Azure Databricks: https://techcommunity.microsoft.com/t5/fasttrack-for-azure/working-with-unity-catalog-in-azure-databricks/ba-p/3693781

- Unity Catalog best practices: `https://learn.microsoft.com/en-us/azure/databricks/data-governance/unity-catalog/best-practices`

- Databricks Unity Catalog: Everything You Need to Know: `https://atlan.com/databricks-unity-catalog/`

- Unity Catalog privileges and securable objects: `https://docs.databricks.com/en/data-governance/unity-catalog/manage-privileges/privileges.html`

Tagging, commenting, and capturing metadata about data and AI assets using Databricks Unity Catalog

Tagging and documenting data and AI assets are important features of Unity Catalog which helps help you to do the following:

- **Improve data discovery**: You can use tags and descriptions to annotate your data and AI assets with meaningful and consistent metadata, such as business terms, data quality, data sensitivity, and data ownership. This makes it easier for data consumers to find the data they need using the Unity Catalog search interface or SQL queries.

- **Improve data quality**: You can use tags and descriptions to document the data lineage, provenance, and transformations of your data and AI assets. This helps you track the origin and history of your data, as well as the dependencies and impacts of any changes.

- **Improve data compliance**: You can use tags and descriptions to classify your data and AI assets according to various regulatory and organizational policies, such as GDPR, HIPAA, and data retention. This helps you enforce the appropriate access controls and auditing requirements for your data.

In this recipe, we will learn how to tag, comment, and document data and AI assets using SQL commands or the Unity Catalog UI.

Getting ready

Before you start tagging and documenting your data and AI assets using Unity Catalog, you need to do the following:

- **Create or access a Databricks workspace**: You need a Databricks workspace to store and access your data and AI assets. You can create a new workspace or use an existing one.

- **Create or access a Unity Catalog catalog**: You need a catalog to organize your data and AI assets into logical groups. You can create a new catalog or use an existing one.

- **Create or access a Unity Catalog schema**: You need a Unity Catalog schema to define the structure and format of your data and AI assets. You can create a new schema or use an existing one.

How to do it...

In this section, we will learn how to add tags and comments to tables:

- Using SQL commands
- Using Unity Catalog UI

Using SQL Commands

1. **Add a comment on a Unity Catalog table**: To add comments on a table, you can use the COMMENT ON TABLE command. For example, suppose you have a table called transactions_table in your de_bookcatalog and credit_card schemas. You can add a description and some tags to this table using the following SQL command:

    ```
    COMMENT ON TABLE de_book.credit_card.transactions_table IS 'This
    table contains transaction information from the credit_card
    database';
    ```

2. **Tag Unity Catalog objects using SQL commands**: To set tags on a table, you can use the ALTER TABLE command. For example, the following code modifies the table named transactions_table in the credit_card schema in the de_book catalog. It sets three tags for the table: business_unit with the value finance, data_sensitivity with the value medium, and data_quality with the value high. These tags can be used to search and filter tables in Unity Catalog:

    ```
    ALTER TABLE
      de_book.credit_card.transactions_table
    SET
      TAGS (
        'business_unit' = 'finance',
        'data_sensitivity' = 'medium',
        'data_quality' = 'high'
      );
    ```

3. **Add comments to table columns**: You can also add descriptions/comments and tags to specific columns of your table using the ALTER COLUMN statements and the COMMENT and SET TAG clauses. For example, suppose you have a column called Transaction_ID in your transactions_table table.

    You can add a description to this column using the following SQL command. The code changes the comment of the column named Transaction_ID in the transactions_table table in the credit_card schema in the catalog de_book. The new comment is a unique identifier for the transaction:

    ```
    ALTER TABLE
      de_book.credit_card.transactions_table
    ```

```
ALTER COLUMN
  Transaction_ID COMMENT 'A unique identifier for the
transaction.';
```

You can add tags to this column using the following SQL command. The code modifies the tags of the column named `Transaction_ID` in the `transactions_table` table in the `credit_card` schema in the `de_book` catalog. It sets three tags for the column: `data_protection` with the value `non-PII`, and `isIdentifier` with the value `true`. These tags can be used to search and filter tables in Unity Catalog:

```
ALTER TABLE
  de_book.credit_card.transactions_table
ALTER COLUMN
  Transaction_ID
SET
  TAGS (
    'data_protection' = 'non-PII',
    'isIdentifier' = 'true'
  );
```

4. **View table comments**: You can view the descriptions using the `DESCRIBE DETAIL` clause. For example, you can view the description of your `transactions_table` table using the following SQL commands:

```
DESCRIBE DETAIL de_book.credit_card.transactions_table;
```

5. **View table tags from the information schema**: You can view all table tags from the information schema of the catalog. `INFORMATION_SCHEMA` has tables that show the objects in the catalog's schema. To get tag information, you can query these tables: `INFORMATION_SCHEMA.CATALOG_TAGS`, `INFORMATION_SCHEMA.COLUMN_TAGS`, `INFORMATION_SCHEMA.SCHEMA_TAGS`, and `INFORMATION_SCHEMA.TABLE_TAGS`. The query returns the tags that apply to the table named `transactions_table` in the `credit_card` schema in the `de_book` catalog:

```
SELECT
  catalog_name,
  schema_name,
  table_name,
  tag_name,
  tag_value
FROM
  de_book.information_schema.table_tags
WHERE
  catalog_name = 'de_book'
  and schema_name = 'credit_card'
```

```
and table_name = 'transactions_table';
```

6. **Remove tags from a table**: You can delete the tags of your data and AI assets using the UNSET TAG clauses. For example, you can modify the description and tags of your transactions_table table using the following SQL command. The code modifies the table named transactions_table in the credit_card schema in the de_book catalog. It unsets two tags from the table: business_unit and data_sensitivity. These tags will no longer be associated with the table in Unity Catalog:

```
ALTER TABLE
  de_book.credit_card.transactions_table UNSET TAGS ('business_
unit', 'data_sensitivity');
```

7. **Remove tags from columns**: You can delete the tags of a specific column of your table using the UNSET TAG clauses. For example, you can modify the description and tags of your email column using the following SQL command. The code modifies the tags of the column named Transaction_ID in the transactions_table table in the credit_card schema in the de_book catalog. It unsets one tag from the column: data_type. This tag will no longer be associated with the column in Unity Catalog:

```
ALTER TABLE
  de_book.credit_card.transactions_table
ALTER COLUMN
  Transaction_ID UNSET TAGS ('data_type');
```

Using the Unity Catalog UI

You can also use the Unity Catalog UI to tag and document your data and AI assets using a graphical interface. To access the Unity Catalog UI, you need to do the following:

1. Go to your Databricks workspace and click on **Catalog** on the left sidebar. You will see a list of catalogs, schemas, tables, views, and volumes in your workspace. You can browse, search, and filter them using the UI.

2. To tag and document your data and AI assets using the Unity Catalog UI, you need to select the data or AI asset that you want to tag or document from the list. You will see a detailed page of the selected asset, with information such as name, type, location, schema, statistics, permissions, lineage, and more:

Figure 10.27 – Selected table details

3.   Databricks AI-generated comments will suggest a descriptive comment for the table. Click on the **Edit** button to edit this suggestion:

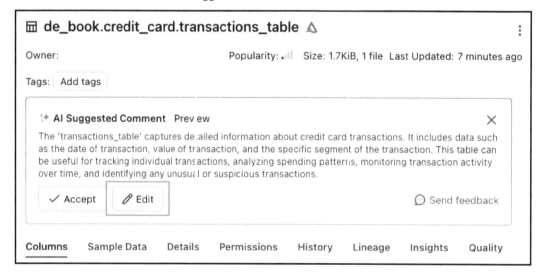

Figure 10.28 – AI-generated table comment

4.   After editing the comment, click on **Save**:

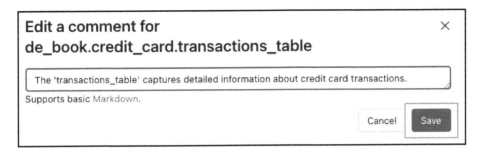

Figure 10.29 – Edit table comments

5. To enter tags, you can click on the **Add tags** button:

Figure 10.30 – Add tags to tables

6. Add the tags in the pop-up window. You can also use the + button to add additional tags. Click on the **Save tags** button to apply the changes:

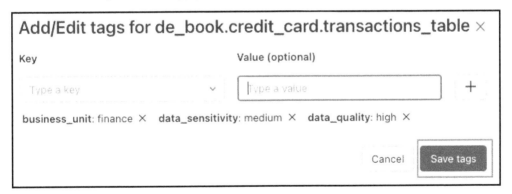

Figure 10.31 – Save table tags

7. You can also tag and comment on specific columns in your table, and Databricks AI-generated comments will suggest a descriptive comment for all the columns. Click on the **AI generate** button to generate suggestions:

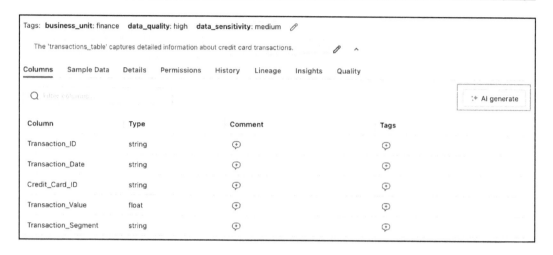

Figure 10.32 – AI-generated column comments

You can accept generated comments for each column:

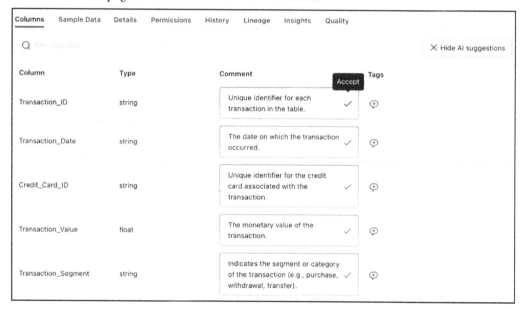

Figure 10.33 – Accept column comments

You can also add tags to the columns by clicking on the button under tags:

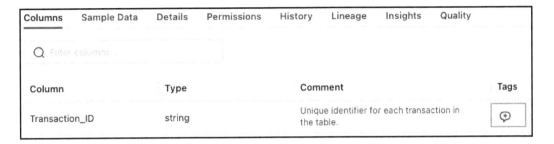

Figure 10.34 – Add tags to columns

8. Click on the **Schema** tab at the bottom of the page. You will see a list of columns in the selected table or view, with information such as name, type, comment, and tags.

9. Add the tags in the pop-up window. You can also use the + button to add additional tags. Click on the **Save tags** button to apply the changes:

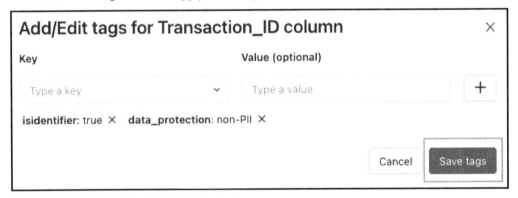

Figure 10.35 – Save tags for a column

See also

- What is Unity Catalog?: https://learn.microsoft.com/en-us/azure/databricks/data-governance/unity-catalog/

- Apply tags: https://docs.databricks.com/en/data-governance/unity-catalog/tags.html

- Announcing Public Preview of AI-Generated Documentation In Databricks Unity Catalog: https://www.databricks.com/blog/announcing-public-preview-ai-generated-documentation-databricks-unity-catalog

Filtering sensitive data with Unity Catalog

One of the key features of Unity Catalog is the ability to filter sensitive data using row filters and column masks. Row filters allow you to apply a filter to a table so that any subsequent queries only return rows, for which the filter predicate evaluates to true. Column masks let you apply masking functions to table columns, such as replacing, hashing, or redacting the original values. Using row filters and column masks enables you to apply fine-grained access controls at the level of individual rows and columns and protect sensitive data from unauthorized access.

In this recipe, you will learn how to use row filters and column masks to filter sensitive data in a table using SQL commands or the Unity Catalog UI. You will also learn how to use dynamic views to create a filtered view of a table.

Getting ready

Before you start, you need to have the following prerequisites:

- An Azure Databricks workspace with Unity Catalog enabled

- A storage credential that provides access to your cloud storage account

- An external location that references your storage credential and a cloud storage path

- A table that contains some sensitive data, such as customer information, stored in your external location

How to do it...

1. **Create a sample table**: You can use the following SQL commands to create a sample table in the de_book catalog and `credit_card` schema:

    ```
    -- Create a sample table with customer information
    USE CATALOG de_book;
    USE SCHEMA credit_card;

    CREATE TABLE customer (
      id INT,
      name STRING,
      email STRING,
      phone STRING,
      ssn STRING,
      country STRING
    );

    -- Insert some sample data into the table
    INSERT INTO customer VALUES
    ```

```
(1, 'Alice', 'alice@example.com', '+1-111-1111', '111--111-
1111','USA'),
(2, 'Bob', 'bob@example.com', '+1-222-2222', '222-222-
2222','USA'),
(3, 'Charlie', 'charlie@example.com', '+1-333-3333', '333-333-
3333','USA'),
(4, 'David', 'david@example.com', '+44-444-4444','444-444-4444',
'UK'),
(5, 'Eve', 'eve@example.com', '+44-555-5555', '+555-555-
5555','UK');
```

Apply row filters

Row filters are functions that can be applied to tables to filter the rows based on some conditions. They can be useful for restricting access to certain data or optimizing queries. In this recipe, you will learn how to create, apply, modify, and delete row filters in SQL:

1. **Create the row filter function**: To create a row filter function, you need to use the CREATE FUNCTION statement with the following syntax. The function name and parameters are user-defined, and the filter clause is a Boolean expression that determines which rows to keep or discard. For example, the following function filters the rows based on the region column:

   ```
   CREATE FUNCTION <function_name> (<parameter_name> <parameter_
   type>, ...)
   RETURN {filter clause whose output must be a boolean};
   ```

   This function returns true for all rows if the user is a member of the admin group; otherwise, it only returns rows where the region is USA:

   ```
   CREATE FUNCTION country_filter(country STRING)
   RETURN IF(IS_ACCOUNT_GROUP_MEMBER('admin'), true,
   country='USA');
   ```

2. **Apply the row filter to a table**: To apply the row filter function to a table, you need to use the ALTER TABLE statement with the following syntax. The table name and column names are user-defined, and the function name is the one you created in the previous step. The column names are the parameters that the function takes as input:

   ```
   ALTER TABLE <table_name> SET ROW FILTER <function_name> ON
   (<column_name>, ...);
   ```

   For example, the following statement applies the country_filter function to the customer table on the region column. This means that any query from the sales table will only return rows that satisfy the country_filter function:

   ```
   ALTER TABLE customer SET ROW FILTER country_filter ON (country);
   ```

   Once the row-level filter is applied, non-admin group members will only see records for USA customers, as shown here:

Figure 10.36 – Table sample data with row filter applied

Users who are part of the admin group will be able to see all the records in the table live, as shown here:

Figure 10.37 – Admin view with row filter applied

3. **Modify the row filter**: To modify the row filter function, you have two options:

    - Run a DROP FUNCTION statement to drop the existing function and then create a new one with the same or different name and logic

    - Use CREATE OR REPLACE FUNCTION to replace the existing function with a new one with the same name but different logic

    For example, the following statement replaces the country_filter function with a new definition. This function returns true for all rows if the user is a member of the admin group, only returns rows for US customers if the user is a member of the US team, or only returns UK customers if the user is from the UK team:

    ```
    CREATE
    OR REPLACE FUNCTION country_filter(country STRING) RETURN IF(
      IS_ACCOUNT_GROUP_MEMBER('admin'),
      true,
      IF(
        IS_ACCOUNT_GROUP_MEMBER('usteam')
        AND country = 'USA',
        true,
        IF(
          IS_ACCOUNT_GROUP_MEMBER('ukteam')
          AND country = 'UK',
          true,
          false
        )
      )
    );
    ```

4. **Delete the row filter**: To delete the row filter function, you need to do two things:

    - Run an ALTER TABLE statement to drop the row filter from the table with the following syntax. The table name is the one that has the row filter applied. For example, the following statement drops the row filter from the sales table:

    ```
    ALTER TABLE <table_name> DROP ROW FILTER;
    ```

    This means that any query from the sales table will return all the rows in the table:

    ```
    ALTER TABLE customer DROP ROW FILTER;
    ```

    - Run a DROP FUNCTION statement to drop the row filter function with the following syntax. The function name is the one you created in the first step. For example, the following statement drops the us_filter function:

    ```
    DROP FUNCTION <function_name>;
    ```

This means that the function is no longer available for use:

```
DROP FUNCTION country_filter;
```

> **Important note**
>
> You must perform the `ALTER TABLE <table_name> DROP ROW FILTER` command before dropping the function or the table will be in an inaccessible state. If the table becomes inaccessible in this way, alter the table and drop the orphaned row filter reference using `ALTER TABLE <table_name> DROP ROW FILTER;`.

Apply column masks

Column masks allow you to control the visibility of sensitive data in a table column. You can use column masks to apply a user-defined function to a column, which returns a masked value based on the identity or role of the user who queries the table:

1. **Create a column mask function**: A column mask function is a user-defined function that takes a column value as an input and returns a masked value as an output. This function can use any of the Databricks built-in runtime functions or call other user-defined functions. Common use cases include inspecting the identity of the user who invokes the running of the function using `current_user()` or the groups they are a member of using `is_account_group_member()`. To create a column mask function, use the following syntax:

    ```
    CREATE FUNCTION <function_name> (<parameter_name> <parameter_
    type>, ...)
    RETURN {expression with the same type as the first parameter};
    ```

    For example, the following function masks the email address column so that only users who are members of the `hr_dept` group can view the original values. Other users will see a masked value of @***:

    ```
    -- Create a UDF that masks the email column by replacing the
    domain part with '***'
    CREATE FUNCTION mask_email (email STRING)
      RETURN CASE WHEN is_account_group_member('hr_dept') THEN email
    ELSE CONCAT(SPLIT(email, '@')[0], '@***') END;
    ```

2. **Apply a column mask to a column in an existing table**: To apply a column mask to a column in an existing table, use the following syntax:

    ```
    ALTER TABLE <table_name> ALTER COLUMN <col_name> SET MASK <mask_
    func_name> [USING COLUMNS <additional_columns>];
    ```

    The `USING COLUMNS` clause is optional and allows you to specify additional columns that the mask function depends on. For example, if your mask function uses the name column to determine the masked value, you can include it in the `USING COLUMNS` clause.

For example, the following command applies the mask_email function to the email column in the customer table:

```
ALTER TABLE customer ALTER COLUMN email SET MASK mask_email;
```

Once applied, users who are not in the hr_dept will see the masked value of the column, as shown here:

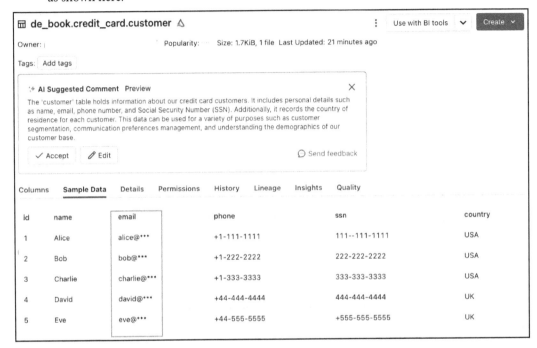

Figure 10.38 – Table sample data with column applied

Alternatively, you can apply the column mask when you create a table using the MASK clause in the CREATE TABLE syntax:

```
CREATE TABLE customer (
  id INT,
  name STRING,
  email STRING MASK mask_email,
  phone STRING,
  ssn STRING,
  country STRING
);
```

3. **Remove a column mask from a column**: To remove a column mask from a column, use the following syntax:

```
ALTER TABLE <table_name> ALTER COLUMN <column where mask is
applied> DROP MASK;
```

For example, the following command removes the column mask from the `email` column in the `customer` table:

```
ALTER TABLE customer ALTER COLUMN email DROP MASK;
```

4. **Modify a column mask function**: To modify a column mask function, you can either drop the existing function and create a new one or use the `CREATE OR REPLACE FUNCTION` syntax to overwrite the existing function. For example, the following command modifies the `mask_email` function to also check if the user is a member of the `finance_dept` group:

```
CREATE
OR REPLACE FUNCTION mask_email (email STRING) RETURN CASE
  WHEN is_account_group_member('hr_dept')
  OR is_account_group_member('finance_dept') THEN email
  ELSE CONCAT(SPLIT(email, '@') [0], '@***')
END;
```

5. **Delete a column mask function**: To delete a column mask function, use the following syntax:

```
DROP FUNCTION <function_name>;
```

However, before deleting a column mask function, you must remove the column mask from any column that uses it. Otherwise, the table will be in an inaccessible state:

```
ALTER TABLE customer ALTER COLUMN email DROP MASK;
DROP FUNCTION mask_email
```

See also

- Filter sensitive table data using row filters and column masks: `https://docs.databricks.com/en/data-governance/unity-catalog/row-and-column-filters.html`

- Unity Catalog best practices: `https://learn.microsoft.com/en-us/azure/databricks/data-governance/unity-catalog/best-practices`

Using Unity Catalogs lineage data for debugging, root cause analysis, and impact assessment

Databricks Unity Catalog is a unified governance solution for all data and AI assets on the Lakehouse platform. It enables users to capture and view data lineage across queries run on Databricks down to the column level. Data lineage describes the transformations and refinements of data from source to insight and includes the metadata and events associated with the data lifecycle.

Some of the benefits of data lineage with Unity Catalog are the following:

- **Impact analysis**: Users can see the downstream consumers of a dataset and understand the potential impact of any data changes

- **Data understanding and transparency**: Users can gain better context and trustworthiness of the data by seeing its source, history, and usage

- **Data provenance and governance**: Users can access lineage data through Catalog Explorer or REST API and apply access control and permissions to the data assets

In this recipe, you will learn how to use Unity Catalog, a unified governance solution for all data and AI assets in your lakehouse on any cloud. Unity Catalog automatically captures user-level audit logs and column-level lineage data that track how data and AI assets are created and used across languages. You will also learn how to query these data using system tables and how to use them for debugging, root cause analysis, and impact assessment.

Getting ready

To use data lineage with Unity Catalog, the following requirements must be met:

- Unity Catalog should be activated in the workspace

- Tables should be enrolled in a Unity Catalog metastore

- Queries should use either the Spark DataFrame or the Databricks SQL interfaces

- Users should have the appropriate rights and permissions to access the lineage data

- Permission to see the lineage and audit data of the tables and views that interest you

How to do it...

1. **Capture data lineage and audit logs**: Unity Catalog automatically captures data lineage and audit logs for every query that reads or writes data from a table or view registered in a Unity Catalog metastore. Lineage data include the source and destination tables and columns, as well as the notebooks, workflows, and dashboards related to the query. The audit logs include the user, timestamp, action, and query details for each access to a data or AI asset. To capture data lineage and audit logs, you just need to run your queries as usual. For example, you can run the following queries in a notebook to read data from a table called `customer` and write it to two tables called `usa_customers` and `uk_customers`:

```
CREATE OR REPLACE TABLE usa_customers AS
SELECT  * FROM customer
WHERE country = 'USA';

CREATE OR REPLACE TABLE uk_customers AS
```

```
SELECT * FROM customer
WHERE country = 'UK';
```

2. **Create a schedules workflow**: Click on **Schedule** in the top bar. In the dialog, select **Manual**, select a cluster with access to Unity Catalog, and then click on **Create**:

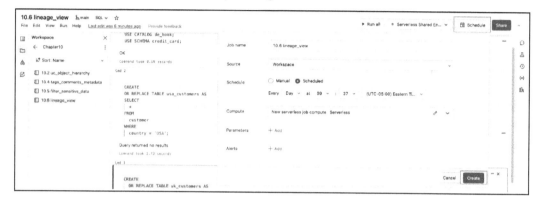

Figure 10.39 – Schedule notebook

Click on **Run now**:

Figure 10.40 – Scheduled notebook Run now button

3. **Use catalog explorer to view the lineage data**:

I.   In the **Search** box in the top bar of the Databricks workspace, enter de_book.
     credit_card.customer and hit *Enter*:

Figure 10.41 – Search table

II.   Under **Tables**, click on the **customer** table:

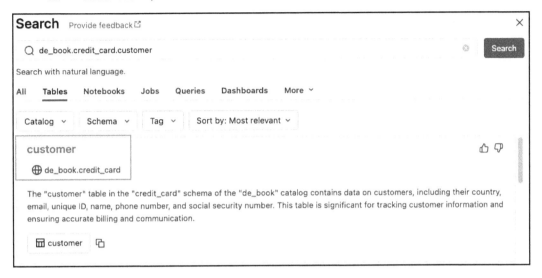

Figure 10.42 – Search results

III.   Select the **Lineage** tab. The lineage panel appears and displays related tables (for this example, it's the `usa_customers` and `uk_customers` tables):

Figure 10.43 – Table details Lineage tab

IV.   To view an interactive graph of the data lineage, click on **See lineage graph**:

Columns	Sample Data	Details	Permissions	History	Lineage	Insights	Quality

Q Filter lineage   All connections   ⌄   ↻   🔗 See lineage graph

▦ Tables

Table name		Lineage direction
de_book.credit_card.uk_customers		Downstream

📓 Notebooks

de_book.credit_card.usa_customers		Downstream

≔ Workflows   Lineage data is captured from the last 90 days

🗘 Pipelines

⊞ Dashboards

Figure 10.44 – The See lineage graph button

V.   The graph shows the first level by default. To see more connections, if any, click on the + icon on a node:

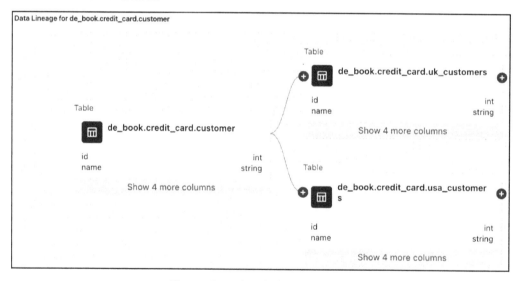

Figure 10.45 – Level 1 lineage graph

VI.   To open the **Lineage Connection** panel, click on the arrow that links nodes in the lineage graph. The connection details, such as source and target, are shown in the **Lineage Connection** panel:

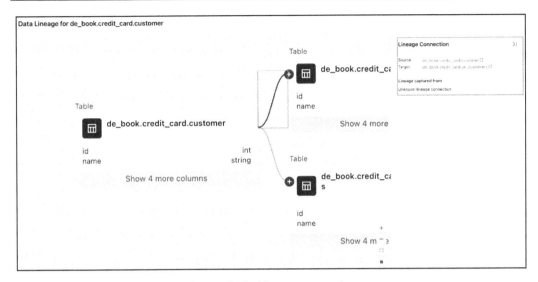

Figure 10.46 – Lineage connection

VII. Click on tables in the lineage graph and then click on **Lineage** on the right panel to see more details on the table's lineage:

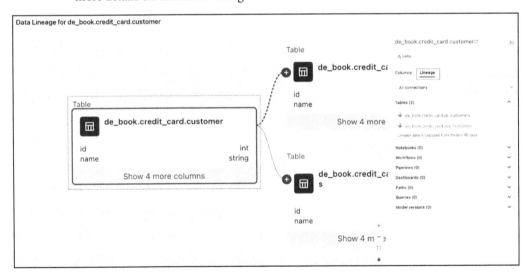

Figure 10.47 – Table lineage view

VIII. To view the column-level lineage, click on a column—in our case, the **ssn** column—in the graph. This will show the links to related columns:

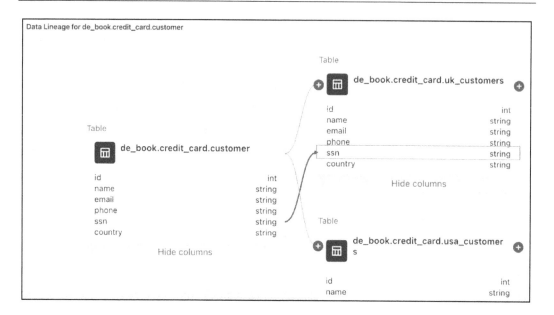

Figure 10.48 – Column lineage

4. **View the workflow lineage**: Select the **Lineage** tab and click on **Workflows**:

Figure 10.49 – Workflow lineage

5. **Understand the lineage permissions**: Lineage graphs follow the same permissions as Unity Catalog. Users need the SELECT privilege on a table to see its lineage. They can also only see notebooks, workflows, and dashboards they can access. For instance, if you run these commands for a non-admin usa_customer_care_team group. A user in the group can see the usa_customer table in the lineage graph for the de_book.credit_card.usa_customer table but not the related tables, such as the upstream de_book.credit_card.customer table. The customer table is a masked node for usa_customer_care_team users, and they cannot see the downstream tables they cannot access:

```
GRANT USE SCHEMA on de_book.credit_card to usa_customer_care_
team;
GRANT SELECT on de_book.credit_card.usa_customer to usa_
customer_care_team;
```

See also

- Capture and view data lineage with Unity Catalog – Databricks: https://docs.databricks.com/en/data-governance/unity-catalog/data-lineage.html

- Announcing the Availability of Data Lineage With Unity Catalog – Databricks: https://www.databricks.com/blog/2022/06/08/announcing-the-availability-of-data-lineage-with-unity-catalog.html

- Announcing General Availability of Data lineage in Unity Catalog: https://www.databricks.com/blog/announcing-general-availability-data-lineage-unity-catalog

Accessing and querying system tables using Unity Catalog

System tables are an analytical store hosted by Databricks that contains your account's operational data. You can use system tables to track the history of your account. System tables enable you to gain operational insights for data and AI on the lakehouse by letting you observe, audit, and report on various aspects of your Databricks usage and billing:

- **Audit logs**: Includes records for all audit events across your Databricks account, such as user actions, cluster events, job runs, and notebook executions

- **Table and column lineage**: Includes records for each read or write event on a Unity Catalog table or path, as well as the source and destination columns involved in the event

- **Billable usage**: Includes records for all billable usage across your account, such as cluster hours, node types, and SKUs

- **Pricing**: Includes a historical log of SKU pricing, which can be used to calculate the cost of your usage

- **Clusters**: Includes the full history of cluster configurations over time for any cluster in your account

- **Node types**: Includes the currently available node types with their basic hardware information

- **Marketplace listing access**: Contains data about consumer info for marketplace listing

- **Predictive optimization**: Contains the history of the predictive optimization feature

Getting ready

To access and query system tables, you need to have the following prerequisites:

- A Databricks account with Unity Catalog enabled

- An account admin role to enable system tables using the Unity Catalog REST API

- A Unity Catalog-enabled workspace to access system tables

- Appropriate permissions to access system tables

How to do it...

1. **Enable system tables using the Unity Catalog REST API**: You can use the following `curl` command to list the available system schemas:

```
curl -v -X GET -H "Authorization: Bearer <PAT Token>"
"https://<workspace>.cloud.databricks.com/api/2.0/unity-catalog/
metastores/<metastore-id>/systemschemas"
```

The response will include a list of system schemas, such as `system.access`, `system.billing`, `system.compute`, and `system.marketplace`. You can enable a specific system schema by using the following `curl` command:

```
curl -v -X POST -H "Authorization: Bearer <PAT Token>"
"https://<workspace>.cloud.databricks.com/api/2.0/unity-catalog/
metastores/<metastore-id>/systemschemas/<schema-name>/enable"
```

For example, to enable the `system.access` schema, which contains the audit logs and lineage tables, you can use the following:

```
curl -v -X POST -H "Authorization: Bearer <PAT Token>"
"https://<workspace>.cloud.databricks.com/api/2.0/unity-catalog/
metastores/<metastore-id>/systemschemas/system.access/enable"
```

2. **Grant access to system tables using the Unity Catalog SQL interface**: You can use the GRANT statement to grant permissions to users or groups to system tables. For example, to grant SELECT permission to the `system.access.audit` table for the analysts group, you can use the following:

```
GRANT SELECT ON TABLE system.access.audit TO GROUP analysts
```

You can also use the REVOKE statement to revoke permissions from users or groups to system tables. For example, to revoke the SELECT permission to the system.billing.usage table from the developers group, you can use the following:

```
REVOKE SELECT ON TABLE system.billing.usage FROM GROUP
developers
```

3. **Query system tables using SQL or Spark APIs**: You can use any SQL or Spark API to query system tables, such as SELECT, JOIN, GROUP BY, ORDER BY, LIMIT, COUNT, SUM, AVG, MIN, and MAX. For example, to query the top 10 users by the number of audit events in the last 30 days, you can use the following:

```
SELECT
  user_identity.email as user_id,
  COUNT(*) AS event_count
FROM system.access.audit
WHERE event_time >= current_date - interval 30 days
GROUP BY user_id
ORDER BY event_count DESC
LIMIT 10
```

- To query the total billable usage and cost by SKU in the last month, you can use the following:

```
SELECT
  b.sku_name,
  SUM(b.usage_quantity) AS usage_hours,
  SUM(b.usage_quantity * p.pricing.default) AS cost
FROM
  system.billing.usage AS b
  JOIN system.billing.list_prices AS p ON b.sku_name = p.sku_
name
  AND b.usage_date BETWEEN p.price_start_time
  AND coalesce(p.price_end_time, current_timestamp())
WHERE
  b.usage_start_time >= date_trunc('month', current_date) -
interval 1 month
  AND b.usage_start_time < date_trunc('month', current_date)
GROUP BY b.sku_name
ORDER BY cost DESC
```

- To query the lineage of a table, you can use the following:

```
SELECT
  source_table_full_name, target_table_full_name, event_time,
created_by, entity_type, entity_id, entity_run_id
FROM system.access.table_lineage
```

```
WHERE target_table_full_name = 'de_book.credit_card.usa_
customers'
ORDER BY event_time DESC
```

See also

- Monitor usage with a system table: `https://learn.microsoft.com/en-us/azure/databricks/data-governance/unity-catalog/`

- Billable usage system table reference: `https://docs.databricks.com/en/administration-guide/system-tables/billing.html`

- Pricing system table reference: `https://docs.databricks.com/en/administration-guide/system-tables/pricing.html`

11

Implementing DataOps and DevOps on Databricks

DataOps and DevOps are two methodologies that aim to improve the quality, speed, and efficiency of data and software development processes. DataOps focuses on the end-to-end orchestration of data pipelines, from data ingestion to analysis and visualization. DevOps focuses on the continuous integration and delivery of software applications, from code development to deployment and monitoring.

Databricks supports both DataOps and DevOps practices by offering various features and tools that enable users to collaborate, automate, and optimize their data and code workflows.

In the chapter, you will learn how to implement DataOps and DevOps on Databricks. We will cover the following recipes:

- Using Databricks Repos to store code in Git
- Automating tasks by using the Databricks **command-line interface** (**CLI**)
- Using the Databricks VSCode extension for local development and testing
- Using **Databricks Asset Bundles** (**DABs**)
- Leveraging GitHub Actions with **Databricks Asset Bundles** (DABs)

By the end of this chapter, you will have a solid understanding of how to leverage the various features and tools that Databricks provides to streamline and improve your data and code workflows.

Technical requirements

To follow along with the examples in this chapter, you will need a Databricks workspace with the Premium plan.

You can find the notebooks and data for this chapter at `https://github.com/PacktPublishing/Data-Engineering-with-Databricks-Cookbook/tree/main/Chapter11`.

Using Databricks Repos to store code in Git

Git is a distributed **version control system** (**VCS**) that allows you to track and manage your code changes across multiple collaborators and branches. Databricks Repos is a feature that integrates Git with Databricks, enabling you to create, clone, and modify Git-backed notebooks and projects within your Databricks workspace. By using Databricks Repos, you can leverage the benefits of Git, such as version control, collaboration, and code reuse, while working with Databricks notebooks and libraries.

In this recipe, you will learn how to use the Databricks Repos UI or the Repos API to link your Git repository to a Databricks workspace and how to sync your code changes between the two.

Getting ready

Before you start using Databricks Repos, you need to have the following prerequisites:

- A Databricks workspace with the Repos feature enabled. You can check if you have access to Repos by clicking the Repos icon on the left sidebar of the Databricks UI.

- A Git account and a remote repository hosted by one of the supported Git providers, such as GitHub, GitLab, or Bitbucket. You can use either a cloud-based or an on-premises Git service, but you need to make sure that your Databricks workspace can access the Git provider's URL. We will be using GitHub.

- A **personal access token** (**PAT**) or SSH key to authenticate with your Git provider. You can generate a token or a key from your Git provider's settings page and copy it to your clipboard. Create a GitHub PAT with the repo scope. This will allow you to access your GitHub repositories from Databricks. Instructions on how to create a token on GitHub are provided here: `https://docs.github.com/en/authentication/keeping-your-account-and-data-secure/managing-your-personal-access-tokens`.

- Configure Git credentials and connect a remote repo to Databricks. Setting up your Git credentials in Databricks allows you to link a remote repo to Databricks Repos. You can follow the instructions here: `https://docs.databricks.com/en/repos/get-access-tokens-from-git-provider.html`.

How to do it...

1. **Fork the book repo**: From the GitHub repo for this book (`https://github.com/PacktPublishing/Data-Engineering-with-Databricks-Cookbook`), fork it for access:

Figure 11.1 – Creating a fork of the book's repo

2. **Clone a remote repository to Databricks**: You need to do the following:

    I.   In the Databricks UI, click on **New** and then click on **Repo** on the left sidebar:

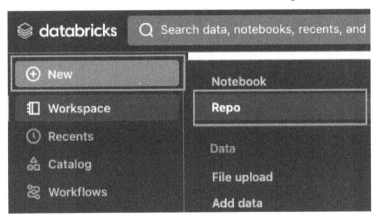

Figure 11.2 – Creating a new repo

    II.   In the **Add Repo** dialog, enter the URL of your remote repository in the **Git repository URL** field. You can copy the URL from your Git provider's web page. We will use this forked GitHub repo, the URL for which will be specific to your account. You can copy it from the GitHub repo page:

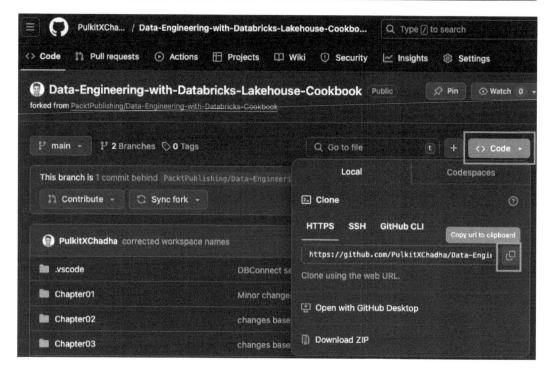

Figure 11.3 – Copying forked repo URL

Optionally, you can change the name of the repo folder that will be created in your Databricks workspace. By default, it will be the same as the name of your remote repository.

III.   Click on the **Create Repo** button to clone the remote repository to Databricks:

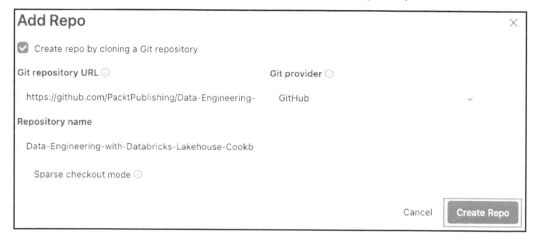

Figure 11.4 – The Add Repo dialog

3.   **Add files to the repo folder**: To create and edit notebooks or other files in the repo folder, you need to do the following:

    I.   In the Databricks UI, navigate to the repo folder that you created in the previous step. You will see the files that are in your remote repository.

    II.   To create a new folder, notebook, or file, click on the **Add** button and select the required option from **Notebook**, **File**, or **Folder**:

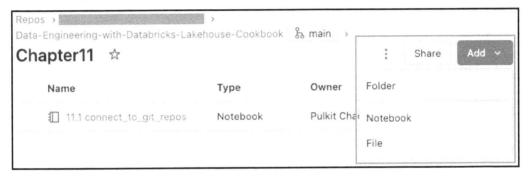

Figure 11.5 – Creating and editing a notebook

    III.   To edit an existing notebook or file, double-click on it to open it in a new tab. You can make changes to the code or text as you wish. You can also run the code cells in the notebook or file.

4.   **Commit and push your changes to the remote repository**: You need to do the following:

    I.   In the Databricks UI, navigate to the repo folder that contains your changes:

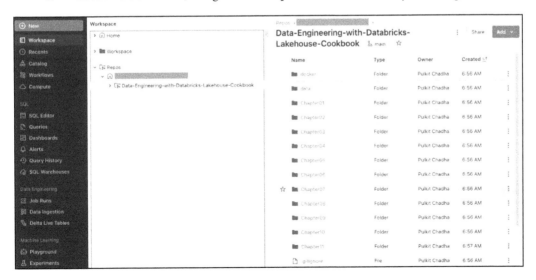

Figure 11.6 – Cloned repo

II.   To commit your changes, click on the branch icon next to the name of the repo:

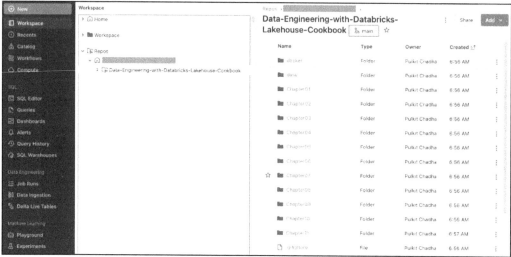

Figure 11.7 – Repo branch icon

III.   In the dialog box, select the files that you want to include in the commit. Enter a commit message that describes your changes and click on the **Commit & Push** button:

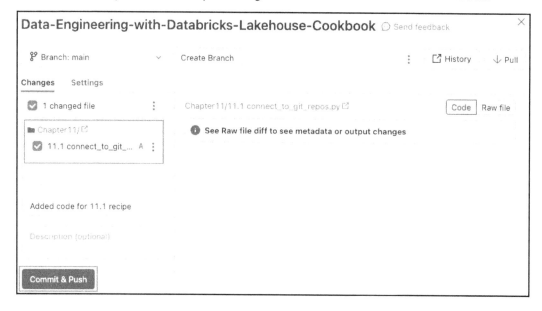

Figure 11.8 – Committing and pushing changes

5. **Pull and merge changes from the remote repository**: You need to do the following:

    I.   In the Databricks UI, navigate to the repo folder that contains your changes. Then, click on the **branch** icon next to the name of the repo.

    II. To pull the changes from the remote repository, click on the **Pull** button. You will see a confirmation message and a link to the remote repository:

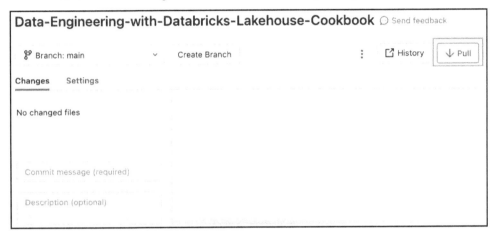

Figure 11.9 – Pulling changes

    III. If there are any conflicts between your local and remote changes, you will see a **Resolve conflicts** button. Click on that button to open the **Conflict resolution** dialog, where you can compare the differences and choose which version to keep for each file. Click on the **Resolve** button to apply your choices and complete the merge.

6. **Create and switch branches for parallel development**: You need to do the following:

    I.   In the Databricks UI, navigate to the repo folder that contains your changes. Then, click on the branch icon next to the name of the repo.

    II. To create a new branch, click on the **Branch** button and then click on the **Create Branch** button:

Figure 11.10 – Creating a branch

III.   Enter a name for the new branch and then click on the **Create** button. You will see a confirmation message and a link to the new branch:

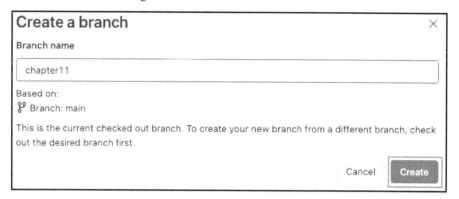

Figure 11.11 – Created branch details

IV.   To switch to an existing branch, click on the **Branch** button and then select the branch name from the drop-down list:

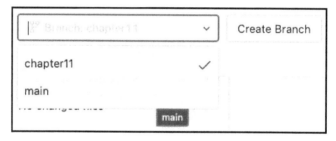

Figure 11.12 – Changing branch

There's more...

Databricks Repos has a few other sets of capabilities that help you work with Git; for example:

- You can use the **History** tab in the Databricks Repos UI to view the commit history of your repo folder. You can also revert to a previous commit or compare two commits in the **History** tab.

- You can use the **Settings** tab in the Databricks Repos UI to manage your repo folder settings, such as changing the name, updating the authentication method, or deleting the repo folder.

- You can use the **Search** feature in the Databricks UI to search for notebooks or other files in your repo folder. You can also filter the search results by repo name, branch name, or file type.

- You can use the **Export** and **Import** features in the Databricks UI to export or import notebooks or other files from or to your repo folder. You can choose the format (DBC or IPYNB) and the destination (local or DBFS) of exported or imported files.

While working with Git providers and Databricks, you may face some common issues, such as the following:

- **Authentication errors**: If you see an **Authentication failed** or **Invalid credentials** error message, it means that your PAT is expired, revoked, or incorrect. You need to generate a new token or key from your Git provider and update it in the Databricks Repos settings.

- **Network errors**: If you see a **Network error** or **Connection timed out** error message, it means that your Databricks workspace cannot access your Git provider's URL. You need to check your network settings and firewall rules and make sure that your Git provider's URL is whitelisted.

- **Merge conflicts**: If you see a **Merge conflict** or **Conflict resolution required** error message, it means that there are conflicting changes between your local and remote branches. You need to resolve the conflicts in the Databricks Repos UI or API before you can push or pull your changes.

- **Invalid Git URL**: You may get an error message that says **Invalid Git URL** when you try to create a repo using the Databricks Repos UI or the Repos API. This means that the URL of your Git repository is not valid or not supported by Databricks. Make sure that you enter the correct URL of your repository, which should start with `https://` and end with `.git`. Also, make sure that your Git platform is one of the supported ones, such as GitHub, GitLab, or Azure DevOps.

- You may get an error message that says **Authentication failed** when you try to create a repo using the Databricks Repos UI or the Repos API. This means that your Git credentials or your access token are not valid or authorized to access your Git repository or your Databricks workspace. Make sure that you enter the correct credentials or token and that they have the required permissions to create and manage repos. For example, if you are using a PAT for GitHub, make sure that it has the **repo** scope enabled. If you are using an access token for Databricks, make sure that it has the **Allow access to Repos** option enabled.

- You may get an error message that says **Repo already exists** when you try to create a repo using the Databricks Repos UI or the Repos API. This means that a repo with the same name or the same Git URL already exists in your Databricks workspace. Make sure that you enter a unique name and URL for your repo, or delete the existing repo before creating a new one.

See also

- *Git integration with Databricks Git folders*: `https://docs.databricks.com/en/repos/index.html`

- *Run Git operations on Databricks Git folders (Repos)*: `https://docs.databricks.com/en/repos/git-operations-with-repos.html`

- *New Support for Conflict Resolution in Repos: Merge, Rebase, and Pull*: `https://www.databricks.com/blog/new-support-conflict-resolution-repos-merge-rebase-and-pull`

Automating tasks by using the Databricks CLI

The Databricks CLI can be used to automate the Databricks platform from your terminal, Command Prompt, or automation scripts. You can use the Databricks CLI to perform various developer tasks, such as creating and managing clusters, jobs, workspaces, libraries, secrets, and notebooks. You can also use the Databricks CLI to perform common operations, such as uploading and downloading files, running commands, and executing notebooks. The Databricks CLI is based on the Databricks REST API, which provides a RESTful interface to interact with the Databricks platform.

In this recipe, you will set up the Databricks CLI on your local machine and learn how to use the CLI to create and run Databricks clusters and jobs.

Getting ready

Before you start using the Databricks CLI, you need the following:

- A Databricks workspace
- You also need to generate a PAT from your Databricks user account, which will be used for authentication
- Depending on your operating system, you may need to install some additional tools, such as Python, `pip`, `curl`, or `zip`

How to do it...

1. **Install the Databricks CLI**: There are different ways to install the Databricks CLI, depending on your operating system and preferences. Here are some examples:

    I. For Linux or macOS, you can install the Databricks CLI using Homebrew by running the following two commands:

    ```
    brew tap databricks/tap
    brew install databricks
    ```

    II. For Windows, you can install the Databricks CLI using **Windows Subsystem for Linux (WSL)** by running the following command:

    ```
    curl -fsSL <https://raw.githubusercontent.com/databricks/setup-cli/main/install.sh> | sh
    ```

2. **Configure the Databricks CLI**: After you install the Databricks CLI, you need to configure it to connect to your Databricks workspace. To do this, you need to provide your workspace URL and your PAT. You can generate a PAT from your Databricks user settings. For instructions on how to create a token, please reference `https://docs.databricks.com/en/dev-tools/auth/pat.html`. To configure the Databricks CLI, run the following command:

```
databricks configure
```

You can also specify a profile name to create multiple configurations for different workspaces or accounts. For example: the following command will create a configuration profile named `dev` that you can use later by adding `--profile dev` to any Databricks CLI command:

```
databricks configure --profile dev
```

This will prompt you to enter your Databricks host and PAT. Here's an example:

```
Databricks Host (should begin with https://): <https://example.
cloud.databricks.com>
Personal Access Token: dapi1234567890abcdef
```

3. **Use the Databricks CLI**: Once you have configured the Databricks CLI, you can use it to perform various Databricks tasks from the command line. The Databricks CLI has several command groups, each corresponding to a different Databricks resource or feature. For example, the `clusters` command group allows you to create and manage clusters, the `jobs` command group allows you to create and run jobs, and the `workspace` command group allows you to import and export notebooks.

- To see a list of available command groups, run the following command:

```
databricks -h
```

- To see a list of commands and options for a specific command group, run the following command with the command group name:

```
databricks <command group> -h
```

- For example, to see the list of commands and options for the `clusters` command group, run the following command:

```
databricks clusters -h
```

- To run a command, you need to specify the command group, the command name, and any required or optional arguments or flags. For example, to create a cluster, you can run the following command:

```
databricks clusters create --profile dev --json '{"cluster_
name": "pulkits_cluster", "spark_version": "13.3.x-scala2.12",
"node_type_id": "i3.xlarge", "num_workers": 2}'
```

- This command will create a cluster named `pulkits_cluster` with Spark version `13.3.x`, node type `i3.xlarge`, and 2 workers. The `--json` flag allows you to pass a JSON string that defines the cluster configuration. You can also use a JSON file instead of a JSON string by using the `--json-file` flag.

- To list all clusters in your workspace, you can run the following command:

```
databricks clusters list --profile dev
```

- This command will display the cluster ID, name, status, and driver node type for each cluster.

- To delete a cluster, you can run the following command with the cluster ID:

```
databricks clusters delete --profile dev 0121-120445-dw3mapwk
```

- This command will delete the cluster with the specified ID.

- Similarly, you can use the `jobs` command group to create and run jobs and the `workspace` command group to import and export notebooks. For example, to create a job that runs a notebook, you can run the following command:

```
databricks jobs create --profile dev --json
'{"name":"chapter_11_cli_job", "tasks":[{"task_key":"10_6_
lineage_view", "run_if":"ALL_SUCCESS", "notebook_
task":{"notebook_path":"/Repos/pulkit.chadha.packt@gmail.
com/Data-Engineering-with-Databricks-Lakehouse-Cookbook/
Chapter10/10.6 lineage_view", "source":"WORKSPACE"}, "job_
cluster_key":"Job_cluster"}], "job_clusters":[{"job_cluster_
key":"Job_cluster", "new_cluster":{"spark_version":"13.3.x-
scala2.12", "node_type_id":"i3.xlarge", "num_workers":2}}]}'
```

- This command will create a job named `chapter_11_cli_job` that runs the notebook at `https://github.com/PacktPublishing/Data-Engineering-with-Databricks-Cookbook/blob/main/Chapter10/10.6%20lineage_view.sql` `lineage_view` on a new cluster with Spark version `13.3.x`, node type `i3.xlarge`, and 1 worker.

- The output will be the job ID:

```
{
    "job_id":601965866717294
}
```

- To run a job, you can run the following command with the job ID:

```
databricks jobs --profile dev run-now 601965866717294
```

- This command will run the job with the specified ID.

There's more...

While working with Databricks CLI, you may face some common issues, such as the following:

- **Authentication errors**: If you get an error message such as **Error: Invalid authentication**, it means that your Databricks host or PAT is incorrect or expired. You can check your configuration by running `databricks configure --token` and verifying your token by logging in to your Databricks workspace. You can also generate a new token from your user settings and reconfigure the CLI with the new token.

- **Connection errors**: If you get an error message such as **Error: Connection error**, it means that your network connection is unstable or blocked. You can check your network connection by running `ping <your Databricks host>` and checking whether you get a response. You can also check whether your firewall or proxy settings are preventing the CLI from accessing the Databricks API. You may need to adjust your settings or use a VPN to bypass the restrictions.

- **JSON errors**: If you get an error message such as **Error: JSON parse error**, it means that your JSON input is invalid or malformed. You can check your JSON syntax by using a JSON validator tool such as JSONLint. You can also use a JSON file instead of a JSON string by using the `-json-file` flag and specifying the file path.

- **Resource errors**: If you get an error message such as **Error: Resource does not exist**, it means that the resource you are trying to access or modify does not exist or has been deleted. You can check the existence and status of the resource by using the appropriate `list` or `get` command. For example, to check if a cluster exists, you can run `databricks clusters list` or `databricks clusters get --cluster-id <cluster-id>`.

- **Command errors**: If you get an error message such as **Error: Invalid command**, it means that your command syntax is not correct or your command arguments are not valid. You can try to check the online help for the command by running `databricks <command> -h`, or refer to the Databricks CLI documentation for more details.

See also

- *Databricks CLI tutorial*: `https://docs.databricks.com/en/dev-tools/cli/tutorial.html`

- *What is the Databricks CLI?*: `https://docs.databricks.com/en/dev-tools/cli/index.html`

Using the Databricks VSCode extension for local development and testing

With the Databricks VSCode extension, you can access your Databricks workspaces and clusters remotely from the Visual Studio Code IDE on your local machine. This lets you do the following:

- Keep your local code in Visual Studio Code in sync with your remote workspace code

- Execute local Python code files on Databricks clusters in your remote workspaces from Visual Studio Code

- Schedule local Python code files and notebooks as Databricks jobs in your remote workspaces from Visual Studio Code

- Take advantage of Databricks features such as remote execution, debugging, testing, and version control to enhance your IDE experience with Databricks

In this recipe, you will learn how to use the Databricks VSCode extension for local development and testing.

Getting ready

Before you start using the Databricks VSCode extension, you need to have the following requirements on your local development machine:

- Visual Studio Code version `1.69.1` or higher.

- Visual Studio Code configured for Python coding, including the availability of a Python interpreter.

- Python version `3.10.0` or higher for Databricks Connect.

- The Databricks extension for Visual Studio Code is installed. Follow the instructions here to install the VSCode extension: `https://docs.databricks.com/en/dev-tools/vscode-ext/install.html`.

- You also need to have at least one Databricks workspace available, and the workspace must meet the requirements detailed here: `https://docs.databricks.com/en/dev-tools/vscode-ext/authentication.html`.

How to do it...

1. **Authenticate the Databricks extension**: The first step is to authenticate the Databricks extension for Visual Studio Code to your Databricks workspace. In this example, we will use OAuth user-to-machine authentication, which is the simplest and most common method. To do this, here are the steps to follow:

    I.   In Visual Studio Code, open the Databricks extension by clicking on the Databricks icon on the sidebar. Then, click on the **Configure Databricks** button:

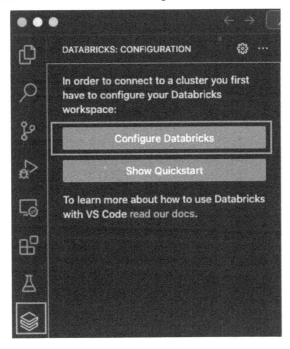

Figure 11.13 – Databricks extension configuration

    II. In the **Add Connection** dialog, enter the URL of your Databricks workspace, such as `https://adb-1234567890123456.12.azuredatabricks.net`, and press *Enter*:

Figure 11.14 – Databricks workspace URL

2. Next, select the **OAuth (user-to-machine)** authentication method:

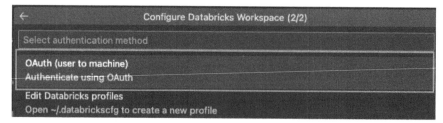

Figure 11.15 – OAuth authentication option

This will configure the Databricks extension to connect to the workspace, and you will see the configuration as shown:

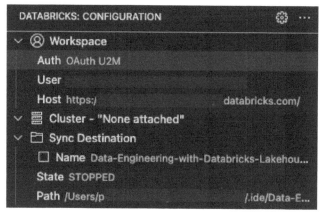

Figure 11.16 – Databricks configuration

3. **Select a cluster**: For the Databricks extension to be able to run local code on Databricks, we will need to select a cluster. Here are the steps to do this:

    I.   In the **Databricks Explorer** view, expand the connection that you created in the previous step, and then click on the **Configure cluster** button:

Figure 11.17 – Configuring a cluster

II.   In the dialog, select the cluster you want to connect to:

Figure 11.18 – Selecting a cluster from the dropdown

4.   **Synchronize your local code project folder with your Databricks workspace directory**: By configuring the sync, any changes you make to your local code files will be reflected in your Databricks workspace directory and vice versa. This allows you to keep your code consistent and up to date across your local and remote environments. Here are the steps to do this:

I.   In Visual Studio Code, click on the Databricks extension that you selected in the previous step, and then click on **Sync Destination**. Select **Config Sync Destination**:

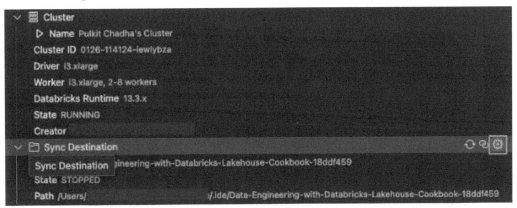

Figure 11.19 – Configuring sync to destination

II.   Pick a remote location:

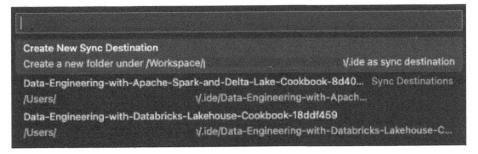

Figure 11.20 – Selecting sync location

III.   To sync changes between your Databricks workspace directory and your local code project folder, you can click on the **Start synchronization** button as shown. This will copy all the files and folders from your local code project folder to your Databricks workspace directory:

Figure 11.21 – Starting sync

5. **Run your local code files from Visual Studio Code on your Databricks cluster**: You can execute your local Python code files on the Databricks cluster that you selected in the previous step and see the results in the Databricks extension output view. This allows you to leverage the power and scalability of Databricks clusters for running your code without leaving your Visual Studio Code IDE. Here are the steps to do this:

I.   In Visual Studio Code, open the Python code file that you want to run. You can open the code associated with this recipe, `Chapter11/11.4 databricks_vscode_ extension.py`, then click on **Upload and Run File on Databricks**. This will run the Python code file on the Databricks cluster and show the output in the Databricks extension output view:

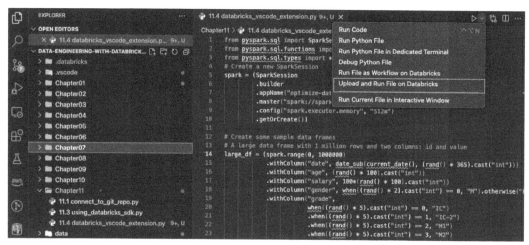

Figure 11.22 – Running file on Databricks

II.   Once the code has run on the Databricks cluster, you will see the output in the console as shown:

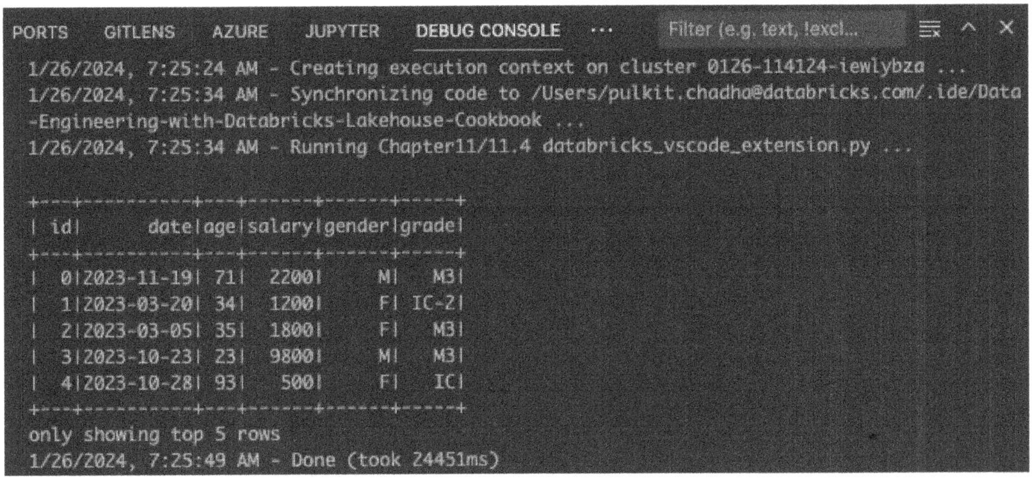

Figure 11.23 – Python file output

See also

- *Software engineering best practices for notebooks*: https://docs.databricks.com/en/notebooks/best-practices.html

- *Install the Databricks extension for Visual Studio Code*: https://docs.databricks.com/en/dev-tools/vscode-ext/install.html

- *What is the Databricks extension for Visual Studio Code?*: https://learn.microsoft.com/en-us/azure/databricks/dev-tools/vscode-ext/

- *Databricks extension for Visual Studio Code reference*: https://docs.databricks.com/en/dev-tools/vscode-ext/index.html

- *Use Azure Databricks in VS Code with the new Databricks extension*: https://techcommunity.microsoft.com/t5/analytics-on-azure-blog/use-azure-databricks-in-vs-code-with-the-new-databricks/ba-p/3741399

Using Databricks Asset Bundles (DABs)

DABs let you use an **infrastructure-as-code** (**IaC**) method to handle your Databricks projects. They enable you to define the infrastructure and resources of your project in a YAML configuration file. By automating your project's tests, deployments, and configuration management with bundles, you can minimize errors while encouraging software best practices throughout your organization with templated projects.

Here are some advantages of using DABs:

- You can handle complex projects that require multiple contributors and automation, as well as **continuous integration and deployment/delivery (CI/CD)**.

- You can work on data, analytics, and **machine learning (ML)** projects in a team-oriented environment. Bundles can assist you in organizing and managing various source files effectively. This guarantees smooth cooperation and simplified processes.

- You can solve ML problems faster. Use ML projects to manage ML pipeline resources (such as training and batch inference jobs) that adhere to production best practices from the start.

- You can establish organizational norms for new projects by creating custom bundle templates that contain default permissions, service principals, and CI/CD configurations.

There are also some challenges associated with DABs:

- You need to learn the YAML syntax and the bundle schema to define your project's metadata

- You need to install and configure the DAB CLI tool to interact with your bundles

- You need to integrate DAB CLI commands with your existing CI/CD tools and workflows

- You need to troubleshoot any errors or issues that may arise during bundle creation, testing, or deployment

In this recipe, you will learn how to use DAB CLI commands in different workflows and scenarios, such as pull requests, branches, and releases. You will also explore the benefits and challenges of using DABs.

Getting ready

Before you start using DABs, you need to have the following prerequisites:

- Access to a Databricks workspace.

- A Databricks CLI that's installed and configured with your workspace credentials. You can follow the instructions to set up the CLI here: `https://docs.databricks.com/en/dev-tools/cli/install.html`.

- Authenticate the Databricks CLI to the workspace you have access to. You can follow the instructions to do so here: `https://docs.databricks.com/en/dev-tools/cli/authentication.html`.

- A Git repository that contains your project source files and metadata. You can use any Git provider of your choice, such as GitHub, Bitbucket, or Azure DevOps.

- A CI/CD tool that can run commands on your Git repository, such as GitHub Actions, Jenkins, or Azure DevOps Pipelines.

- A local development environment with Python 3.6 or higher and `pip` installed.

How to do it...

1. **Create a DAB project**: The databricks bundle init command is used to create a new DAB project from a template. When you run the databricks bundle init command, you will be prompted to enter some information:

   - **Template to use**: We will choose default-python
   - **Unique name for the project**: We will use de_book_dabs_example
   - **Include a sample notebook**: Select **Yes**
   - **Include sample Delta Live Table**: Select **Yes**
   - **Include sample Python package**: Select **No**

   The command will then generate a project directory with the source files and metadata of your bundle project. You can then modify the project files as needed and use other DAB CLI commands to test and deploy your bundle project. The project folder structure is shown in the following screenshot:

   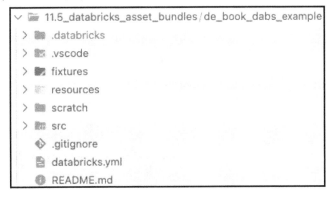

   Figure 11.24 – DAB project outline

2. **Modify the bundle configuration file**: The databricks.yml file is the main configuration file that defines your bundle project. It contains information such as the name, version, description, dependencies, tests, resources, and environments of your project.

   The code specifies the following information:

   - The name of the bundle is de_book_dabs_example.
   - The files to include in the bundle are all the YAML files in the resources folder.
   - The targets for deploying the bundle are dev and prod. Each target has a different mode, workspace, and run_as settings:

     - The dev target is for development purposes, and it uses the development mode, which prefixes the deployed resources with [dev my_user_name], pauses any job schedules and triggers, and uses the development mode for **Delta Live Tables** (**DLT**) pipelines. The dev target is also the default target, which means it will be used if no other target is specified. The dev target deploys the bundle to the workspace.

- The prod target is for production deployment, and it uses the production mode, which enables strict verification of the settings. The prod target deploys the bundle to the same workspace as the dev target but with a different root_path value.

The YAML definition for the bundle is as follows:

```
# This is a Databricks asset bundle definition for de_book_
dabs_example.
# See <https://docs.databricks.com/en/dev-tools/bundles/
index.html> for documentation.
bundle:
  name: de_book_dabs_example


include:
  - resources/*.yml


targets:
  # The 'dev' target, for development purposes. This target is
the default.
  dev:
    # We use 'mode: development' to indicate this is a
personal development copy:
    # - Deployed resources get prefixed with '[dev my_user_
name]'
    # - Any job schedules and triggers are paused by default
    # - The 'development' mode is used for Delta Live Tables
pipelines
    mode: development
    default: true
    workspace:
      host: <https://adb-7934447987890827.7.azuredatabricks.
net/>


  ## Optionally, there could be a 'staging' target here.
  ## (See Databricks docs on CI/CD at <https://docs.
databricks.com/dev-tools/bundles/index.html>.)
  #
  # staging:
  #   workspace:
  #     host: <https://adb-7934447987890827.7.azuredatabricks.
net/>


  # The 'prod' target, used for production deployment.
  prod:
```

```
    # We use 'mode: production' to indicate this is a
production deployment.
    # Doing so enables strict verification of the settings
below.
    mode: production
    workspace:
      host: <https://adb-7934447987890827.7.azuredatabricks.
net/>
      # We always use /Users/pulkit.chadha.packt@gmail.com for
all resources to make sure we only have a single copy.
      # If this path results in an error, please make sure you
have a recent version of the CLI installed.
      root_path: /Users/pulkit.chadha.packt@gmail.com/.
bundle/${bundle.name}/${bundle.target}
    run_as:
      # This runs as pulkit.chadha.packt@gmail.com in
production. We could also use a service principal here,
      # see <https://docs.databricks.com/dev-tools/bundles/
permissions.html>.
      user_name: pulkit.chadha.packt@gmail.com
```

- Within the resources folder, we have a YAML definition of the workflow and a delta live table.

  The de_book_dabs_example_job.yml YAML file defines the job for the DAB. The code specifies the following information:

  - The name of the job is de_book_dabs_example_job.

  - The schedule of the job is every day at 8:37 A.M. in the Europe/Amsterdam time zone.

  - Email notifications for the job are sent to pulkit.chadha@databricks.com on failure.

  The tasks of the job are a sequence of steps that the job executes. The tasks are the following:

  - notebook_task: This task runs a notebook located at ../src/notebook.ipynb using a cluster specified by job_cluster_key.

  - refresh_pipeline: This task depends on the completion of the notebook_task task and refreshes a pipeline with the ${resources.pipelines.de_book_dabs_example_pipeline.id} ID.

  - main_task: This task depends on the completion of the refresh_pipeline task and runs a Python wheel package named de_book_dabs_example with the main entry point. This task also uses the same cluster as notebook_task and includes the wheel file generated for the package as a library dependency.

The job clusters are the following:

- `job_cluster`: This cluster has a Spark version of `13.3.x-scala2.12`, a node type of `i3.xlarge`, and an `autoscale` setting that allows the cluster to scale from `1` to 4 workers.

- The `de_book_dabs_example_pipeline.yml` YAML files defined the pipeline for the DAB.

The code specifies the following information:

- The name of the pipeline is `de_book_dabs_example_pipeline`.

- The target of the pipeline is `de_book_dabs_example_${bundle.environment}`. This is a variable that depends on the environment of the bundle, such as `dev` or `prod`.

- The libraries of the pipeline are the dependencies that the pipeline needs to run. In this case, the pipeline uses a notebook located at `../src/dlt_pipeline.ipynb` as a library.

- The configuration of the pipeline is the setting that the pipeline uses during execution. In this case, the pipeline sets `bundle.sourcePath` to `/Workspace/${workspace.file_path}/src`, which is the path to the source code of the bundle in the Databricks workspace.

To learn more about the syntax and options of this file, reference the syntax here: `https://docs.databricks.com/en/dev-tools/bundles/settings.html`.

3. **Check the bundle configuration files**: Before you run any job or pipeline or deploy any artifacts, make sure your bundle configuration files do not have any syntax errors. To check this, execute the `bundle validate` command in the bundle root folder, which is where the bundle configuration file is located:

```
databricks bundle validate
```

This will be a JSON definition of all the assets that will be deployed from the bundle. If there is an error, the `validate` command will call it out.

4. **Deploy the bundle**: You can transfer local artifacts to the remote workspace by running the `bundle deploy` command in the folder that contains the bundle configuration file, also known as the bundle root folder. The Databricks CLI will use the default target from the bundle configuration files if you do not provide any command options:

```
databricks bundle deploy
```

To specify a different target, use the `-t` (or `--target`) option and the name of the target from the bundle configuration files. For instance, for a target named `dev`, you can run this command:

```
databricks bundle deploy -t dev
```

Here is the output of the `deploy` command:

```
● Pulkit.Chadha@              1 de_book_dabs_example % databricks bundle deploy -t dev
  Uploading bundle files to /Users/                        /.bundle/de_book_dabs_example/dev/files...
  Deploying resources...
  Updating deployment state...
  Deployment complete!
○ Pulkit.Chadha@              1 de_book_dabs_example % █
```

Figure 11.25 – Output of the deploy command

Within the workspace, the bundle has been deployed in the specified workspace directory:

Figure 11.26 – DAB deployed in the workspace

5. **Run the bundle**: To execute a particular job or pipeline, run the `bundle run` command in the bundle root folder. You need to indicate the job or pipeline that is defined in the bundle configuration files. If you don't use the `-t` option, the default target from the bundle configuration files is applied. For example, to run a job in our `de_book_dabs_example_job` bundle with the default target, the following command is used:

    ```
    databricks bundle run de_book_dabs_example_job
    ```

    To run a job called `hello_job` with a target that has the name `dev`, run the following command:

    ```
    databricks bundle run -t dev de_book_dabs_example_job
    ```

    Here is the output of the `run` command:

```
● Pulkit.Chadha@          de_book_dabs_example % databricks bundle run -t dev de_book_dabs_example_job
  Run URL: https://e2-demo-field-eng.cloud.databricks.com/?o=1444828305810485#job/308658145688130/run/699365181750201

  2024-02-05 07:54:46 "[dev pulkit_chadha] de_book_dabs_example_job" TERMINATED SUCCESS
○ Pulkit.Chadha@          de_book_dabs_example % ▯
```

Figure 11.27 – DAB job run output

Within the workspace, the bundle has been deployed in the specified workspace directory:

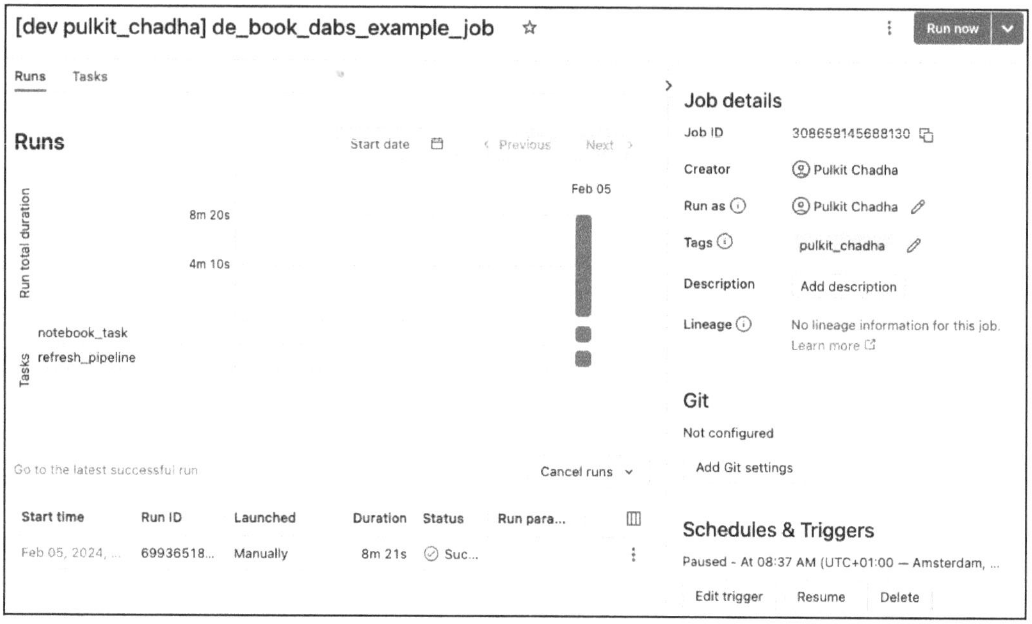

Figure 11.28 – DAB job run details

See also

- *What are Databricks Asset Bundles?*: https://docs.databricks.com/en/dev-tools/bundles/index.html

- *Databricks Asset Bundles*: https://www.databricks.com/resources/demos/tours/data-engineering/databricks-asset-bundles

Leveraging GitHub Actions with Databricks Asset Bundles (DABs)

CI/CD are software development practices that aim to automate the building, testing, and deployment of code on various platforms. CI/CD can help developers deliver high-quality software faster and more reliably, as well as reduce the risks and costs associated with manual and error-prone processes. Using CI/CD with Databricks can enable developers to leverage the power and flexibility of the platform while ensuring that their code is consistent, tested, and deployable across different environments and teams. CI/CD can also help developers integrate their Databricks workflows with other tools and services, such as source control, issue tracking, monitoring, and security. By using DABs, you can leverage popular CI/CD tools, such as GitHub Actions, Azure DevOps, Jenkins, and CircleCI, to automate the testing and deployment of your bundle projects. This way, you can reduce errors, promote software best practices, and iterate faster on your Databricks workflows.

In this recipe, we will show you how to use DABs and some of the CI/CD tools and frameworks to implement industry-standard CI/CD best practices for Databricks. We will provide examples of how to use these tools and frameworks to perform common operations, such as building, deploying, and monitoring Databricks applications and pipelines. We will also discuss best practices and considerations for using CI/CD tools and frameworks with Databricks, such as version control, code quality, security, and automation.

Getting ready

The prerequisites for this recipe are as follows:

- A Databricks workspace with a PAT that has sufficient permissions to create and manage bundles.

- A CI/CD tool of your choice that supports running custom commands or scripts. In this recipe, we will use GitHub Actions as an example, but you can adapt the steps to other tools as well.

- A Databricks CLI that's installed and configured with your workspace credentials. You can follow the instructions to set up the CLI here: `https://docs.databricks.com/en/dev-tools/cli/install.html`.

- Authenticate the Databricks CLI to the workspace you have access to. You can follow the instructions to do so here: `https://docs.databricks.com/en/dev-tools/cli/authentication.html`.

- GitHub CLI installed. Follow the instructions here: `https://github.com/cli/cli#installation`.

- A Git repository that contains your project source files and metadata. You can use any Git provider of your choice, such as GitHub, Bitbucket, or Azure DevOps.

How to do it...

1. **Create a GitHub repo**: You can follow the steps here to create a GitHub repo using the UI: `https://docs.github.com/en/repositories/creating-and-managing-repositories/quickstart-for-repositories`.

   We will name the GitHub repo `dabs_cicd_example`. If you have the GitHub CLI set up, you can run the following command on a terminal:

   ```
   gh repo create dabs_cicd_example --private
   ```

2. **Create a DAB**: Refer to *step 1* of the *Using Databricks Asset Bundles (DABs)* recipe.

3. **Push the DABs project to GitHub**: The following commands are used to create and upload a local repository to a remote server, such as GitHub:

- `git init` initializes a new **Git repository** in the current directory.

- `git add .` adds all the files and folders in the current directory to the **staging area**, which is a temporary area where changes are prepared before being committed.

- `git commit -m "Initial Commit"` creates a new commit with the `"Initial Commit"` message.

- `git remote add origin <https://github.com/PulkitXChadha/dabs_cicd_example.git>` adds a new remote named `origin` that points to the `https://github.com/PulkitXChadha/dabs_cicd_example.git` URL. A remote is a reference to another repository that can be used to synchronize changes.

- `git branch -M main` renames the current branch to `main`. A branch is a parallel version of a repository that can be used to work on different features or bug fixes.

- `git push -u origin main` uploads the `main` branch to the `origin` remote and sets it as the `upstream` branch. This means that future `git pull` and `git push` commands will use this branch as the default.

The following set of commands will push the changes to the `main` branch of the remote Git repo:

```
cd dabs_cicd_example
git init
git add .
git commit -m "Initial Commit"
git remote add origin <https://github.com/PulkitXChadha/dabs_cicd_example.git>
git branch -M main
git push -u origin main
```

4. **Create a development branch**: The following command creates and switches to a new branch called `dev`. A branch is a parallel version of a repository that allows you to work on different features or tasks without affecting the `main` branch. The `-b` option is a shortcut that tells Git to run `git branch dev` before running `git checkout dev`. The `git branch dev` command creates a new branch called `dev` based on the current branch. The `git checkout dev` command switches to the `dev` branch and updates the working directory accordingly:

```
git checkout -b dev
```

5. **Create a GitHub Actions workflow**: A GitHub Actions workflow is a YAML file that defines steps and triggers for your CI/CD pipeline. You can create a workflow file in your GitHub repository under the `.github/workflows` directory. The steps to do so are as follows:

I. Create a `.github/workflows` directory:

```
mkdir .github
cd .github
mkdir workflows
cd workflows
```

II. Create a workflow for QA deployment and name it `deploy_to_qa_CI.yml`. Please reference the file here: `https://github.com/PacktPublishing/Data-Engineering-with-Databricks-Cookbook/blob/main/Chapter11/11.5_databricks_asset_bundles/dabs_cicd_example/.github/workflows/deploy_to_qa_CI.yml`. The workflow automates the process of validating, deploying, and running a DAB on the QA Databricks workspace:

- The workflow has the name `QA deployment` and uses the concurrency setting to ensure that only one instance of the workflow runs at a time.

- The workflow is triggered by pull request events on the `main` branch of the repository, such as when a pull request is opened or updated.

- The workflow has two jobs: `Deploy bundle` and `Run Job`. These jobs run on Ubuntu virtual machines provided by GitHub.

- The `Deploy bundle` job performs the following steps:

  - It checks out the repository using the `actions/checkout` action.

  - It downloads the Databricks CLI using the `databricks/setup-cli` action.

  - It deploys the bundle to the qa target, which is a pre-production environment defined in the bundle's settings file. The deployment uses the `databricks bundle deploy` command and requires the Databricks token and bundle environment as environment variables. The deployment also validates the bundle by checking its syntax and dependencies. If the validation fails, the workflow fails.

  - The `Run Job` job depends on the `Deploy bundle` job and performs the following steps:

  - It checks out the repository using the `actions/checkout` action.

  - It uses the downloaded Databricks CLI using the `databricks/setup-cli` action.

  - It runs the Databricks workflow named `de_book_dabs_example_job` using the `databricks bundle run` command. This workflow is defined in the bundle that was just deployed. The command also refreshes all resources in the bundle, such as jobs and pipelines. The command requires the Databricks token and bundle environment as environment variables.

III. **Create a workflow for production deployment and name it deploy_to_prod_CD.yml**: The production deployment workflow is very similar to the qa deployment one, except that it is for **production deployment**. This means that the code will deploy and run the bundle on the **production target** named prod, which is the final environment where the project is intended to run. The code also has a different trigger condition: it will run whenever a pull request is **pushed** to the main branch of the repository, instead of when a pull request is opened or updated. This ensures that the production deployment only happens after the code has been reviewed and approved. The rest of the code is same as before, except for environment variables that specify the target and bundle environment. Please reference the file here: https://github.com/PacktPublishing/Data-Engineering-with-Databricks-Cookbook/blob/main/Chapter11/11.5_databricks_asset_bundles/dabs_cicd_example/.github/workflows/deploy_to_prod_CD.yml.

6. **Create a secret for the Databricks PAT**: Use the GitHub CLI tool to create or update a secret named DATABRICKS_TOKEN in the current repository. A secret is a piece of sensitive information, such as a password or an API key, that you do not want to expose in your code or logs. The value of the secret is provided by the user through a prompt or a file. The DATABRICKS_TOKEN secret is used to authenticate to the Databricks CLI, which requires a PAT. A PAT is a credential that allows a user or a service principal to access Databricks resources and operations.

   By using the gh secret set command, you can store the DATABRICKS_TOKEN value securely in GitHub and then use it in your GitHub Actions workflows or other GitHub integrations. This way, you can avoid hardcoding or exposing your Databricks PAT in your code or configuration files:

```
# Set a secret value for the current repository in an
interactive prompt
gh secret set DATABRICKS_TOKEN
```

7. **Push your bundle project to the development branch**: Once you have created your bundle project, you need to push it to a GitHub repository that you own or have access to. You can use any branch or tag that you want, but we recommend using a main branch for your stable code, and feature branches for your development work. For example, you can push your bundle project to GitHub like this:

```
git add .
git commit -m "Added Github Action Workflows"
git push -u origin dev
```

8. **Trigger GitHub actions for QA deployment**: Once you have created the GitHub Actions workflow file, you can trigger it by pushing or creating a pull request to the main branch by going to the GitHub repo and creating a pull request. If the workflow succeeds, you should see a green check mark next to the workflow name. If the workflow fails, you should see a red cross mark and an error message.

Figure 11.29 – QA GitHub action running

9. **Trigger GitHub actions for production deployment**: You can trigger the production deployment by merging the pull request into the `main` branch:

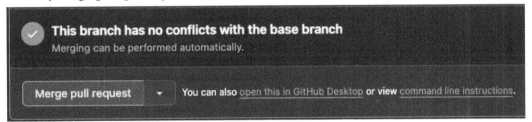

Figure 11.30 – Merging request into main

If the workflow succeeds, you should see a green check mark next to the workflow name. If the workflow fails, you should see a red cross mark and an error message.

Figure 11.31 – Production deployment GitHub action

See also

- *Run a CI/CD workflow with a Databricks Asset Bundle and GitHub Actions*: `https://docs.databricks.com/en/dev-tools/bundles/ci-cd.html`

- *CI/CD techniques with Git and Databricks Git folders (Repos)*: `https://docs.databricks.com/en/repos/ci-cd-techniques-with-repos.html`

Index

packtpub.com

Subscribe to our online digital library for full access to over 7,000 books and videos, as well as industry leading tools to help you plan your personal development and advance your career. For more information, please visit our website.

Why subscribe?

- Spend less time learning and more time coding with practical eBooks and Videos from over 4,000 industry professionals

- Improve your learning with Skill Plans built especially for you

- Get a free eBook or video every month

- Fully searchable for easy access to vital information

- Copy and paste, print, and bookmark content

Did you know that Packt offers eBook versions of every book published, with PDF and ePub files available? You can upgrade to the eBook version at packtpub.com and as a print book customer, you are entitled to a discount on the eBook copy. Get in touch with us at customercare@packtpub.com for more details.

At www.packtpub.com, you can also read a collection of free technical articles, sign up for a range of free newsletters, and receive exclusive discounts and offers on Packt books and eBooks.

Other Books You May Enjoy

If you enjoyed this book, you may be interested in these other books by Packt:

Business Intelligence with Databricks SQL

Vihag Gupta

ISBN: 978-1-80323-533-2

- Understand how Databricks SQL fits into the Databricks Lakehouse Platform
- Perform everyday analytics with Databricks SQL Workbench and business intelligence tools
- Organize and catalog your data assets
- Program the data security model to protect and govern your data
- Tune SQL warehouses (computing clusters) for optimal query experience
- Tune the Delta Lake storage format for maximum query performance
- Deliver extreme performance with the Photon query execution engine
- Implement advanced data ingestion patterns with Databricks SQL

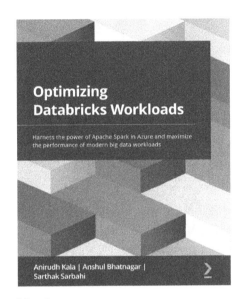

Optimizing Databricks Workloads

Anirudh Kala, Anshul Bhatnagar, Sarthak Sarbahi

ISBN: 978-1-80181-907-7

- Get to grips with Spark fundamentals and the Databricks platform
- Process big data using the Spark DataFrame API with Delta Lake
- Analyze data using graph processing in Databricks
- Use MLflow to manage machine learning life cycles in Databricks
- Find out how to choose the right cluster configuration for your workloads
- Explore file compaction and clustering methods to tune Delta tables
- Discover advanced optimization techniques to speed up Spark jobs

Packt is searching for authors like you

If you're interested in becoming an author for Packt, please visit authors.packtpub.com and apply today. We have worked with thousands of developers and tech professionals, just like you, to help them share their insight with the global tech community. You can make a general application, apply for a specific hot topic that we are recruiting an author for, or submit your own idea.

Share Your Thoughts

Now you've finished *Data Engineering with Databricks Cookbook*, we'd love to hear your thoughts! Scan the QR code below to go straight to the Amazon review page for this book and share your feedback or leave a review on the site that you purchased it from.

https://packt.link/r/1-837-63335-5

Your review is important to us and the tech community and will help us make sure we're delivering excellent quality content.

Download a free PDF copy of this book

Thanks for purchasing this book!

Do you like to read on the go but are unable to carry your print books everywhere?

Is your eBook purchase not compatible with the device of your choice?

Don't worry, now with every Packt book you get a DRM-free PDF version of that book at no cost.

Read anywhere, any place, on any device. Search, copy, and paste code from your favorite technical books directly into your application.

The perks don't stop there, you can get exclusive access to discounts, newsletters, and great free content in your inbox daily

Follow these simple steps to get the benefits:

1. Scan the QR code or visit the link below

https://packt.link/free-ebook/9781837633357

2. Submit your proof of purchase
3. That's it! We'll send your free PDF and other benefits to your email directly

www.ingramcontent.com/pod-product-compliance
Lightning Source LLC
Chambersburg PA
CBHW060110090326
40690CB00064B/4550